# Television Awareness Training

## The Viewer's Guide for Family & Community

edited by Ben Logan

ASSISTANT EDITOR, KATE MOODY

marc

A MEDIA ACTION RESEARCH CENTER RESOURCE

Published by
Abingdon/Nashville

*Creators:*
Stewart M. Hoover
Carolyn K. Lindekugel
Ben Logan
Nelson Price

*Coordinator:*
Shirley Whipple-Struchen

*Consultants:*
Eli A. Rubinstein
George C. Conklin
Joyce Sprafkin
Patricia Kowalski

*Visual Training Materials:*
Produced by Jeffrey Weber
Stanley Nelson


Library of Congress Catalog Card Number: 76-395-89
ISBN 0-687-41200-5
(Previously published by MARC under ISBN 0-918084-02-4)
Printed in the United States of America

# CONTENTS

## Part I

## Part II

# HOW TO USE THIS BOOK

The relationship each of us has with television is very personal, and any study of that relationship must be equally personal. Another person can't move into our minds, tell us how we now use TV, and what changes will make television a more creative part of our lives.

No one can do that for us, and apparently, most of us are not very good at doing that by ourselves either because it is a rare person who has developed any real guidelines for viewing. TV is simply there, and we join it for an average of three to four hours a day, perhaps occasionally feeling doubts and asking ourselves questions: Do I want to give almost a fourth of my waking hours to television? What are the positive and negative messages received over the years, and how, if at all, do those accumulating messages change how I (or my children) feel, think and believe? And if I want to make changes, how do I do that?

Again, it is each of us as an individual who can best find the answers—*if we will take some time and energy to study our relationship with TV.*

For those who want to do that, this new edition of the *Television Awareness Training* book has been redesigned to make it more complete and self-helpful to individuals, families and groups.

Experience with the earlier edition suggests ways this book becomes valuable in studying television.

Groups use it in informal sessions or in more formal Television Awareness Training Workshops where a trained leader helps a group move through a series of readings, viewing of on-film program excerpts, discussions and exercises. Participants share ideas, learn more about their relationship with TV, learn how the television system works and then make decisions about appropriate changes, both in how they use TV and in the system.

Individuals and families use the book in a similar way, but in a more self-help approach. The TV provides the viewing experiences. The book provides selective reading on the most common areas of interest and concerns, plus dozens of exercises that help one move, step by step, into a deepening awareness about television and opportunities for change.

Families find the suggested activities can open up communication on practically every subject introduced by the screen, with discussions moving very naturally to the questions of what kinds of values and behaviors programs seem to suggest as appropriate.

The goal of Television Awareness Training is to make possible a self-determined trip of exploration and discovery into ourselves, the people we live with and the real world around us. Television has its own personality and identity—at least we give it those characteristics —and it becomes almost a separate country and culture. Exploring that exotic culture and our relationship to it can be an exciting journey filled with ups and downs, interesting insights, startling surprises. The trip can be both light and heavy, sobering and fun-filled as it turns into an intriguing game of coming to terms with this electronic "Being" that keeps tumbling so willy-nilly into our homes and lives.

The earlier edition of this book is also used in college courses, high school social studies classes, adult education and continuing education for teachers. Since Television Awareness Training uses a values approach in looking at television, it provides a good back-up and expansion of the U.S. Office of Education (HEW) television curriculum which takes a critical viewing approach.

For most of us, television is a taken-for-granted presence, and asking us to understand it is like asking goldfish to understand water. Yet, groups and individuals who have used the earlier edition report that the book has been a valuable aid in developing new awareness and beginning to use TV in a way that more closely represents who they are and who they want to be.

The result is rarely just a decision to watch or not to watch TV. Almost always it is a decision to use TV differently, to stay in command of the set, watching specific programs instead of just watching TV, looking for positive and negative values, and placing the television experience in perspective, so that it does not become more important than one's real life self, other persons and the issues that confront our communities and the world.

STEWART M. HOOVER
CAROLYN K. LINDEKUGEL
BEN LOGAN
NELSON PRICE

## PUBLISHER'S NOTE

Worksheets No. 1 on pages 12, 24, 38, 49, 71, 91, and 107 are designed for use with the film excerpts used in T-A-T presentations and/or workshops.

# Chapter 1

# TELEVISION: INTRODUCTION AND OVERVIEW

## They Started The Revolution Without Me

*Ben Logan*

*BEN LOGAN, an editor of this book, is a broadcast producer with United Methodist Communications, a member of the board of directors of Media Action Research Center, and a free-lance writer. He has written drama for TV, written and produced TV and film documentaries and public service spot campaigns. He is author of* The Land Remembers—the Story of a Farm and Its People, *Viking Press 1975, Avon paperback 1976.*

I went to live in Mexico in 1950, aware that a new plaything called television was beginning to flicker here and there. I had seen one program, a half-hour drama by a friend who told me writing for television was going to be difficult "because the picture will never be good enough so we can see facial expressions clearly."

My response to that first viewing experience was a response to an interesting toy.

When I returned to the U. S. in 1955, I began to realize a revolution had started without me. Almost half of the homes in the country now had the flickering screen. The picture was improved. I *could* see facial expressions. I didn't know, however, what was happening until New Year's day, 1956. I had invited a friend for a mid-afternoon dinner. He came in, said hello, and began pacing around the apartment, obviously looking for something.

"Where's the TV?" he finally asked.

I told him there was none.

He couldn't have looked more stunned if I had shot him.

"But how are we going to watch the game?"

I didn't even know that "watch the game" was code for viewing the Rose Bowl.

Stunned again by my confession of that, he reluctantly stayed for dinner, then left, obviously hoping to catch the end of the game somewhere. He admitted as he went out that he would not have come if he had known we did not have TV.

It was my turn to be stunned. I had considered him a close friend. It was my first real knowledge that the revolution was a lot more than a technological one.

Not long after, I returned to Mexico. When I came back in 1959, the revolution was nearing completion. Eighty-six per cent of U. S. households now had TV. At the home of a favorite friend in Wisconsin, the set was never turned off. My good days of searching talk with that man were over, except for a time or two when I found him in the kitchen and quickly sat down there, ignoring the flickering light and strange voices from the next room.

I never fully recovered that friend from the TV screen. Our relationship was a casualty of the revolution.

At my favorite bar, where I had always gone to hear good stories, the story-tellers were silent. A large screen TV had become the story-teller with no one willing to compete with this new window on the world.

I had been studying anthropology in Mexico, focusing in on what makes up a culture, identifying those rituals and behaviors of everyday life that people center around. My textbooks had told me such traits were extremely persistent, going on from generation to generation.

The books were wrong. The groups I had been studying had never been confronted with anything like television. In the few years of my looking closely at other cultures, TV had become an instrument of incredible change in my own. If, overnight, a stone-age society had developed a complex language, learned to read and write, started printing books, and accumulated a vast library, I think the impact would not have equaled that of television.

The changes were everywhere. Most significant of all for me were those touching the family, that basic cultural grouping where we live closely with one another and teach children. I found startling changes in how people ate their meals, in bedtime hours, child routines, letter writing, conversation, reading, meeting and church attendance. The language itself was changed, filled with new code words and phrases, new jingles and slogans, new expressions that had entered people's lives from the screen. Sports were already being altered to more nicely fit the needs of television. There was even a story going around about city water systems being redesigned to cope with the sudden loss of pressure when millions of people used a commercial break for a trip to the bathroom.

The most startling fact of all was the quietness of the revolution. Almost no one was questioning the changes. In my background, gathering around a table at mealtime, with all its talk and laughter, had been the very heart of being a family. Yet people were casually giving that up, as though the mealtime had no purpose other than getting food into the body.

TV was the new hearth, with the warmth and voices coming from the magic box, taking the place of talk with each other, taking the place, it seemed to me, of family itself.

Sometimes disturbed by what we were seeing, some times simply unimpressed, my wife and I resisted television, feeling a bit like aliens.

Then, in July 1961, we joined the revolution. We brought home a TV and spent almost all of a long weekend watching program after program. We called it "getting acquainted," and television quickly began to change our lives, too, though it also confirmed our fears and we soon set up strict rules. The first rule was that no child was allowed to turn on TV without specific permission.

I think I was lucky. The objectivity I brought to the TV experience was an accident of my being out of the country. I had the advantage of seeing the accomplished revolution instead of having it creep over me in a way that never raised defenses.

I was lucky and even so I had a hell of a time fighting back. The rules we made at the beginning were mainly for the children. They didn't apply to me.

I found at times that my appetite for TV was too great to be satisfied by the delivery system. One of my approaches was to watch two or three movies at once, changing channels often enough to get the basic storylines of all. You haven't really lived the TV experience until you watch two John Wayne westerns at the same time, though I think my ultimate achievement was a combined viewing of *The Scarlet Pimpernel* and *The Return Of The Scarlet Pimpernel.* The chronology gaps didn't matter. The contradictions didn't matter. Even the fact that it didn't make sense didn't seem to matter. What did seem to matter was that I keep my eyes on that screen, letting each scene become its own reality that reached out to wrap itself around me.

In case it still isn't clear, let me say it. For a time I was hooked. I was an addict. The biggest clues to that addiction were these:

—I was watching television, not specific programs.

—I was letting TV crowd out of my life a great deal that I had always *said* was important to me, including people.

—I kept turning the knob, looking for higher levels of excitement.

—I was doing the equivalent of the alcoholic's "blacking out," having to work hard the next day to remember what I had watched the night before.

That first binge didn't last very long. Helped by my still fresh memory of the pre-TV world and the pre-TV me, I extricated myself from the screen, feeling as though I had been put through a cultural homogenizer. That was the beginning of my eighteen-year worry affair about my relationship with TV.

I am a rather well-disciplined person and have tended to live out my values, but television was more than I was prepared to handle. It is the most demanding, tempting, seductive medium of change I have ever faced. It seems to have been designed with all my needs, interests, vulnerabilities, and weaknesses in mind. It seems to know all about my tendencies to be passive and lazy, my willingness to suspend judgment while I stare at the screen and frantically construct images out of dots, too busy doing that to question. I am hooked on fingertip availability of a big, exciting, exotic world. I am beguiled by a romantic never-never land where today's behavior has no tomorrow consequences and where things/products from this Alladin's lamp reach out to me, promising to make me sexier, happier, wiser, pain free, and more successful at everything from being a great lover to keeping my barber from ever seeing a ring around my collar.

**"My own identity and reality can never again be fully separated from the myths and facts of the television universe."**

Television all too easily becomes my modern fairy tale book and I a child again when I view it, disappearing into nothing, looking for happy times, easy solutions and wanting them to come from outside myself and save me from myself, do my thinking for me, tell me who I am and what I'm supposed to do.

We are irretrievably linked, me and my TV. (Sing to same tune as "Me and my R.C.") Even if I never viewed again, and I do and will because I find it valuable and interesting, I am still swimming in its universe, where it has become the most pervasive and persuasive of all values teachers. My own identity and reality can never again be fully separated from the myths and facts of the television universe. More than 98 percent of U.S. homes have TV now and the images have become the new reality.

As I became more concerned about the impact of television on me, my family and my culture, I found other voices speaking to the issue. A lot of us were naive. We expected too much of TV, wanted it to be "pro-social" and "culturally enriching." It was some of that, but mostly it was something else entirely, and we stayed confused because we had not yet clearly identified its purpose.

While commercial television is clearly in the entertainment business more than any other, it is not in business to entertain. Its purpose is the same as that of Ford Motor Company, General Electric, and IBM—to make money. It does that by selling us, the viewers, to the advertisers at so much a head. The more of us who view, the higher the rate charged the advertisers.

Television is very good at making money. Networks operate at high profit levels. It is a rare station, usually one in a very small market area, that is not making a margin of profit far above most other types of businesses. The manager of a station in a large city once told me, "Owning a TV license is the closest thing there is to having permission to print money."

There are two big reasons for TV's financial successes. It is to some degree a monopoly business because of the limited number of channels. That makes it a seller's market, with more advertisers wanting on the air than there is air time available. The other reason is that the television industry knows us very well, gives us what we will watch in large numbers, and we happily deliver ourselves up to receive the sales messages.

That means we, you and I, are key

participants in what TV is and does.

Over the years I have listened to a lot of arguments about what the purpose of commercial television should be. There are usually two poles-apart positions—and some compromises in between.

The reformer view is that all TV is educational, even when it just sets out to be entertaining, and that puts a special burden of responsibility on the industry. Also, TV uses scarce airwaves that belong to the public. Not everyone can have a channel so choices must be made in issuing licenses. The obvious way of selecting one party over another in granting and renewing a license is to base it on which licensee will best serve the public interest. Just what "public interest" means is difficult to pin down. For one person it is more of *Three's Company,* for another it is hard-hitting documentaries, for another it is serious drama.

A variation of the reformer view is that any business which distributes information must take great responsibility for accuracy, completeness of information, and even take some responsibility for how people, especially children, use the information. This argument often ends with the assertion that we are not necessarily being culturally responsible when we give people what they want. Trying to find what is meant by "we," "people," and "they" in that sentence can become a lifetime task.

The opposite point of view comes from those who say commercial television is just another business, one which makes its money by helping sell products, a valuable stimulator of the gross national product. It is unfair to ask the television industry to be different from other businesses. There should be less control, not more, and networks aren't controlled by anyone anyway. It is a mass medium that sets out to program to the largest possible audience, a General Motors that produces for the general public, not a business that custom designs programming to fit every group and individual. Also, viewers have choices; they can turn it off; they can change the channel. In the end, these persons say, the public gets what it wants because programming decisions are based on ratings which accurately tell us what people view in the largest numbers.

Of course television does live in a system of control. The Federal Communications Commission (FCC) sets requirements calling for stations to operate in the public interest, and it grants and renews licenses. The Federal Trade Commission (FTC) has power over commercials, requiring them to be honest and to avoid modeling or suggesting behavior that could be dangerous.

In both cases, that "control" is quite theoretical. Neither organization is sufficiently staffed to do the job that could be done, and political realities limit their effectiveness. Most of the licenses were issued long ago and a renewal by the FCC is largely a rubber stamping of the present licensee. Refusal to renew is so rare that each denial becomes a landmark case. Even though applications for a license could be seen as competitive bids for the monopolistic right to use a very valuable frequency, if the present licensee does the mountain of paperwork required by the FCC and makes the right promises, it will get its renewal in almost every case.

In stating that TV's purpose is to make money, and that control often consists of more promise than performance, I do not mean that as a put-down. Television lives and acts within its own context of reality, just as other businesses do, and given the practicalities of now, I accept that. What I do mean to say is that we need to be fully aware that decisions in the TV industry are economic decisions and that we are kidding ourselves into a dangerous apathy if we think television is a carefully controlled medium where some mythical "they" looks out for our interests.

**"Television is very good at making money."**

The problem seems to be that we early saw a special vision of human possibility in television, and now that vision appears to have been pushed aside in an orgy of self-indulgence and business as usual. Disappointed and frustrated, I sometimes find myself seeking an enemy when what I really want is to find some new opportunity to use this miracle in a better way.

In the past, I think I make the mistake of wanting the television industry to take more responsibility than we have been willing to take ourselves. I now see that as unrealistic. We viewers are full participants in the television system, as responsible as the industry itself for how things are, and in a better position to break free and cause change than is any other part of the system.

I have been naive about other aspects of television. Like most people, I assumed it was free. It isn't. In the United States we spend about 11 billion dollars a year buying sets, repairing them, and buying the electricity to run them. We also pick up the annual tab for about eight billion in advertising costs when we buy the products. Only as individuals can we estimate how much we spend each year when we allow TV to talk us into buying products we don't need. I find that knowing I'm not getting an entirely free ride makes me more willing to bring some critical judgment to my viewing.

From the beginning I have been bothered by the content of programming, particularly about stereotyping, the way sexuality is presented as something to be laughed at, the way violence is so often shown as an effective way of dealing with conflict. One visual image stays disturbingly clear in my mind. At the Media Action Research Center we interviewed a seven-year-old boy after he had viewed violent programming, asking why he thought there was so much shooting and killing on television. He said, "Because that's what's happening outside in the world."

*Is* violence that common out there? Is TV what some people tell me it is, simply a mirror that reflects real life back to me? I don't think so. I think the images are more like those in a funhouse mirror, recognizable but very distorted.

Despite my concern with content areas such as violence, I now see a danger in studying my relationship with television too much category by category without putting it all together into the larger context of TV as a persistent, repetitive way of looking at human behaviors and possibilities.

There is, of course, still a great deal of physical violence on the screen. There is also a more subtle violence, such as put-downs and ridicule. But the problem is larger than that. Violence, in its broader sense, is for me linked with the word "violate." To perpetuate stereotypes is to do violence to the individual or group being stereotyped. To show the sexual side of us as primarily shallow, predatory and performance-related, does violence to our potential as human beings. To constantly dramatize the aggressive, the highly competitive, the violent side of people does violence to our vision of being able to relate cooperatively and reasonably with each other.

When I examine my relationship with TV in this larger arena, I find linkages I did not see before. I now believe that most of the content problems come out of television's tendency to sensationalize, a situation fed by the competitive need of each network to beat out the other networks, and fed, also, by our own willingness, even eagerness, to watch increasingly sensationalized programming.

It is a hydraulic situation. Push down violence, as has been done slightly, and up pops a new form of sensationalism—hyped-up sexually related material, for example. If we push down the sexual material, we can be sure some new form of sensationalism, some other shallow way of looking at human behavior, will pop up next.

Again, it is an illustration of how we viewers are in a position to cause change. The networks, locked into a battle to put the top income-producing programs on the air, are not going to back away from sensationalism while great numbers of us go on watching it.

I don't want to minimize my concern with content on the screen, which for me adds up to a very unreal view of ourselves and our society, but I think many of us have become so preoccupied with what's on the screen that we have ignored an equally critical concern. I'm talking about television as a presence in our lives, what happens to us not because of *what's* on but because *it is* on.

Here the study of television gets personal, indeed. No matter how great the programming, how much is too much? What am I *not* doing while I give three to four hours a day to television? How does it separate me from other persons, from physical activity, from exploring diverse new arenas?

During the past three years, a lot of my energies have gone into helping create and introduce Television Awareness Training. I have worked as a group leader with hundreds of persons who have come to workshops and presentations, wanting to explore the role of TV in their lives.

Some of those participants have let television just happen to them and have no guidelines for its use. Some have carefully and thoughtfully made a place for TV in their lives, preventing it from becoming a substitute for life. Some don't watch TV and are surprised to find they are caught up in television's universe even so. Some openly admit addiction. They, like me when I first got TV, can't remember what they viewed last night. They just watch television, like the Connecticut boy whose parents discovered he was getting up before any programing went on in the morning and was staring for a half hour at the test pattern.

Some persons at workshops are so hostile to any suggestion that TV could be used more creatively that a hidden addiction is suggested. They seem to be saying, "I'm OK, but I want to learn how other people should use TV." Or they want to learn how their children should use it and are surprised to find there is no way to help your children use TV differently without using it differently yourself.

T-A-T groups often get tangled up in another difficult issue—just how appropriate is television's extremely persuasive sales approach when it stresses things as more important than people and encourages consumerism when we are already plundering the planet of its dwindling supply of resources? The arguments can be long and heavy as the gross national product advocates and the conservationists meet head-on.

A lot of stored frustrations with the television experience come rushing chaotically to the surface at workshops, but as people look deeper, a clearer picture of television's role in our lives emerges.

It becomes obvious that we have a love/hate relationship with TV. We want it. We would miss it terribly if it vanished. Where else would I find so conveniently packaged the latest news, entertainment, sports, escape, babysitter, a way to learn about people and places I will never visit any other way, and those sudden insights that come from drama and from hearing exciting people talk about exciting ideas?

Yet, may of us are confused, have nagging doubts and are angry at ourselves for the way we use and allow ourselves to be used by TV.

It would be a miracle if most of us were not confused. We are constantly handicapped by the instant nature of TV. We have little opportunity to know in advance what a program is going to be like, what values it will suggest, how appropriate it will be for children, how appropriate to our own tastes. We are dealing with a mass medium where if only twenty million of us love a series, it will go off the air for lack of viewer interest. Or, even if a series is getting enough viewers, it may be taken off anyway because it is not reaching an age group willing to spend lots of money for the advertised products.

Television Awareness Training sessions have helped me realize how little we can prescribe for each other in coming to terms with television. We must each sort out our relationship to TV in a way that represents who we are and what we want to do with the ticking-away moments of our lives.

I once took a mirror off a wall and leaned it against a chair. The dog came past, saw his reflection and slid to a halt. After a barking fit, he came slowly closer until he touched his nose against the nose of the dog in the mirror. Then, he did the intelligent obvious. He looked behind the mirror, trying to find the real dog.

Something about it reminded me of my television watching and suggested a game I've been playing ever since. I look behind the images on the screen by enlarging the scene in my mind and seeing more of the real life reality TV is part of. In a news show I widen the screen to include the cameras, and the hairdresser who makes sure the anchor person's hair is just right. I extend a reported event and try to see what came before and after the minute or so of action I am given. I look over an editor's shoulder and imagine the stories that got crowded out or weren't used because no good visual was available.

In an action-adventure or situation comedy series I include the director who is telling everyone exactly what to do, the stunt person who pretends to fall off a building, the actors delivering lines they have memorized from a script written by a writer who is fantasizing out of his or her own view of the world. I watch the laughter and applause being edited in or made to happen by cue cards held up in front of a live audience.

I go even farther behind the screen to an advertising agency and listen in on conversations about whether or not an actor can drink a dark-colored liquid on a show carrying commercials for a non-cola. I watch a commercial being designed to convince me I have a problem and simultaneously sell me the product that will take care of the problem.

It is a great game, this peeking behind the screen, and the possibilities are endless. I don't play it all the time, just often enough to keep TV in a clear perspective, to remind myself that it is made up. It is fantasy. Its characters don't exist anywhere else.

I find a lot of "theys" behind and around the screen. "They" are always there and "they" make the decisions that control the partly real, partly unreal world called television.

Just as there are always "theys" behind the screen, there is always a me in front of it. I bring a unique me to my viewing. I bring a past, a set of values, a need to be entertained and informed, a mind that is always learning and storing whether I want it to or not. I, too, have a larger context than just the immediate viewing experience. That knowledge leads me into another game. I try to step outside myself, then turn around and confront with questions the me that is staring at the screen. "Why are you watching a program filled with gratuitous violence when you say you are a non-violent person? How come you approve when the good guy kills the bad guy? Why are you laughing at a sexist joke when you try to be nonsexist?

People program TV. I all too often let TV program me. I don't have to guess at that. I know it and here's an example.

For a time there was a rash of motorcycle-riding "creeps" on the TV screen, doing almost exclusively things that made me angry and afraid. They were brutal and irresponsible. Their standard uniform was helmets, black jackets, metal studs and hair. Now here's the problem. I have never known a real-life, motorcycle-riding, leather-jacket wearing person. I lack what I call a "reality check" to test my TV experience against. So when I see a real-life motorcycle rider who fits the stereotyped description, I am suspicious, uneasy and even frightened. My first reaction is to get out of sight. Intellectually, I know my reaction is silly. Emotionally, I have allowed my TV experience to push me into a stereotypic reaction, generalizing all such persons into TV's negative mold.

**"I'm talking about television as a presence in our lives. . ."**

In this situation I know what is happening and can be on guard. But there are a lot of areas where I lack "reality checks" so I may not know about other cases where I have let myself become programmed.

Despite all this, I don't think television is a monster. TV is a screen full of dots. I don't think the people who program TV are monsters, either, though I plan to go on yelling at them to give me dots that add up to better programs. That leaves us, the viewers, and I don't think we're monsters, either. But I think if I join the screen carelessly, non-selectively, blindly, then a monster is being created. That relationship between an unthinking screen and a mindless me is monster enough.

This revolution called television that started without me, this odyessey into the unknown for which we had no leader and no map, is now thirty years old. We can no longer play the game of innocence. We have only to step back, look objectively at our relationship with television to know a lot about what happened when we bought into this 20th century pied piper that seemed to promise so much and has left so many of us feeling hollow.

There is no way to avoid the truth that we are the ones who did it, both the good and the bad of the television experience. Surely then, there is no longer any way to avoid the knowledge that we are also the ones who will have to undo what needs undoing, change what needs changing, start over again where that is what's required.

Those who want to do that, will find in the rest of this book a lot of different kinds of clues, starting points and insights from the minds of a lot of different persons.

## LOG OF PROGRAM EXCERPTS

News—Local news show

Sports—Basketball game

Talk Show—Mike Douglas

Game Show—$1.98 Beauty Show

Dallas—Husband and wife kiss and make up

Three's Company—"I know what we can do."

Three Hungry Wives—Man pays woman in bed

Wonder Woman—Men shoot at Wonder Woman

Barnaby Jones—Man grabs woman

Vegas—Snipers shoot at old woman

Commercial—-Pine Sol

Syndicated Show—Untouchables

Movie—What did you do in the war, Daddy?

Special—Happy Birthday, Las Vegas

Lifeline—Brought to you by Xerox

News—McNeil/Lehrer

Evening at Symphony

Sesame Street—"This is a T"

ABC Out of School Special

Scooby Doo, Where Are You?

Commercial—Fruity and Coco Pebbles Commercial

Introduction & Overview/Worksheet No. 2

## THE TELEVISION EXPERIENCE

1. In a word, phrase or sentence, describe positive and/or negative responses you have to your personal or family use of television.

2. Talk with another person, describing your most negative and positive feelings about television. Ask the other person to do the same.

## Introduction & Overview/Worksheet No. 3

Complete the sentence: I saw the following values portrayed in programming I just watched.

## Introduction & Overview/Worksheet No. 4

## HOMEWORK

1. Read Chapter I. Use this new information and awareness as you view and use television.
2. Monitor your own viewing patterns for a one-week period. Learning about your own relationship to television is an important discovery process. In systematically monitoring viewing patterns, most persons find some surprises about themselves.

## VIEWING LOG

| | Mon. | | Tues. | | Wed. | | Thurs. | | Fri. | | Sat. | | Sun. | |
|---|---|---|---|---|---|---|---|---|---|---|---|---|---|---|
| *Program title* | 1 | 2 | 1 | 2 | 1 | 2 | 1 | 2 | 1 | 2 | 1 | 2 | 1 | 2 |
| | | | | | | | | | | | | | | |
| | | | | | | | | | | | | | | |
| | | | | | | | | | | | | | | |
| | | | | | | | | | | | | | | |
| | | | | | | | | | | | | | | |
| | | | | | | | | | | | | | | |
| | | | | | | | | | | | | | | |
| | | | | | | | | | | | | | | |
| | | | | | | | | | | | | | | |
| | | | | | | | | | | | | | | |
| | | | | | | | | | | | | | | |
| | | | | | | | | | | | | | | |
| | | | | | | | | | | | | | | |
| | | | | | | | | | | | | | | |
| | | | | | | | | | | | | | | |
| | | | | | | | | | | | | | | |
| | | | | | | | | | | | | | | |
| | | | | | | | | | | | | | | |
| | | | | | | | | | | | | | | |
| | | | | | | | | | | | | | | |

Column 1—record the time of day the program played.
Column 2—record the amount of time you viewed the program.

## HOMEWORK

Create a profile and analysis of your television viewing for the week of____________________

a. Total number of hours viewed: __________

b. Total hours viewed from 6 a.m. to noon: __________

c. Total hours viewed from noon to 6 p.m.: __________

d. Total hours viewed from 6 p.m. to midnight: __________

e. Total hours viewed from midnight to 6 a.m.: __________

f. Total viewing time for the following program types:

Variety.................... __________

Sports.................... __________

Action/adventure.................... __________

Situation comedy.................... __________

Drama.................... __________

Soap opera.................... __________

News.................... __________

Documentary.................... __________

Game.................... __________

Talk.................... __________

Other.................... __________

g. Figuring 16 commercials per hour, how many times were you exposed to commercial messages? __________

h. ______________________________ __________

i. ______________________________ __________

j. ______________________________ __________

(Add other information which will be of help or interest to you.)

## HOMEWORK

**Evaluate why you view television. Get away by yourself and take some reflection time. Probe your motives honestly and deeply. The purpose of this exercise is to provide you with additional insight regarding the psychological needs TV may be fulfilling for you—as well as the more obvious needs. On a scale of 1-5 (with 1 being the least weighty and 5 the most weighty), assign a value to TV viewing reasons where applicable for you.**

| | | |
|---|---|---|
| a. Information (News, weather, documentaries) | ______ | ______ |
| b. Entertainment (To relax, have fun, laugh) | ______ | ______ |
| c. Instruction (To learn how to do something, such as cooking, auto repair, formal courses, etc.) | ______ | ______ |
| d. Learn about human relationships (To see how people relate in different situations and places) | ______ | ______ |
| e. To get away from problems (Work, office, debt, relationships) | | |
| f. To get away from people (Friends, fellow workers, family) | ______ | ______ |
| g. To be with people (Friends, family) | ______ | ______ |
| h. Conversation topics (To be able to talk with friends, family, co-workers) | ______ | ______ |
| i. To avoid loneliness (To have companionship when alone) | ______ | ______ |
| j. To pass time (To help time go by more quickly) | ______ | ______ |
| k. Commercials | ______ | ______ |
| l. ______________________ | ______ | ______ |
| m. ______________________ | ______ | ______ |
| n. ______________________ | ______ | ______ |
| o. ______________________ | ______ | ______ |

**In the second column, place values on the reasons for viewing television as you would like to change them during the next two months. What precise actions and decisions will you commit to—what day this week will you begin to change?**

## HOMEWORK

A fantasy exercise about TV as a presence; imagine that all TV has gone off the air for one week. (Electronic termites ate up the signals!) You suddenly find yourself with more than twenty hours (or as many hours as you now view each week) of new free time. Write a schedule of what you would like most to do with all those free hours during that week.

| | *Morning* | *Afternoon* | *Evening* |
|---|---|---|---|
| **MONDAY** | | | |
| **TUESDAY** | | | |
| **WEDNESDAY** | | | |
| **THURSDAY** | | | |
| **FRIDAY** | | | |
| **SATURDAY** | | | |
| **SUNDAY** | | | |

Chapter 2

# TELEVISION AND VIOLENCE

*by Larry Gross*

*LARRY GROSS is Associate Professor of Communication at the University of Pennsylvania. He is co-principal investigator of the CULTURAL INDICATORS PROJECT, and co-editor of* Studies in the Anthropology of Visual Communication. *His research is in the cultural determinants of symbolic behavior.*

"Don't you understand, I replied, that we begin by telling children stories, which, taken as a whole, are fiction, though they contain some truth? Such storytelling begins at an earlier age than physical training; that is why I said we should start with the mind."

"You are right."

"And the beginning, as you know, is always the most important part, especially in dealing with anything young and tender. That is the time when the character is being molded and easily takes any impress one may wish to stamp on it."

"Then shall we simply allow our children to listen to any stories that anyone happens to make up, and so receive into their minds ideas often the very opposite of those we shall think they ought to have when they grow up?" (Plato, *The Republic*).[1]

Plato probably was not the first to articulate a concern over the effects of "fictional" storytelling on young minds; he certainly wasn't the last. Parents have always been understandably wary of those who may wish to entertain or educate their children.

Traditionally, the only accepted extra-familial storytellers were those certified by religious institutions. The growth of educational institutions, also originally religious, gradually intruded a new group of specialists into the relationship between children and the world they grow up into.

The emergence of the mass media of communications fundamentally altered the picture: children were increasingly open to influence which parents, priests and teachers could not monitor or control. Beginning with the widespread availability of printed materials to the literate, increasing with the invention of media less dependent upon reading (movies, radio), and culminating with television's omnipresence, children have become independent consumers of mass-produced stories. Plato's ancient question reverberates very vividly for today's parents as they confront the television set which so fascinates their children.

The televised stories that generate the most concern seem to be those that contain scenes of violence. Why should this be? For one thing, acts of physical aggression, even when committed by the "good guys" in the name of law and order (as the majority of TV violence is) are intuitively suspected of instigating similar acts by impressionable viewers.

A second reason for the focus on the violent content of television is the incredible prevalence of aggressive acts depicted in television drama, particularly in programs aimed specifically at children. It has often been noted that by the time the average American child graduates from high school he will have seen more than 13,000 violent deaths, all of them on television. The sheer amount of exposure to television violence is itself a source of concern: we worry that children will become jaded, desensitized and inured to violence—not only on television but in real life as well.

In the thirty years that we have lived with television, public concern with the medium's prediction for violence has been reflected in at least eight separate congressional hearings, a special report to the National Commission on the Causes and Prevention of Violence in 1969, and a massive study of television and social behavior commissioned by the Surgeon General, which focused largely on the prevalence and effects of televised violence and culminated in a five volume report. Since then, the flow of research titled, *Television and Social Behavior—A Technical Report to the Surgeon General's Scientific Advisory Committee on Television and Social Behavior*[2] and debate has continued unabated. While scientific caution requires us to proceed carefully in drawing conclusions from the wealth of data and evidence that has been accumulated, many patterns seem well established.

## How Much Violence

The first question for which we have a clear answer is: how much violence is there on television? Since 1968 my colleague, George Gerbner, and I, along with our associates, have been conducting a multi-faceted research project called Cultural Indicators. This project involves a continuing study of the "world" of television drama and of the conceptions of social reality that television viewing cultivates. The first part of the project is the yearly recording and analysis of a sample of prime-time and weekend morning network television drama. The analysis, based on the work of teams of specially trained coders, provides systematic, cumulative, and reliable information on the composition and structure of the symbolic world of television—its geography, demography, thematic and action structure. At this writing, 1,437 programs, 4,106 major characters and 10,429 minor characters have been analysed.

Among the aspects of program content which we analyse is the portrayal of violence. We define as the overt expression of physical force, with or without a weapon, against self or other; compelling action against one's will on pain of being hurt or killed; or actually hurting or killing.

In the latest "Violence Profile" report of our findings, we noted that the percentage of programs containing violence has ranged in the nine years of our study from 80 to 90 percent. In the fall of 1977 it was 75.5 percent. The rate of violent episodes per hour rose to a record high of 9.5 in 1976; in 1977 it was a typical 6.7 episodes per hour. The percentage of characters involved in violence has ranged from 56 to 75 percent, and averages at 63 percent over the last nine years.

Weekend morning programs, the period referred to in the industry as the "kidvid ghetto," contain the highest rate of violent actions. In 1977 there were scenes of violence in nine out of ten programs at a rate of 16 per hour.

### Effects of Violence

Having established that violence is a frequent feature of television drama—a finding that most of us find surprising only in its consistency over nine years, despite continued network claims that the incidence of violence is being reduced—we turn to the more difficult question of its effects on viewers.

Here I would identify two main hypotheses. The first, alluded to above, is that viewing television violence will make people, children in particular, more likely to commit acts of violence themselves. At the time of the Surgeon General's report in 1972, about 50 experimental studies supported the view that viewing violence increases the likelihood of children engaging in agressive behavior, at least in the short-term context of the laboratory. However, we know that findings based on laboratory studies have their shortcomings because many factors are not present in the laboratory.

The significance of the experimental studies covered in the Surgeon General's report was stengthened by the addition of surveys which found positive correlations between everyday violence viewing and aggression among adolescents in real life (Chafee, 1972). Moreover, these relationships were not accountable for by other factors—socio-economic status, sex, school achievement—which often prove quite powerful in explaining adolescent behavior.

In two surveys conducted by the Cultural Indicators project we asked children, "How often is it all right to hit someone if you are mad at them?" We found that heavy viewers of television, more often than light viewers, respond that it is "almost always all right" to hit someone.[3]

We can conclude, then, that the first hypothesis reflects a justifiable fear about the influence on children of viewing violence on television. Is this really, however, the most disturbing and widespread consequence of television's preoccupation with mayhem?

**"'How often is it all right to hit someone if you are mad at them?' We found that heavy viewers of television, more often than light viewers, respond that it is 'almost always all right to hit someone.'"**

The second hypothesis about the effects of television violence begins by expressing serious reservations about the first hypothesis. Can we really argue that violence on television consistently or dangerously instigates acts of violence among viewers? Surely if this were the case we wouldn't need elaborate research studies; the evidence would be piled up in front of our eyes (and our television screens). If the average child witnesses thousands of homicides and lesser acts of violence, the average sibling, parent, and teacher should be reeling from the blows of television-stimulated aggression.

Clearly this isn't the case. We may all be aware of frequent but low-level imitative-aggression among children. We also know about the very rare, if widely publicized, cases of serious violence which seem to be influenced by television programs or movies. But, on the whole, *most* viewers, children as well as adults, do not seem to respond to the violence they see on television by committing acts of violence. Are there, then, no real grounds for concern?

In order to answer this question, we must begin with a fuller understanding of the total phenomenon of television.

### The System Is The Message

All societies have ways of explaining the world to themselves and to their children. Socially constructed "reality" gives a coherent picture of what exists, what is important, how things are related, and what is right. The constant cultivation of such "realities" is the task of rituals and mythologies. They legitimize actions along lines which are conventionally acceptable and functional for the status quo.

The economic, political and social integration of modern industrial societies allows few communities or individuals to maintain an independent integrity. We are parts of a Leviathan, like it or not, and its nervous system is telecommunications. Our knowledge of the "wide world" is what this nervous system transmits to us.

Television is the central cultural core of this society. It is an agency of the established order and as such serves primarily to maintain, stabilize and reinforce conventional values, beliefs and behaviors. Media messages are items manufactured for sale by the institutional establishment. The goal of greatest audience appeal at the least cost demands that these messages follow conventional social morality.

Violence plays a key role in television's celebration of the status quo. It is the simplest and cheapest dramatic means to demonstrate the rules of the game. Encounters with physical violence in the real world are rare, more sickening than thrilling. But in the symbolic world, violence is dramatic, exciting and entertaining. Television violence is a *show of power* that tells us who are the aggressors and who are the victims. It demonstrates who has power and who will have to acquiesce to that power. Television violence may achieve the immediate goal of real violence—producing fear. Fear, and the power that can be achieved because of it, may be the main dangers of television violence to all viewers, children and adults alike.

Our second hypothesis states that the important effect of television violence is the result of the selective and sterotypical portrayals of power and people in society. The Cultural Indicators project has addressed this hypothesis by charting the world of television drama and by determining the extent to which exposure to this *symbolic* world cultivates among viewers conceptions about the *real* world.

### Fact Or Fiction?

Television drama conforms to the traditional narrative conventions of representational realism. However contrived their plots might be, TV stories appear to take place against a backdrop of the real world. The basic "reality" of the world as shown on most TV shows is unquestioned; and it is highly "informative." It offers to the unsuspecting viewer a continuous stream of "facts" and impressions about the ways of the world, about the constancies and vagaries of human nature, and about the consequences of actions.

Normal adult viewers, to be sure, are aware of the fictiousness and vicarious nature of TV drama: no viewer calls the police when a character on a TV program is shot. But one may still wonder how often and to what extent viewers suspend their disbelief in the persuasive realism of the symbolic world. Even the most sophisticated among us can find many facets of our 'knowledge' of the real world which are derived wholly or in part from fictional representations.

How many of us have ever been in an operating room (and been awake), a criminal courtroom, a police station, jail, or a corporate boardroom? How much of our general 'knowledge' about these and other spheres of activity, about how various jobs are carried out—how much of our real world has been learned from fictional worlds? To the extent that viewers see TV drama—the foreground plot or the background world—as naturalistic, they may derive a wealth of incidental knowledge from the progams they watch.

In real life much is hidden from our eyes—motives are obscure, outcome ambiguous, personalities complex—the truth is never pure and rarely simple. But television presents life differently. Problems are not left hanging, the rules of the game are clear and rarely change, rewards and punishments are present and accounted for. TV drama shows us the normally hidden workings of many important and fascinating institutions—medicine, law enforcement and justice, big business—and it shows us the people who fill important and exciting roles. Television provides the broadest common background of assumptions about what things are, how they work (or should work), and why.

**"Weekend morning programs, the period referred to in the industry as the 'kidvid ghetto', contain the highest rate of violent actions."**

The symbolic world of television drama is a mixture of truth and falsehood, of accuracy and distortion. This is an inevitable concommitant of naturalistic story-telling; should it be a cause for concern?

### Fear and Mistrust

In March of 1977, *The Morning Herald* of Hagerstown, Maryland, conducted a poll among residents of the city. In this survey it was found that, "almost two-thirds of the people polled fear for their safety after dark—despite the fact that they know that crime rates (in Hagerstown) are relatively low." In fact, the paper reported, "while more than a third of the people polled thought they only had an even chance of escaping violent crime, police statistics show that their chances of being attacked are less than 1 in 100.[5]" Why should the residents of this peaceful small town exhibit such unreasonable fear for their safety?

*The Morning Herald* also asked their respondents about their television viewing habits. They found that "the heavy TV viewers"—those who watch more than four hours a day—"tended to overestimate all the crime figures they were asked to offer opinions on."[6] This finding parallels the results obtained by the Cultural Indicators project in analysing a series of survey questions asked by ourselves and others to samples of children and adult viewers. In case after case, heavy viewers of television are more likely than are light viewers to answer questions about the *real* world in terms which seem to reflect the facts of life in the *symbolic* world of television drama.

In instance after instance, adults who watch more television are more likely to overestimate their chances of encountering violence, to overestimate the percentage of crimes that are violent, and to overestimate the percentage of men employed in law enforcement and crime detection. They are more likely to reflect interpersonal mistrust, suggesting that "most people" just look out for themselves, take advantage of others, and cannot be trusted.

In addition, our analysis of data obtained in the 1976 American National Election Study showed that adults who report that they "frequently" watch evening police and crime programs also report that they have obtained dogs, guns, and locks for purposes of protection in greater proportions than those respondents who "sometimes" or "rarely/never" watch crime and police programs.[7]

The obvious question that arises is whether factors other than television are responsible for these differences. We know that heavy and light viewers are different in many ways prior to—and aside from—television. Heavy viewing is part and parcel of a complex of characteristics which includes lower education, lower income and other class, age, and sex-related attributes. However, these powerful social and psychological variables do not explain our results: the differences we find still show up when we control for each and all of these characteristics. Within each group—the college educated, for example, or women—the heavy viewers are more likely than are the light viewers to give what we call "television answers" to questions about the real world.

It should come as no surprise that children are, if anything, more likely than adults to reveal conceptions of the world which seem to derive in large part from television's stories. The patterns of adult responses outlined above are all characteristic of children's responses as well. Controlling for age, sex, socioeconomic status, and IQ, we find that children who watch a lot of television are more likely than are light viewers to give "television" answers to questions about violence, crime and law enforcement in society. As with adults, but even more so, heavy viewing children are likely to say they would be afraid to walk alone at night—even those who live in a notably safe suburban community.

In parallel with our findings, the Foundation for Child Development reported on a survey of 2,200 children aged seven to eleven and of their parents, in which, "children who are reported to be heavy TV-watchers—whose parents say they watch four or more hours of television on the average weekday—were twice as likely as other children to report that they 'get scared often.'"

## Kojak Vs. The Cops

Two law professors recently took a careful look at the behavior of police officers as depicted on television over an 18-month period, "to see whether the behavior legitimized by television drama is the same as that required by the Constitution that the man on the beat is sworn to uphold and protect."[8] What might we expect such a study to reveal?

Mr. Telly Savalas, TV's Lt. Kojak, sees his show as an accurate reflction of the real police world: "The attitudes and the accent of these shows are accurate. Everything is on the nose except for the cases. They're fiction. The public thinks *Kojak* is the way the police actually operate, and they're right."[9]

Captain Thomas Gallagher of the New York police force has a different view: "The police shows I've seen do violate people's constitutional rights. They show police officers circumventing established methods of getting information by brutalizing people or performing illegal searches and seizures of property. You wouldn't see that in real police work—Unlike TV policemen, we don't always come up with the bad guy. We have a lot of restrictions, many of them constitutional."

**"We define violence as the overt expression of physical force, with or without a weapon, against self or other; compelling action against one's will on pain of being hurt or killed; or actually hurting or killing."**

Who was right, Mr. Savalas or Captain Gallagher? Unfortunately, they both were. The television shows analysed were rife with illegalities and procedural violations by the police, as Captain Gallagher claimed. "The overall image the shows project is clearly one that is alien to the Constitution. The facts would horrify the average judge if they were brought into court as real cases."

However, Professors Arons and Katsh also confirmed Telly Savalas' confidence in the credulity of viewers: the programs were believed to show the normal and appropriate conduct of police officers. "As we have discovered, many people are unaware of the police violations they see on the screen until after considerable discussion."[10]

Is this really a problem? Arons and Katsh think so: "It may not be a very great leap from the dim awareness of constitutional rights to the failure to recognize or react to constitutional violations perpetuated by police in everyday life. . . . (If) crime-show violations of the Constitution always turn out to be a good thing, then these TV morality plays may amount to nothing more than reactionary propaganda."

## Power Scenario

We have seen that children and adults who spend much of their time in the violent world of television drama express a heightened sense of fear and interpersonal mistrust. They may also come to be more accepting of police practices which violate Constitutional protections in the name of law and order. Why is there no public outcry about *these* effects of television violence?

I mentioned earlier that parents' concern over television violence is a characteristic fear on the part of an established group about the subversion of a subservient class. Popular media have historically raised the specter of unrest and misbehavior among various "lower orders" —poor people, ethnic and racial minorities, women and children. "Whether the suspect and controversial media are newspapers, novels, and theater, as in the nineteenth century, or movies, radio, comic books, and television as in the twentieth, concern tends to focus on the possibilities of disruption that threaten the established norms of belief, behavior, and morality."[11]

What these concerns have in common is the danger of the *ruled* attempting to emulate the behavior of the *rulers*. In contemporary society, the privileges of power most jealously guarded are those of violence and sex. In the public realm the State claims the sole legal prerogative to commit violence (in defense of law, order and national security), and to regulate the commission and depiction of sexual acts (in defense of "decency"). In each case the State asserts substantial powers of determining (through various agencies—legislatures, courts, police, schools) who can do (or show) what to whom, and with what consequences for disobedience.

In the private realm, parents generally assert the same prerogatives over their children—the power to determine the range of permissable and forbidden behavior. In the case of violence and sex the policy is usually clear: these are not acceptable forms of activity for children to engage in, although parents may do so when the need (for discipline) or the urge (for sex) moves them.

It is not really surprising, therefore, that the representatives of the established social order—both congressional committees and parents—are primarily worried about television violence as a threat to their monopoly over physical coercion—however limited that threat might be. In the larger perspective I have been attempting to delineate, we may all have more reason to worry about television's potential to influence our relationships with each other and with our public institutions.

In 1776, John Adams wrote that fear is the foundation of most governments. By depicting a social power hierarchy, television drama may cultivate assumptions that tend to maintain this hierarchy. This is not to claim that television alone is responsible or decisive; only that it makes a contribution in that direction.

The policies that govern television programming are made on economic grounds; and they may be said to achieve their immediate goals. However, they may fall short of achieving long-range social and moral goals of justice and the development of healthy attitudes about power and trust. Given the evidence that TV may sometimes cause subsequent aggressive behavior, and given our findings that it promotes a climate of fear and suspicion, we might ask again in modern language what Plato said . . . shall we simply broadcast to our people any stories that anyone happens to make up—and to receive into their minds ideas often the very opposite of those we think they ought to have. . . ?

### *REFERENCES*

1. Cornford, F. M., (trans. and ed.) *The Republic of Plato,* London: Oxford University Press, 1941.
2. Comstock, G. A., E. A. Rubinstein, and J. P. Murray, (eds.), *Television and Social Behavior,* five volumes, Washington, D.C.: U. S. Government Printing Office, 1972.
3. Gerbner, G. L. Gross, M. Jackson-Beeck, S. Jeffries-Fox and N. Signorielli, "Cultural Indicators: Violence Profile No. 9," *Journal of Communication* 28:3, 1978, pp. 176-207.
4. Chafee S., "Television and Adolescent Aggressiveness (an overview)," in G. A. Comstock and E. A. Rubinstein (eds.), *Television and Social Behavior,* vol. III: *Television and Adolescent Aggressiveness,* Washington, D.C.: Government Printing Office, 1972.
5. Bertorelli, P., "A City Afraid?" *The Morning Herald,* Hagerstown, Maryland, March 22, 1977, p. 1.
6. Ibid.
7. Gerbner, et al., 1978. Op. cit.
8. Arons, S. and E. Katsh, "How TV Cops Flout the Law," *Saturday Review,* March 19, 1977, pp. 11-18.
9. Ibid.
10. Ibid.
11. Gerbner, G. and L. Gross, "Living with Television: The Violence Profile," *Journal of Communication,* 26:2, 1976, pp. 173-199.

Violence/Worksheet No. 1

## LOG OF PROGRAM EXCERPTS

1. **Murders and Assaults**
   Barnaby Jones—Man grabs woman
   Charlie's Angels—Woman electrocuted in bathtub
   Starsky and Hutch—Man is shot at a party
   Starsky and Hutch—Woman hit by car
   Hawaii Five-O—Man is killed by a bomb
   Wonder Woman—Two men shoot at Wonder Woman
   Eddie Capra Mysteries—Dead woman is discovered in a bathtub
   Eddie Capra Mysteries—Fist Fight
   Movie—Woman shoots ambulance driver
   Sword of Justice—Man is shot by paid assassin
   Vegas—Men terrorize women with rattle snake

2. **Period Violence**
   Little House on the Prairie—Five boys gang up on a sixth
   Centennial—Soldiers insult man and his family—fight ensues
   Centennial—Man is attacked by Indians

3. **Humor/Violence**
   Love Boat—Men fight over a woman
   Soap—Family slap and cuff one another
   Saturday Night Live Teaser—"Knoogies"
   Movie—Women wrestlers
   Soap—Woman attacks her daughter-in-law

Violence/Worksheet No. 2

## GROUP WORK

Participants are divided into four groups. It is the groups' responsibility **to develop action models and coping techniques to counteract the anti-social messages of television.** The four groups are:

**Individual and personal action**
**Family action**
**Community action**
**National and governmental action**

This is a brainstorming session in which ideas will be freely proposed. There may be limited time after the brainstorming to flesh out the ideas. The purpose at this point is to begin to explore possible options for action.

1. Allow individual persons to read Sections I, II and III below.
2. In Section IV, select your group area. Propose new ideas; suggest ways of implementing or expanding on those ideas which are suggested.
3. Report your ideas back to the total group.

## COPING WITH THE ANTI-SOCIAL MESSAGES OF TELEVISION

### I. THE NEW CONDITION

**A. Television is different from other media.**

1. It is more pervasive.
2. It is available to children as no other previous medium.
3. It is non-discriminatory in selecting viewers—for example, children see adult programs.
4. It goes into the home and is readily accessible at any time to children and adults alike.
5. Children are exposed to more television than to any previous medium.
6. Television teaches by means of observation, instruction, and example.
7. Television presents massive quantities of anti-social messages and lessons.

**B. Television teaches anti-social messages.**

1. Violence is presented often as an appropriate and desirable means of solving problems.
2. Weapons of violence are normal parts of American everyday life.
3. Happiness is achieved through consumption of things.
4. Competition—creating winners and losers is desirable and normal.
5. Women, men, racial groups and others are stereotyped so that viewers are taught mis-information about other groups and persons.
6. The world's problems often are presented superficially.
7. Viewers are taught to solve problems through quick, easy and superficial solutions.
8. Human relationships and human sexuality are presented in exploitive ways.

*(Worksheet No. 2 Continued)*

## II. SOCIETY CONTROLS FREEDOM OF SPEECH IN SEVERAL WAYS.

A. "False advertising . . .
B. "Speaking a prayer, reading from the Bible, or giving instruction in religious matters . . . in public schools . . .
C. "Libel, slander, defamation of character. . . .
D. "Saying words which amount to a conspiracy or an obstruction of justice. . . .
E. "Sedition. . . .
F. "Words that tend to create a 'clear and present danger,' such as yelling 'fire' in a crowded theatre.
G. "Using words that constitute offering a bribe . . .
H. "Words that threaten social harm because they advocate illegal acts.
I. "Words (from a loudspeaker) at 3:00 a.m. in a residential neighborhood. disturbing the peace.
J. "A public address in the middle of Main Street at high noon, which as a consequence interferes with the orderly movement of traffic. . . .
K. "Being in contempt of court. . . .
L. "Committing perjury under oath. . . .
M. "Television cigarette advertisements.
N. "Saying words or giving information which have been classified (e.g., secret) by the government.
O. "Obscenity. . . .
P. "Copyright violations.
Q. "Pretrial publicity which might interfere with a defendant's opportunity to secure a fair trial by his peers.
R. "U.S. Government employees engaging in political speech or activity (prohibited by the Hatch Act, 1939, 1940)."

## III. TELEVISION ALREADY IS PARTIALLY CONTROLLED.

**A. By Government**

1. Federal Communications Commission
2. Federal Trade Commission
3. Congress
4. The Fairness Doctrine (FCC)
5. Equal Time requirements (FCC)
6. Cigarette advertising (Congress)
7. Truth in advertising (FTC)

•

**B. By the Television Industry**

1. By program production and selection.
2. By programming schedule.
3. By editing and news selection.
4. By NAB Code.
5. By internal censors.

**C. By Advertisers and big business**

1. By what they will sponsor.
2. By what controversy they will allow and disallow.
3. By direct proscriptions on content.
4. By the amount of advertising and its affect on program content (more commercials, less program).
5. By audience size (largest number of viewers possible).

(The quoted excerpts above are taken from the book, **Where Do You Draw the Line,** by Victor B. Cline, Brigham Young University Press, Provo, Utah, pp. 7-9 & 355.)

*(Worksheet No. 2 Continued)*

**D. By the people**

1. By what they view (to a limited degree).
2. Through pressure by citizens groups.
3. Through occasional legal action on station license renewal.
4. Through letters to networks, advertisers, FCC, FTC.

## IV. COPING/ACTION OPTIONS

**A. Individuals**

1. Decide to watch specific programs instead of watching TV.
2. Determine how much television is using you, how much you are using television.
3. Evaluate **why** you view television (see homework).
4. Become active in a consumer community group or help to organize a group.
5. Write letters of concern to stations, networks, advertisers, FCC, FTC, Congresspersons.

**B. Family Action**

1. Throw the TV out.
2. View with children and help counteract anti-social messages.
3. Discuss programs: violence, problem solving, commercials, issues, values.
4. Join protest and consumer groups.
5. Write letters of protest to stations, networks, advertisers, FCC, FTC, Congress.
6. Don't use TV as a baby sitter or pacifier.

**C. Community Action**

1. Organize consumer groups.
2. Confront stations on antisocial message issues.
3. Challenge license renewals of stations.
4. Boycott advertisers products.
5. Introduce Television Viewing Literacy courses in all levels of the school system.
6. Create recreational and growth programs for children to experience real relationships and to develop new skills.
7. Use television programs as curriculum in church schools for children and adults.
8. Use television in the church as a moral/ethical/theological laboratory for preaching and teaching.
9. Use television production in church and public schools to demythologize the medium.

*(Worksheet No. 2 Continued)*

**D. National and governmental action**

1. The issues of censorship
   a. Television already is being censored. The question is, who should censor and to what degree?
   b. Should the majority determine what the minority can see—or the minority determine what the majority will see?
   c. Who will decide what is appropriate violence and sex on television?
2. React to the following statements:
   a. Violence on television is creating a "clear and present danger" similar to the yelling of "fire" in a crowded theatre. The teaching of violence as an appropriate way of life to our young people can only end in anarchy and disaster.
   b. It is ludicrous to use our most powerful and pervasive medium to primarily sell goods and to unintentionally teach anti-social and destructive behavior. The basis for program decision-making must move to a higher morality than simply what will secure the largest audience for advertisers.
3. Alternatives
   a. Create laws defining excessive violence and sex. Allow the jury system to determine what is objectionable.
   b. License stations on a part-time basis so that the community can create programs which are more directly prosocial and useful.
   c. Create a National Television Program Board and Community Television Program Boards to participate in program decision-making.
   d. Develop others.

Violence/Worksheet No. 3

## HOMEWORK

**Select three programs you normally watch in the action/adventure, children or dramatic formats and record the violent or aggressive acts using the following definition:**

***"The overt expression of physical force against others or self, or the compelling of action against one's will on pain of being hurt or killed, or actually hurting or killing." (Gerbner)***

**1. Program ______________________ Day __________ Time __________**

**Violent or aggressive acts ______________________**

**Number of shootings, killings, deaths of violence ______________________**

**Victims: Male ________ Female ________ Minority ________ Over 50 ________**

**Perpetrators: Male ________ Female ________ Minority ________ Over 50 ________**

**2. Program ______________________ Day __________ Time __________**

**Violent or aggressive acts ______________________**

**Number of shootings, killings, deaths of violence ______________________**

**Victims: Male ________ Female ________ Minority ________ Over 50 ________**

**Perpetrators: Male ________ Female ________ Minority ________ Over 50 ________**

**3. Program ______________________ Day __________ Time __________**

**Violent or aggressive acts ______________________**

**Number of shootings, killings, deaths of violence ______________________**

**Victims: Male ________ Female ________ Minority ________ Over 50 ________**

**Perpetrators: Male ________ Female ________ Minority ________ Over 50 ________**

**Observations/personal notes/possible actions:**

## Violence/Worksheet No. 4

### HOMEWORK

Psychological aggression is subtle but also is damaging to relationships and to the development of a secure and happy person. It is those situations in which one person discounts another by the put-down, sarcasm, discrediting and belittling. The discount says to the other, "You are not important to me, you do not exist for me." Select two situational comedies. Tabulate the number of discounts or put-downs. Become aware of the way in which humor often is at the expense of the personhood of another. Listen especially for cues from the laugh track or live audience. This is an interesting exercise, too, for the game shows and the soaps.

1. Program ______________________ Day __________ Time __________

   Number of discounts ______________________

   Notes:

   New discoveries:

2. Program ______________________ Day __________ Time __________

   Number of discounts ______________________

   Notes:

   New discoveries or surprises:

   Possible actions or changes for me:

## Violence/Worksheet No. 5

### HOMEWORK

1. The following exercise may help to increase your awareness of the role of television in your life.

   a. Do you automatically turn your TV on when you get up or enter your home?
   Almost always ________ Often ________ Rarely ________

   b. When you are home, is your TV on . . .
   Almost always ________ Half time ________ ¼ of time ________

   c. Do you have favorite television programs which you will not miss except in the most critical circumstances?
   Often ________ Once in a while ________ Rarely if ever ________

   d. Do you find yourself preferring to watch TV rather than go out or be with people?
   Often ________ Once in a while ________ Rarely if ever ________

   e. Do you find your conversation with people centers around TV programs, TV personalities and the problems they are facing?
   Usually ________ Often ________ Rarely ________

   If your answers are weighted to the left, you may want to consider whether you want television to be that much a part of your life.

It is not intellectually smart
to ignore the most significant force in our society.
*Nicholas Johnson*

Television (and other media) presents men and women
in stereotypic roles that tend to limit
their capabilities for knowing themselves—
and for knowing others.
Stereotypes limit self-knowledge
and limit a knowledge of the world and people in the world.
Stereotyping attributes to an individual
the general characteristics of a group or sex.
*Nelson Price-Journal of Current Social Issues*

Heavy TV viewers under the age of 30 are more fearful of life's dangers
than their elders.
*George Gerbner*

From the vantage of any culture, entertainment of almost any
other culture can be seen as propaganda.
Erik Barnouw—
*The Sponsor*

In a faraway land, a very sensitive young woman was speaking to me about the opulence, the grandeur, and the generosity of the United States. Abruptly, however, she added, "but of course I shall never return there." Her reason was that she was far too alarmed and terrified by the violence of our society. "You have ten million guns in your homes, you can't walk in your parks, you bomb children in churches, and as long as I can remember you've been at war all over the earth. Your TV programs are also violent. Football is the worst thing of all for it seems more like serious, non-nonsense violence than it does a game."
Chester M. Pierce, M.D.
*Medical Opinion & Review*

They became what they beheld.
*William Blake*

Chapter 3

# STEREOTYPES ON TELEVISION

*Joyce N. Sprafkin*

*JOYCE N. SPRAFKIN received her Ph.D in Clinical Psychology at SUNY at Stony Brook. She is now a Senior Research Scientist at Long Island Research Institute and a member of the Department of Psychiatry at SUNY at Stony Brook. She has published research reports on the prosocial effects of television, children's reactions to sex role portrayals on TV, and the impact of children's TV commercials.*

For most of us, television provides a timeless "eye on the world" through which we can see current and past happenings, people, places, and even physical events we could not otherwise experience. We are exposed to life styles very different from our own—that of the millionaire, the drug addict, the prostitute, or the religious missionary. Situation comedies, soap operas, family dramas, and crime adventure programs show us marital quarrels, medical consultations, and financial hassles, to which we can all relate. Our TV brings into our living room individuals who, over a time, become so familiar that we just know how they would handle certain situations. Overall, TV appears to be a somewhat dramatized, but accurate representation of the non-TV world. However, like a funhouse mirror, while television actually reflects characteristics which are real, it distorts them.

For most of us, the TV world becomes our world three or four hours each day, seven days a week, throughout most of our lives. Further, the television world appears to be so realistic that we passively accept the people and situations presented and rarely question them. What exactly do we accept that could have negative effects on us?

Any society, including the one portrayed on television, has informal social rules suggesting how males and females of various ages, racial/ethnic groups, and social classes should live. The rules indicate how we should act, where we should live, who we should socialize with, how we should support ourselves, and generally how much wealth, status, and happiness we should have. In the real world, there is flexibility in these rules, with a variety of lifestyles and behaviors available to individuals. However, the TV producer condenses these rules, projecting a simpler picture of society in an attempt to reach an extremely diverse audience. Producers feel the characters and stories must be instantly recognizable and believable to the Madison Avenue businessman, the Southern farmer, the Midwest housewife, the Berkeley college professor. This is most easily accomplished by exaggerating the more commonly accepted societal rules. The result is a rigid and distorted image and research tells us that the distortions influence how we feel about non-TV people and what we expect of them.

## Distorted Images of People

The distortions are most critical for those who have had limited life experiences. For children and adults who are entrenched in homogeneous communities all their lives, television provides practically the only source of information about people from different walks of life. Exposure to biased presentations can, then, influence such a viewer's behavior towards unfamiliar people. This can be direct, as in actual face-to-face encounters. It can be indirect, through children who adopt parents' views, or it can be expressed in political behavior. For viewers having more diversified life experiences, the biased TV portrayals can still have a negative influence by solidifying stereotypic views about people different from us. Most insidious of all are the effects of such distortions on our feelings about ourselves. At some level, we judge our own success, happiness, and status in comparison to TV characters who are similar to us.

These subtle influences have caused much public concern. Parents, educators, and special interest groups are becoming more conscious of how TV projects informal rules or "should's" for males and females of various ages, racial/ethnic groups, and social classes. While television is by no means the only influence on our attitudes and expectations of people, with Americans spending approximately one of every six waking hours in front of the TV, its potential effect is considerable. It becomes important to ask: How accurate is the barrage of information we are exposed to daily?

## Recognition and Respect

Researchers have explored the televised portrayals of men and women in the various racial/ethnic groups. It has been suggested that the way the TV viewer develops attitudes toward social groups involves a two-step process.[1] First, the group has to gain recognition through sheer representation on television. If we see a particular social group very frequently on television, we presume they are at least a relevant segment of society. A group never (or hardly ever) presented leads us to presume that the group must be insignificant.

If a group is represented, our impression is further influenced by the degree of respect the group is given. Respect for a group is shown by social class/occupational roles it is assigned as well as the personalities its members project. A group seen frequently on TV can still be relegated to an inferior status if its members occupy a low socioeconomic class or show undesirable characteristics. In short, recognition addresses how often a group is represented and respect, how it is portrayed.

It is within this framework of recognition and respect that I will summarize what TV researchers have learned about portrayals of men and women in the various racial/ethnic groups and the unintended effects these portrayals have on the viewer.

## Status of TV Males and Females

One would expect the TV world to reflect the real world's sex distribution—approximately 50% male/50% females. However, in the world of TV, males have occupied between 66% and 75% of all television roles for the past 25 years.[2] Clearly, females are grossly underrepresented. Networks apparently assume that audience size is maximized by showing mostly males. However, studies on both children and adults suggest that female viewers prefer to watch female characters.[3]

Let's look at how males and females are portrayed on television. Even though women make up about 40% of the national labor force, only about 20% of TV roles having a definite occupational activity are held by women. Most TV women are assigned marital, romantic, or family roles, and when females have jobs, they are rarely prestigious ones.[4]

Examining Saturday morning cartoons, one investigator found that females have a far more restricted range of occupational roles than do males; females are pretty teenagers in adolescence and housewives in adulthood; they are always defined in status by their relationship to males.[5] In contrast, two-thirds of TV males are employed, and they occupy a variety of highly prestigious occupational roles such as physicians, lawyers, and law enforcement officials.[6]

## Male and Female Roles

Aside from their formal occupational/social status, what are males and females like on TV? First, the nature of the roles given to females and males is quite different. Females are more often cast in light or comic roles. Almost 75% of television female characters are found in comedies or similar shows, while more than 50% of the male characters are in crime, western, and action-adventure shows.[7] Within these roles, females are portrayed as more attractive, fair, sociable, warm, happy, peaceful, and youthful than males. In a sense, they are portrayed as "nicer" people, being more likely to help, share, and cooperate with others, to sympathize and explain their feelings to others, to repair damage caused to others, and to resist the temptation to break societal rules.[8]

**"Males have occupied between 66% and 75% of all television roles for the past 25 years."**

Females, however, are less likely than males to accomplish tasks. They are generally passive, deferent, and very likely punished if they are highly active. They are impulsive. They have great difficulty delaying gratification and persisting at a task.[9] Further, females are very vulnerable on TV, far more likely than males to be the victims of violence,[10] to be patients in a hospital and less likely to recover. Seventy percent of medical interventions are successful with male patients, whereas only 23% are successful with female patients![11]

Objections to females' TV portrayals have recently focused on presentations of the female as a sex object. The number of programs featuring sexy women has been steadily increasing (*e.g., Charlie's Angels, Flying High, Three's Company, American Girls*), and influential women have taken a stand. Witness the statement by Margita White, an FCC Commissioner:

> *"I'm not sure whether I'm more outraged because the medium has missed the message of the antiviolence campaign, or more offended because women are to be battered through a new low in sexploitation. But I do know that the networks will be hearing a new chorus of 'I'm mad as hell and I won't take it any more.' "*[12]

Indeed, emotionalism aside, objective content analyses have shown that females are more likely than males to be involved in such behaviors as kissing, hugging, affectionate touching, and suggestive behaviors.[13]

Commercials have also come under attack for sexist portrayals. In 1975, the National Advertising Review Board issued a report based on an analysis of commercial portrayals of females. They concluded that women were inaccurately presented as sex objects and rarely as professionals.[14] Several of the airline commercials illustrate their finding. Recall National Airlines' "Fly me" campaign and Continental Airlines' "We really move our tail for you."

Television males are portrayed as more powerful, smart, rational, and stable than females. Overall, males are presented as very callous, that is, aggressive, nonaltruistic, and unable to resist the temptation to break societal rules, unable to repair damage caused to others or to express their feelings in resolving conflict. However, they have characteristics that help them succeed: they are very planful, constructive, and able to delay gratification and persist at tasks—all important for accomplishing goals. Further, their activity is generally rewarded.[15]

## Portrayal of Minorities

Over the last 10 years, whites have been given between 70% to 90% of all scripted parts on television, leaving between 10 to 30% of the roles for black and nonblack minority members.[16] Due to pressure from black groups, the current proportion of blacks on TV approximates their number in the U.S. population. But nonblack minority members are still underrepresented. (Also see chapter on *Minorities*)

Even for blacks the representation leaves much to be desired. Though we now see a number of black characters, they are most likely cast in minor roles. A recent examination of 133 evening network programs found only eight featured a black character as a regular. In five of these, the black shared the spotlight with several whites. Further, blacks were shown significantly more than whites in bit parts and minor roles.[17]

The formal social-occupational status of blacks is quite improved since the early days of television when they were cast as lovable buffoons (Amos and Andy), servants (Steppinfetchit), or entertainers. Now blacks are presented as "regulators" of society in positions such as law enforcement officers or teachers.[18] (Many take a cynical view of this new occupational emphasis in that television blacks are presented as supporters of the status quo rather than as social reformers.)

### Blacks Shown as Powerless

Despite recent changes in frequency of appearance and status of roles, the black character portrayals within roles convey and perpetuate stereotypes. Blacks are portrayed as good and likeable but neither forceful nor powerful. In an examination of biracial regularly broadcast programs, it was found that black men, unlike their white counterparts, are nonaggressive, altruistic, and likely to make amends for damage caused to others. Black females are seen quite often trying to resolve conflicts by expressing feelings.[19] Other studies reveal that television blacks are generally industrious, competent, and law-abiding.[20] Positive attributes are associated with blacks in children's programs as well.[21] However, blacks are still portrayed as less powerful than whites. Even in prestigious occupations, blacks invariably have a white supervisor.[22] In commercials, they are most often seen in public service announcements, typically do not speak or hold the product, and usually appear with whites.[23]

The rare appearance of nonblack minority members in itself indicates the low respect given this group. Furthermore, non-Americans are presented in a negative manner. In children's programs, good characters speak standard English and more than 50% of the bad characters have foreign accents.[24] In prime-time programming, non-Americans are more often cast as villains and as the victims of violence.[25]

**"Whites have been given between 70 to 90% of all scripted parts on television. . ."**

### The Effects of Stereotypic Portrayals

Clearly men and women from the diverse racial/ethnic groups are presented in very restricted and biased ways on television. The critical issue now is—how influential are these portrayals? To date, there is ample evidence that sex role, racial, and occupational stereotypes are internalized by viewers and accepted as models for their own behavior in proportion to the frequency of television viewing.

The acceptance of stereotypic sex role portrayals was explored by two investigators, Frueh and McGhee,[26] who reasoned that if children do accept TV's sex role messages, frequent TV viewers should have more stereotypic beliefs (such as girls should play with dolls, dishes, and dresses and boys should play with trucks, guns, and tools) than do infrequent TV viewers. They found that for boys and girls in kindergarten, second, fourth, and sixth grades, heavy watchers showed significantly greater identification with the traditional sex role associated with their own sex.

Entertainment television's potential influence on racial attitudes was shown most dramatically by Sheryl Graves[27] who studied how positive and negative portrayals of black characters in cartoons affected white and black children's attitudes towards blacks. She selected eight previously broadcast cartoons which varied in the portrayal of blacks (positive or negative). The positive portrayals showed blacks as competent, trustworthy, and hardworking. The negative presentations showed them as inept, destructive, lazy, and powerless. Graves found that there was a positive attitude change for the black children who saw either positive or negative portrayals. For the white children who saw the positive portrayals there was also a positive attitude change. White children exposed to the negative portrayals changed the most and in a negative direction. Overall, these results suggest that while the mere presence of black TV characters may have a positive impact on black children, the type of characterization of blacks is critical in terms of the potential negative impact on white children. These findings are particularly striking because the attitude changes were found after only a single program exposure and for children who live in racially integrated neighborhoods. In other words, one would expect to find even stronger effects for white children having little direct contact with blacks.

Research also indicates that adults are not immune to being influenced by TV's portrayal of life. Controlling for a variety of factors (age, socioeconomic status, education, etc.), Gerbner and Gross[28] found that heavy TV users (four or more hours per day) had a more sinister and distrustful view of people than did light TV users (two hours or fewer per day).

What about the humorous presentations of prejudiced characters such as Archie Bunker in *All in the Family?* Does the series reduce the viewer's stereotypes (as its producer Norman Lear intended) or does it perpetuate them? A recent study conducted in the U.S. and Canada suggests that a large percentage of people don't "get the joke" and see nothing wrong with Archie's use of racial and ethnic slurs. The researchers found that prejudiced persons identify with Archie, think he makes better sense than Mike (the liberal son-in-law), and see Archie as generally winning in the end.[29] It is, then, difficult to imagine how such "satires" can reduce viewer's prejudices. If anything, the showing of bigots such as Archie on prime-time television gives such individuals a degree of acknowledgment and acceptance.

Clearly not all viewers are equally influenced by TV's stereotypic portrayals. The influence would be expected to be greatest for those who have limited direct experience with or information about the social groups portrayed. This was the case in the Graves study in which negative portrayals of blacks negatively affected white children but not black children. Certainly black children have more information available about blacks than do white children. In fact, in another study it was found that white children who had the least amount of contact with blacks named television much more often as a major source of information about this group than those white children who lived near blacks.[30]

Overall, one can conclude that the restricted portrayal of people and life on television has an impact on both the child and adult viewer—he or she comes to see a narrow range of behaviors, feelings, life styles—for both himself and others. In a sense, the value system in the TV world puts a premium on stereotyped functioning. The ideal situation would be for TV to portray the widest possible range of behaviors, to emancipate all people from conforming to one set of standards and values. If TV were even just an accurate reflection of the real world, we should see men crying, women being competent, and minority members leading fulfilling lives. This is a time of rapid social change—the women's liberation and the ethnic awareness movements have taken major steps in opening life alternatives to people. Television can either help or hurt these developments; perpetuating stereotypes can discourage change for the individuals themselves and cause other society members to be unreceptive to any attempts at change.

## Positive Changes

At the same time it should be said that TV has potential to influence positive attitudes toward people. All of the evidence comes from research on educational programs which were designed to facilitate acceptance of others. *Sesame Street* with its positive portrayal of blacks and Hispanics was found to improve children's racial attitudes toward these groups.[31] Exposure to the *Big Blue Marble,* which was designed to encourage international awareness, seems to have caused American youngsters in the fourth through sixth grades to perceive people around the world as being more similar to one another and children in other countries as happier, healthier, and better off than they had before viewing the program.[32] *Vegetable Soup,* which presents a positive picture of many ethnic groups, was found to make child viewers between 6 and 10 years of age more accepting of children of different races and to more strongly identify with their own racial/ethnic group.[33] Finally, *Freestyle* is a very recent series that was carefully researched and designed to combat sex-role and ethnic stereotyping in the career-related attitudes of 9- through 12-year-old children.

**"Showing of bigots such as Archie on prime-time television gives such individuals a degree of acknowledgment and acceptance."**

One possible solution to the problem of sex role and racial stereotypes on television is to include more women and minorities in script writing and broadcasting station decision-making positions. Currently, the overwhelming majority of these positions are held by white males.[34] It is reasonable to question whether any one group can present the diversity of human experience. In 1977, the U.S. Commission on Civil Rights completed a study of employment practices of broadcasters and TV portrayals of women and minorities, and in its conclusions it recommended that production companies and network programming executives "actively solicit" scripts from minority and female writers and train new minority and female writers, while the FCC should defer renewing broadcast licenses to stations that do not include an equitable proportion of women and minority group members in its employment.[35]

Also, we viewers do not have to passively accept television the way it is. We can work for changes in how different kinds of people are presented. (You will find suggestions on how to do this elsewhere in this book.) We can be more careful in our program selections, particularly when we are viewing as a family. Television is a much more convincing teacher when parents seem to accept or affirm what is happening on the screen.

Perhaps most important of all we can think about what kind of view of the world and its people we would like our children to have. If people are presented in ways that seem biased and stereotypic to us, we can talk to our childen about that and help them better understand that the television world often presents a very distorted view of people.

***REFERENCES***

1. Clark, C. C., Television and social control: Some observations on the portrayal of ethnic minorities. *Television Quarterly,* 1969, *8,* 18-22.
2. Gerbner, G., Violence in television drama: Trends and symbolic functions. In G. A. Comstock & E. A. Rubinstein (Eds.), *Television and social behavior. Vol. 1. Media content and control.* Washington, D.C.: Government Printing Office, 1972.

   Head, S., Content analysis of television drama programs. *Film Quarterly,* 1954, *9,* 175-194.

   Tedesco, N. S., Patterns in prime-time. *Journal of Communication,* 1974, *24,* 118-124.
3. Sprafkin, J. N., and Liebert, R. M., Sex-typing and children's television

preferences. In G. Tuchman, A. K. Daniels, and J. Benet (Eds.), *Home and hearth: Images of women in the mass media.* New York: Oxford University Press, 1978.
Maccoby, E. E., & Wilson, W. C., Identification and observational learning from films. *Journal of Abnormal and Social Psychology,* 1957, *55,* 76-87.
4. DeFleur, M., Occupational roles as portrayed on television. *Public Opinion Quarterly,* 1964, *28,* 54-74.
Gerbner, 1972
Tedesco, 1974
Harvey, S. E., Sprafkin, J. N., and Rubinstein, E. A., "Prime-time television: A profile of aggressive and prosocial behaviors." *Journal of Broadcasting,* in press.
Long, M., & Simon, R., The roles and statuses of women on children's and family TV programs. *Journalism Quarterly,* 1974, *51,* 107-110.
5. Levinson, R. M., From Olive Oyl to Sweet Polly Purebread: Sex role stereotypes and televised cartoons. *Journal of Popular Culture,* 1973, *9,* 561-572.
6. Gerbner, 1972
Tedesco, 1974
Harvey, et al., in press
7. Tedesco, 1974
Long and Simon, 1974
8. Tedesco, 1974
Donagher, P., Poulos, R., Liebert, R., & Davidson, E., Sex, race and social example: An analysis of character portrayals on interracial television entertainment. *Psychological Reports,* 1976, *38,* 3-14.
9. Sternglanz, S., & Serbin, L., Sex-role stereotyping in children's television programs. *Developmental Psychology,* 1974, *10,* 710-713.
Donagher, et al., 1976
10. Gerbner, 1972
11. McLaughlin, J., The doctor shows. *Journal of Communication,* 1975, *25,* 182-184.
12. White, M. E., "Mom, Why's the TV Set Sweating?" *New York Times,* March 29, 1978.
13. Silverman, L. T., Sprafkin, J. N., and Rubinstein, E. A., "Physical contact and sexual behavior on prime-time TV." *Journal of Communications,* 1979, in press.
14. *Window dressing on the set: Women and minorities in television.* A report of the U.S. Commission on Civil Rights. August. 1977.
15. Donagher, et al., 1976
Sternglanz and Serbin, 1974
16. Smythe, D. W., Reality as presented on Television. *Public Opinion Quarterly,* 1954, *18,* 143-156.
Gerbner, 1972
Harvey, et al., 1979
17. Hinton, J., Seggar, J., Northcott, H., & Fontes, B. Tokenism and improving the imagery of blacks in TV drama and comedy. *Journal of Broadcasting,* 1973, *18,* 423-432.
18. Clark, 1969
Roberts, C. The portrayal of blacks on network television. *Journal of Broadcasting,* 1970-71, *15,* 45-53.
19. Donagher et al., 1976
20. Hinton, Seggar, Northcott, and Fontes, 1973
21. Mendelson, G., & Young, M. *Network children's programming: A content analysis of black and minority treatment on children's television.* Boston: Action for Children's Television, 1972.
22. Hinton, et al., 1973
23. Dominick, J., & Greenberg, B. Three seasons of black on television. *Journal of Advertising Research,* 1970, *10,* 21-27.
24. Mendelson and Young, 1972
25. Gerbner, 1972
26. Freuh, T., & McGhee, P. Traditional sex-role development and amount of time spent watching television. *Developmental Psychology,* 1974, *11,* 109.
27. Graves, S. B., *How to encourage positive racial attitudes.* Paper presented at the Society for Research in Child Development, 1975, Denver, Colorado.
28. Gerbner, G., & Gross, L. Living with television: The violence profile. *Journal of Communication,* 1976, *26,* 173-199.
29. Vidman, N. & Rokeach, M. Archie Bunker's bigotry: A study in selective perception and exposure. *Journal of Communications,* 1974, *24*(1), 36-47.
30. Greenberg, B. S., Children's reactions to TV blacks. *Journalism Quarterly,* 1972, *50,* 5-14.
31. Bogatz, G. A., & Ball, S. *The second year of "Sesame Street": A continuing evaluation.* Princeton, N.J.: Educational Testing Service, 1971.
32. Roberts, D. F., Herold, C., Hornby, M., King, S., Sterne, D., Whiteley, S., & Silverman, T., *"Earth's a Big Blue Marble": A report of the impact of a children's television series on children's opinions.* Unpublished manuscript, Stanford University, 1974.
33. Mays, L., Henderson, E. H., Seidman, S. K., & Steiner, V. S., *An evaluation report on "Vegetable Soup": The effects of a multi-ethnic children's television series on intergroup attitudes of children.* Unpublished manuscript, New York State Department of Education, 1975.
34. *Window dressing on the set,* 1977.
35. Ibid.

## Stereotypes/Worksheet No. 1

The television excerpts you are about to see represent program and commercial selections from one week of programming. They were selected to show how different kinds of people are portrayed—especially the types of life roles that seem to be suggested as appropriate. Most of the excerpts are very short. They are in this order:

1. Images of Women Negative

   Man the smartest animal
   Man tells women about Mr. Coffee
   Cascade—Woman delighted to have glasses free of spots
   Carol Burnett—"Last chance to catch a man"
   Joe Forester—Woman accepts her husband beating her because he loves her
   Harry-O—Sex object in bikini
   Medical Center—Female doctor was bad mother

2. Images of Women Positive

   Petrocelli—Manager a **she,** and a **he**
   Happy Days—Girl flips boy
   Waltons—Mom and Dad have non-sexist attitudes (Note: so hard to find counter examples we had to go to another week to find this example)

3. Images of Old People (Note: Old People are hardly ever shown on TV)
   Waltons—Stubborn old woman
   Joe Forester—Guy has been beating up old people (Note: Old people are likely victims on TV)

4. Minorities

   On the Rocks—Slur on Puerto Ricans
   $6 Million Man—Contemptuous statement about Japanese
   Barbary Coast—Black asks, "How's that, Boss?"
   Police Story—Black criminal element
   Joe Forester—Good black to bad black, "I'm not your brother"
   Kojak—Hispanic woman a drunk and a loser
   Hawaii Five-O—Oriental hit man
   $6 Million Man—Japanese plot to regain military power
   Ellery Queen—Russian defector

5. Positive Example of Blacks

   Emergency—Competent black doctor
   Movin' On—Successful black woman

6. Stereotyping Males

   Sony Ad—Man had to compromise pleasure for women
   Jeffersons—Man incompetent in kitchen
   Ellery Queen—Man incompetent in kitchen
   All in the Family—Archie demanding dinner, rejecting affection

## Stereotypes/Worksheet No. 2

Every TV commercial has many different messages, some very direct, some very subtle. Recall the commercial in which a woman says, "Gentlemen prefer Hanes." Write for a couple of minutes, listing as many as possible of the different messages you saw in that commercial:

# Stereotypes/Worksheet No. 3

Now assume that **you are a Martian** who has had no contact with Americans other than your TV experience. Using that TV experience write a summary of what individual Americans are like:

## Stereotypes/Worksheet No. 4

Write down four or five guidelines you think will help you be more aware of stereotyped portrayals of persons as you watch television.

## Stereotypes/Worksheet No. 5

## HOMEWORK

1. Review the guidelines for watching TV that you developed in this session and earlier sessions. Then watch television for two hours continuously with one or more other persons. A family group would be ideal. Play a game of being Martians. What you are seeing on the screen is the only information you have about Americans. Spend the first hour watching two half-hour shows or two half-hour portions of longer shows. Spend the second hour switching the channel selector every five minutes to sample programming at random.

2. Discuss the experience with the others. Questions that you, as a Martian, might be asking the TV screen could include:

—What are women like?
—What are men like?
—What are children like?
—What do people do most of the time?
—What is the goal of most Americans?
—What do Americans value highly?
—What do they believe in?
—What are Blacks like? Hispanics? Asian Americans? Native Americans?
—What do people do for a living?
—What do they do for pleasure?

3. Now switch your identity from Martian observers to the Americans being observed. Discuss the experience. How does it feel to be seen and described the way the Martians did that?

4. Talk with each other about ways of using television differently.

Chapter 4

# TELEVISION ADVERTISING AND VALUES

*Diane E. Liebert*

*DIANE E. LIEBERT earned her Ph.D. degree in educational psychology at Hofstra University. A professional teacher and researcher, she is currently finishing a book on child development.*

Television advertising touches all of us and our children. In exchange for entertainment of a wide variety, news, and children's programs, we allow ourselves to be exposed to commerical messages, all trying to persuade us with a variety of sophisticated methods to buy certain products. Some argue that TV is not really free since the dollars earned by the networks, stations, and advertisers once belonged to us, the consumers; we pay for the commercials and programs when we buy the products.

Though we are aware of commercials and often complain about their frequency, it is surprising to learn that an average TV-watching child sees more than 22,000 commercials a year.[1] No wonder parents hear, "Mom, can I have. . ." so often.

Public and government concern about the effects of television advertising on children has recently led to government inquiry. Both the Federal Trade Commission (FTC) and the Federal Communication Commission (FCC) are investigating advertising practices directed to children which could result in a reduction or ban on such advertising.

Such actions are considered by broadcasters and advertisers as an attack on the heart of network television. The commercial is indeed the basis of life for the industry. In order to understand the important issues involved in this battle between public and commercial interests, and to be aware of the effects of advertising on all of us, it is necessary to examine the purpose of television and the role of the advertisers as well as public and government concerns about the effect of television commercials.

## The Purpose of Television

Some people assume that television's purpose is to entertain. It isn't. The purpose of commercial television is to make money for the networks—and it has achieved remarkable success. What the networks and stations sell, of course, is advertising time. And what they promise the advertiser is an audience.

**"Moreover, young children trust commercials and believe they always tell the truth."**

Television advertising grew from a 300 million dollar business in 1952 to $1.8 billion in 1964, to six billion dollars in 1976 and a projected 10 million in 1981. In one three-month period in 1970, almost 100 advertisers spent more than one million dollars each on TV spot commercials.[2]

Networks and advertisers also found that children's programming can be a gold mine. Heavy advertising campaigns directed at children began in 1955, the year that Mattel tapped into the buying influence of children when it committed a half-million dollar budget to the sale of Barbie dolls. Mattel's advertising propelled it from a $500,000 business to one of 12 million dollars.[3] The lesson was not lost on other advertisers. In 1976, Dr. Allan Pearce of the FCC staff estimated that a half billion dollars a year are spent on advertising in children's television. That advertising has paid off well for the toy and cereal companies that dominate the children's market. The toy industry is greatly indebted to television for its $3.4 billion dollar yearly sales.[4]

Since advertisers are eager to reach the largest possible number of people with their commercials, they are willing to pay more money to have their commericals shown on popular programs. The networks and advertisers learn the number of people watching given programs from the rating services. The Neilson ratings, for instance, are based on what 1,200 representative households are watching. The dollar value of commercial minutes on programs is then based on the presumed size of the audience. Both the networks and the advertisers are interested in the programs with the best ratings, so these ratings often determine whether our favorite shows will stay on the air.

## Commercials and Program Quality

Since the price paid by the advertiser is determined by the size of the audience and not the cost or quality of the program, it is in the network's commercial interest to sacrifice quality for revenue as long as it can attract and hold the audience it is selling to the advertiser. The *type* of programs offered is also determined by the profit motives. The game shows are particularly inexpensive to produce and allow a type of double advertising—brand name prize offerings, plus regular commercials.

The commercial structure of television also results in a lack of innovation and experimentation, and the avoidance of controversial topics. Programs which do not sell well are given limited time. To attract the largest possible audience for the advertising dollar, networks do what is safest and imitate successful programs already on the air. This results in seasons oversaturated with police, doctors, or detectives, and the feeling that you've seen the programs many times before.

The quality of children's television is harmed by excessive use of cartoons which can be re-run year after year, reducing their per play cost. Children's programming is particularly profitable for the networks. Advertisers are willing to invest a lot of money for this audience and the networks can provide cheap programs. Bugs Bunny and Road Runner alone grossed more than two and one-half million dollars in one 12-month period. It is easier to understand network and station reluctance to change programming or to provide better quality when one realizes such huge profits are involved. However, advertisers and decision-makers are not insensitive to public criticism and have begun to make changes. The ABC *Children's Afternoon Specials* are quality and thoughtful programs although they have not appeared on a regular weekly schedule.

On the positive side, advertisers can use their buying power to veto programs they feel are not in the public interest. Some advertisers withdrew all money from programs with violence saying they had received thousands of letters supporting this position. A vice-president of the advertising firm of J. Walter Thompson said that violent programs on television could have a negative effect on the sales of products advertised on such programs. He cited evidence from a study in which 8% of adults said they had refused to buy a product because it was advertised in violent programming; two out of five people said they avoided overly violent programs; a third of the women said they prevented their children from watching such shows. The agency executive also indicated his fear of consumer action which could lead to letters of protest and product boycotts.

Direct public pressure on the advertiser offers one potential for changing undesirable programming.

### The Question of Health

The greatest demonstration that commercial television can be changed was the removal of all cigarette commercials from TV in the interest of public health. The battle started in 1967 when the FCC ruled that the Fairness Doctrine (equal time for both sides of a controversial topic) applied to cigarette advertising. Therefore, anti-smoking spots had to be given free time. Then in 1969, Congress opened special hearings to consider a total ban of cigarette ads. Cigarette companies and the networks—who took in 200 million dollars each year in revenue—fought long and hard. Despite such pressures, the Public Health Cigarette Smoking Act was passed by Congress and signed into law on April 1, 1970. Cigarette advertising stopped the following January.

Today, a major concern regarding television commercials and good health is the fact that children are being sold products that are not good for them. A very high percentage of children's commercials are for food, ranging from 50% in the pre-Christmas season to somewhere between 67 and 85% at other times of the year. Most of the foods advertised are high in sugar and low in nutrition, with the main items being sugared drinks. The overwhelming amount of advertisements for sweet foods are charged with contributing to the national problems of obesity, improper nutrition, and tooth decay.

**A disclaimer for fruit punch boasts, "10% real fruit juice!"**

The absence of ads for fruit, vegetables, and other valuable nutritional foods intensifies this problem.

Television commercials have also been attacked for advertising medicine. Action for Children's Television (ACT) is particularly concerned when vitamins are advertised as being just like candy or as something that will make you big and strong. Such arguments can be dangerous. Four-year-old Erin Shelton took 40 Pal vitamins after seeing such an ad because he thought he would "grow big and strong real fast." Erin was lucky and recovered—after having his stomach pumped and spending two days in intensive care. HEW's Clearinghouse for Poison Control Centers documented 5,146 cases in 1973 of children under 5 who took overdoses of vitamins.[5]

An ACT petition to the Federal Trade Commission resulted in a major victory in 1972. Three major drug companies withdrew televised vitamin advertisements directed to children. But since the companies acted voluntarily the FTC did not prohibit vitamin ads. Then ACT had to fight the same battle against a different company, the Hudson Pharmaceutical Corporation, to stop a one-million dollar campaign for Spider-Man Vitamins.

Surely there is something wrong when our children tell us what kind of pills they want.

### Unintended Effects

Commercials sell more than products. One commerical, in stressing how natural and good the cereal was, also communicated that wild plants are good to eat. After complaints were made to the FTC, researchers were asked to investigate. They found that children were indeed more likely to eat harmful plants after viewing the commercial.[6] The cereal company then voluntarily removed the ad.

There are potential dangers when children watch commercials that are primarily aimed at adults as on family programs. Drugs, detergents, razor blades, cigarette lighters, lawn mowers and many other products are frequently advertised in front of child audiences. The products are made to seem exciting and important with little suggestion that they can be extremely dangerous in the hands of the child. Behavioral scientists say that even if warnings were added, they may only stimulate children's curiosity. It would appear that parents must watch carefully to counter these dangerous possibilities.

## Stereotypes in Commercials

Another effect is the stereotype roles that present prejudicial or distorted views of the world. Women particularly are presented in limited occupational roles. In 1972, a study of over 1,200 commercials by the National Organization for Women (NOW) found that 42.6% of the women in commercials were portrayed in household tasks, 37.5% were presented as domestic subordinates to men, and 16.7% were shown as sex objects, leaving only 0.3% in autonomous roles.[7] Another study by the National Advertising Review Board in 1975 unanimously concluded that the portrayal of women was not only unfair, inaccurate, and out of date, but that it was also bad marketing. The panel made these comments in regard to the portrayal of women as sex objects: "Compared to a vibrant living person with a variety of interests, talents, and normal human characteristics, the woman portrayed as a sex object is like a mannequin, with only the outer shell of a body, however beautiful. Many women have stated their resentment at the use of the female body as a mere decoration or as an attention-getting device in advertising."[8]

Today more commercials seem to be portraying women in business and other jobs but there are still many commercials that are offensive to women, particularly those that show women being saved from "difficult" problems by the Man from Glad, Mr. Clean, or the well-dressed man who appears like magic at the laundromat.

Blacks also are presented in limited roles, and they appear in only 10% of commercials.[9] Children's commercials are also dominated by white males.[10] One investigator found that ads on Saturday morning had three times more males than females.[11] Ads showing only females were mostly for dolls. The voice-over "expert" on children's products was male 90% of the time. When a non-stereotype ad appears, such as the Wannabee's commercial showing a girl saying, "I want to be a doctor," its uniqueness is striking.

## Creation of Unnecessary Markets

Television commercials often create unnecessary markets or direct the consumer to more expensive but not necessarily better products. For example, some highly advertised feminine hygiene products are said not only to be unnecessary but possibly harmful.

Children's toys have been completely changed by TV and the holidays, too. Toys that look exciting on TV are what sell. Toy companies now make toys with the commercial in mind. Since rubber ducks look dead on TV, it is almost impossible to find one in a toy store anymore. A child no longer wants only a doll for Christmas, but wants a doll that walks, talks, and dances. What the child doesn't realize is that the more complex the toy, the more likely it will break, or not perform as seen in the TV commercial.

Toy companies increase their advertising by 600% in the pre-Christmas season and they have been very effective. Children today give their parents long lists of *specific* toys that they want—and parents are reluctant to disappoint the children. Advertising executive, Mel Helitzer, says: "Children can be very successful naggers. By and large parents readily purchase products urged upon them by their youngsters . . . it was found that a parent will pay 20% more for an advertised product with child appeal—even when a less expensive, non-advertised product is no different."[12] Unfortunately, paying more for advertised toys is not the only problem. Many children are quickly disappointed when expensive toys break easily, are difficult to repair, and won't do all the things advertised. Perhaps parents should make lists too—of the toys they buy each holiday, whether it was advertised on TV or not, its price, how much the children play with the toy, and how long the toy lasts. Parents should also assume more responsibility for the selection of their child's toys if they want to use their money in the best interests of their child.

## The Question of Values

Other deep concerns about the effects of commercials are that advertising encourages materialism and the need to consume, that *things* are made more important than *people,* and that advertising does not teach rational consumer decision-making but depicts instead vanity and flashiness as a reason for choosing one product over another. Children are pressuring parents for the many things they see constantly on television and some parents feel inadequate if they don't provide them. It is the mass nature of advertising that gives it such an impact. A single commercial for candy will not corrupt a child's perspective, but 30 advertisements for sweets per day has to have some effect. Obviously, the advertisers would not spend so much money for the commericals if they did not have evidence that this will sell their products.

Research reports do indicate a definite correlation between exposure to TV advertising and materialism, with lower status children being influenced the most strongly. Younger children were more materialistic than older children and most often expressed the attitude that physical possessions and money are important to personal happiness. In spite of growing cynicism among the older children, they still want what they see advertised.

The advertisers say commercials only reflect society's values, not make them. But it is difficult to separate "society's values" from the media which help to shape them. However, parents have power and they can increase or decrease the influence of television advertising. Research indicates that the more the mother approves television viewing by watching television herself, the greater the number of purchase requests and the more likely the requests will be met. Comments by parents about commercials could help to reduce the desire for certain kinds of products. If parents said, "That sweet cereal is bad for your teeth" or "I wouldn't buy you any toys that have to be put together since they fall apart so easily," children's expectations and requests for getting such things might decrease.

### Deceptive Messages

Advertising practices that cause a lot of concern are those that are unfair, misleading, or deceptive. Such practices are specifically forbidden by the government and action can be taken by the FTC against violators. The FTC, for example, restrained a major automobile manufacturer from using an ad showing the clarity of its window glass because the scenes were actually shot with the windows rolled down. Others have been prevented from making extravagant claims which they cannot meet; Geritol was stopped from advertising its product as a cure for "tired blood." Listerine no longer claims it can prevent colds.

Advertisers have protected themselves from regulation by including words which qualify or clarify the message. (e.g., "of doctors surveyed, four out of five . . .; As part of a complete dietary program . . . ; Your own mileage will vary according to driving habits and road conditions.") Such qualifying statements are called disclaimers. By correcting misimpressions which the commerical might otherwise suggest, they legally legitimatize the broadcasting of the message. However, sometimes this disclaimer message is presented in a way that is difficult to understand.

Children's toy commercials often have disclaimers saying "partial assembly required" or "sold separately" which sound like wording suggested by lawyers, not wording intended to communicate to children.

Research funded by the Better Business Bureau found that young children who heard the message "some assembly required" following a toy commercial had no more understanding that the toy had to be put together than those who heard no disclaimer. Yet, over 50% of the young children who heard a revised message, "It has to be put together," did understand the meaning of the disclaimer, showing that such messages are better understood by children if the wording is clear and appropriate for their age.[13]

**Surely there is something wrong when our children tell us what kind of pills they want.**

Understanding such messages is important. Imagine a child's disappointment on getting a much wanted toy to find that all the accessories shown with the toy on TV are sold separately. Imagine also the parents' anger and frustration at now being asked for all these expensive and unanticipated extras. Equally frustrating to parents and children are the hours spent on Christmas morning putting toys together (which were shown already assembled) only to find parts missing.

Children are not the only ones deceived by disclaimers. A disclaimer for fruit punch boasts, "10% real fruit juice!" as if this were a bonus. The real message is that 90% of the "fruit punch" is not fruit juice but just sweetened water.

Techniques used for selling products may also be unfair or deceptive. Special photography and editing make the product appear larger or more exciting. There are special camera angles, close-ups, increased speeds, lighting, laughter, and musical tracks. That such advertisements are misleading has been confirmed by actual research studies.[14]

Of concern in both adults' and children's commercials are: offering a free item with a purchase, use of hero figures and program characters to endorse or sell the product, presentation of ambiguous information and undefinable words, and omission of significant information on price and nutritional value. Criticism has resulted in some changes. Captain Kangaroo no longer delivers sales messages.

### Special Limitations of Children

Though concerns about television advertising apply to both children and adults, most people feel that children are more vulnerable and therefore require more protection.

The most important difference between young children and adults is that they do not understand the selling purpose of commercials. The famous developmental psychologist, Jean Piaget, has explained that children progress through a series of "developmental stages," each stage possessing different ways of thinking or solving the same problem. From approximately 2 to 7 years of age, children have special limitations in their way of thinking that prevent full understanding of certain kinds of information. Research studies show that the majority of young children lack the ability to see commercials for what they are. When asked, "What are commercials for?" a typical kindergarten child's response would be, "So you'll know about things, how to buy things. So if somebody washes your clothes, they'll know what to use." Second graders usually have a better idea concerning the selling motive, but it is not until fourth grade that most children have a clear recognition of the profit-seeking motives of the advertiser.[15] Young children are also not able to separate the commercial from the program being watched. ABC uses a clown insertion to help separate the commercials from its children's programs.

Moreover, young children trust commercials and believe they always tell the truth. As children grow older they become increasingly skeptical. About 40% of the 9-12 year-olds believed the products sold on TV were often "not like the ads say."[16] However, it appears this skepticism may be limited to things the children have had experience with, such as toys, cereal premiums, and children's products. Researchers find that 10- to 13-year-old children generally accept as true the messages for pain relievers, personal hygiene and other health-related products.[17]

Child advocates argue that it is unfair to advertise to young children, particularly those under 8, because of their inability to understand commercials. Others respond that advertising prepares the child for his or her role as a consumer. *(See Children: A Special Audience)*

## Public and Government Action

The 1970s have seen a dramatic increase in public and government concern about the effects of advertising on children. Government agencies have issued policy statements noting the vulnerability of children and have provided guidelines for their protection. They are engaged in investigations that may lead to a reduction or outright ban on children's advertising.

Much credit for this goes to ACT (Action for Children's Television) founded in 1968 by a group of concerned parents. In 1970, ACT petitioned the FCC to eliminate all commercial advertising for children and made efforts to inform the public and the government of the harmful effects of children's commercials. Nearly 100,000 people wrote to the Commission in support of the petition. The FCC decided that the broadcast industry should regulate itself and it turned back the ACT petition.

ACT claimed that self-regulation has not been a satisfactory solution and petitioned the FCC again in 1978. The FCC unanimously voted to reopen an inquiry into children's programming and advertising practices. Its current investigation will determine whether the broadcast licensees are in compliance with the Commission's 1974 guidelines and will evaluate the effectiveness of self-regulation. They will also consider whether to recommend a reduction in the amount of permissible commercial time on children's programs and whether to require that stations increase the amount of children's programming.

Meanwhile, the Federal Trade Commission, which has the responsibility for taking action against commercials that are deceptive or unfair, has suggested that television advertising of any product to children who are too young to understand its selling purpose may be unfair. It is therefore considering a proposed ban on all advertising directed to children under 8. Also under consideration is a ban on all advertising directed to 8- to 11-year-olds for sugared foods or a requirement that advertisers who promote sugared food balance their commercials with information about nutrition and proper dental health. Consumers have sent letters 16 to 1 in favor of such rules. Experts from both sides presented evidence at formal public hearings. Advertisers claim that First Amendment rights and the free enterprise system are on their side.

No doubt the debate over advertising issues will be long and hard. Lobbyists representing the networks, the advertisers and consumer groups will all try to influence the outcome of the decisions in Washington. The advertisers' "two million dollar war chest" is a powerful plus for their side.[18] They have already shown their strength in an earlier threat by Congress's Senate Appropriations Committee to cut off all FTC funding if the Commission did not abandon its children's inquiry. That effort did not succeed but advertisers were successful in having the FTC Chairman, Michael Pertschuk, disqualified from participating in the FTC rulemaking on the grounds that his outspoken remarks on children's advertising indicated he was "biased" in favor of the public interest!

**. . . children are being sold products that are not good for them.**

## Conclusion

Because television advertising clearly influences all of us and our children, it is important that we be aware of its purpose and its effects. Advertisers spend billions of dollars each year on television for one reason—because commercials cause people to buy things. Though we criticize commercials, we are often persuaded by them rather than our own judgment of what is best. An awareness of possible deception in commercials and of potential harmful effects can make us better consumers. An awareness of the special limitations of children can make us more responsible parents. That public opinion can result in government action is also an important lesson. The concerned parents who started ACT in 1968 have indeed provided a valuable example. Despite the rich networks, powerful advertisers, and big government, significant changes have been made and more are likely to follow.

***REFERENCES***

1. Choate, R. B., In *Broadcast advertising and children: Hearings of the House Interstate and Foreign Commerce Committee,* 1975. (Serial No. 94-53)
2. *Broadcasting,* April 19, 1971.
3. Jennings, R., *Programming and advertising practices in television directed to children.* Boston: Action for Children's Television, 1970.
4. Johnson, J. Toywars., *Re: Act,* Action for Children's Television News Magazine, Fall, 1978, Vol. *8,* pp. 12-13.
5. 1973 statistics, HEW's National Clearinghouse for Poison Control Centers. In *Broadcast advertising and children: Hearings of the House Interstate and Foreign Commerce Committee,* 1975. (Serial No. 94-53)
6. Poulos, R., Unintentional negative effects of food commericals on children: A case study. In *Broadcast advertising and children: Hearings of the House Interstate and Foreign Commerce Committee,* 1975. (Serial No. 95-53)
7. Hennessee, J. A., & Nicholson, J., N.O.W. says: TV commercials insult women. *New York Times Magazine,* May 28, 1972.
8. *Advertising and women, a report on advertising portraying or directed to women.* Prepared by a consultative panel of the National Advertising Review Board, March, 1975.
9. Roberts, C., The portrayal of blacks on network television. *Journal of Broadcasting,* 1970-71, *15,* 45-53.
10. Schuetz, S., & Sprafkin, J. N., Spot messages appearing within Saturday morning television programs: A content analysis. In *Broadcast Advertising and children: Hearings of the House Interstate and Foreign Commerce Committee,* 1975. (Serial No. 94-53)

As a psychological matter, children are cognitively incapable of understanding all television commercials directed to them, and no amount of consumer education can do much to improve this natural, age-based limitation.

As a dental matter, highly sugared products cause dental caries, and are much more cariogenic than other food products.

As a medical matter, highly sugared products have been shown to be related to other child health problems, such as obesity.

As a nutritional matter, a diet which includes highly sugared products in significant quantities may be lacking in essential nutrients.

As a factual matter, highly sugared foods and toys are the products most often advertised directly to children on television, and the average child spends an inordinate amount of time watching television programs and commercials targeted to him.

As a practical matter, it is unrealistic to expect parents always to mediate between their children and television commercials, especially when the commercials are directed to the children.

As a legal matter, it is unfair and deceptive to direct television commercials to children who are by nature unable to understand the inherent bias in advertising and the underlying selling intent of commercials, and who are unable to evaluate competing product claims and bring outside information to bear on product and brand choices; and this deception is patently harmful to children."

Peggy Charren
President, Action for Children's Television
Testimony before the Federal Trade Commission March, 1979

11. Barcus, F. E., Saturday children's television: A report of TV programming and advertising on Boston commerical television. *Educational Resources Information Center,* 1972, *7*(2).
12. Helitzer, M., *The Youth Market.* New York: Media Books, 1970, p. 32.
13. Liebert, D. E., Sprafkin, J. N., Liebert, R. M., and Rubinstein, E. A., Effects of television on commercial disclaimers on the product expectations of children. *Journal of Communication,* 1977, *27,* 118-124.
14. Barcus, F. E., Saturday children's television: A report of TV programming and advertising on Boston commercial television. *Educational Resources Information Center,* 1972, *7*(2).
15. Ward, S., & Wackman, D., Television advertising and intrafamily influence: Children's purchase influence attempts and parental yielding. In E. A. Rubinstein, G. A. Comstock, and J. P. Murray (eds.). *Television and social behavior. Vol. IV: Television in day-to-day life: Patterns of use.* Washington, D.C.: U.S. Government Printing Office, 1972, pp. 516-525.
16. Ward, S., Reale, G., & Levinson. D., Children's perceptions, explanations, and judgments of television advertising: A further exploration. In E. A. Rubinstein, G. A. Comstock, and J. P. Murray (eds.). *Television and social behavior. Vol. IV: Television in day-to-day life: Patterns of use.* Washington, D.C.: U.S. Government Printing Office. 1972, pp. 468-490.
17. Lewis, C. E., & Lewis, M. A., The impact of television commericals on health-related beliefs and behaviors of children. *Pediatrics,* 1974, *53,* 431-35.
18. Apropos the FTC. *Re: Act,* Action for Children's Television News Magazine, Fall, 1978, *8,* 3.

## Advertising/Worksheet No. 1

# LOG OF COMMERCIAL EXCERPTS

**1. Institutional Ads**—Company wants you to believe everything they do is in your interest too.

Union Carbide—"Something we do will touch your life."

**2. Use our product and you will be sexy.**

Rango
Coppertone
Close-Up
Ultra-Brite

**3. Use of sex objects to sell.**
Joey Heatherton
Nair

**4. Whenever you feel bad, take a pill.**
Arthritis pain formula
Anacin
No-Doz
Sominex
Contact
Geritol

**5. Selling Techniques**
A—Use of celebrities to sell
Roger Miller
Ed McMahon
Angie Dickinson

B—Celebrities emphasizing past TV roles
Karl Malden—wears hat to suggest role in Streets of San Francisco
Hugh Downs—playing old newscaster role

C—Deceptive wording
David Jansen for Excedrin
Anacin has 23% more pain reliever (plain aspirin) and adds an "extra ingredient" (unnamed)

*(Worksheet No. 1, Continued)*

**6. Children's Commercials**

A. Deceptive commercials for kids—even harder for them to see the deception, understand the qualifiers.

Big Jim — Spliced shots of vehicles make them look like they go longer than they do

Evil Knievel

Haunted Mansion—Spooky music and closeup shots make product look better than it is. Qualifier at end: "Partial Assembly"

Ghost Gun—"filmed with special light."

Big Jim—"separate purchase of Kung Fu gear includes everything you see here."

B. Premiums.

Sugar Crisp yo-yo
Buster Brown shoes boogle bag

C. Sex typing in ads

Barbie's beauty parlor
Big Jim's pack
Miss America game

D. The commercials taken from one-half hour of children's programming—all sweetened foods and toys.

Fig Newtons
Baby Christie
Nestle's Crunch
Chip Ahoy
Hershey's Chocolate
Sugar Crisp
Magic Shot Game
Punch Crunch

# Advertising/Worksheet No. 2

Analysis of selling techniques of TV commercials:

| | **Commercial #1**<br>**Product Name:** | **Commercial #2**<br>**Product Name:** |
|---|---|---|
| Technical techniques: | | |
| Color | | |
| Documentary | | |
| Mini-drama | | |
| Demonstration | | |
| Animation | | |
| Sound effects | | |
| Music | | |
| Kaleidoscopic | | |
| Editing | | |
| Style techniques: | | |
| Humor | | |
| Sex appeal | | |
| Romantic setting | | |
| Macho | | |
| Slice-of-life | | |
| Excitement | | |
| Fun | | |
| Irritation | | |
| Repetition | | |
| Propaganda techniques: | | |
| Big names (Buy it 'cause I said so) | | |
| Bad names (You're stupid if you don't) | | |
| Glad names (You can be a good guy/gal) | | |
| Transfer (You can be like this too) | | |
| Testimonial (It worked for me) | | |
| Plain folks (Just like us) | | |
| Stacking the cards (No other way) | | |
| Bandwagon (Everybody's doing it) | | |
| Other techniques: | | |

List major techniques in each category for each commercial.

## Advertising/Worksheet No. 3

| Values TV commercials communicate | Human Values you think are important |
|---|---|
| | |

## Advertising/Worksheet No. 4

Commercials appeal to the viewer on the basis of meeting real needs (food, clothing, housing, transportation, etc.) and upon meeting psychological needs (love, self-esteem, security, acceptance, etc.). In your small group session or with family and friends, propose a product to complete each sentence. Be specific in naming a product and a need which it promises to meet.

1. If I want to be healthy, I will buy ____________ so that I ____________________________ .
2. If I want to be loved, I will buy ____________ so that I ____________________________ .
3. If I want to be successful, I will buy ____________ so that I ____________________________ .
4. If I want to show my love, I will buy ____________ so that I ____________________________ .
5. If I want to be accepted, I will buy ____________ so that I ____________________________ .
6. If I want to feel safe, I will buy ____________ so that I ____________________________ .
7. If I want to be attractive, I will buy ____________ so that I ____________________________ .
8. If I want to be sexually appealing, I will buy ____________ so that I ____________________ .
9. If I want to be feminine, I will buy ____________ so that I ____________________________ .
10. If I want to be masculine, I will buy ____________ so that I ____________________________ .
11. If I want to love myself, I will buy ____________ so that I ____________________________ .
12. If I want to think highly of myself, I will buy ____________ so that I ____________________ .
13. If I want to take good care of myself, I will buy ____________ so that I ________________ .
14. If I want to be sensual, I will buy ____________ so that I ____________________________ .

## Advertising/Worksheet No. 5

## HOMEWORK

1. Watch at least two hours of television during the next week giving special attention to what the commercials are saying to you. Use the form provided (worksheet #6), keeping these questions in mind:

   —What is the commercial's view of me as a person?

   —What is it asking me to do/believe?

   —Do I need the service or product being advertised?

   —How will I feel if I buy/use the product or service?

   —Will it be life-supporting and health-giving?

   —If the commercial really intrigues me or tempts me to buy, how did it do that?

2. Analyze the list of products to learn what products/services dominate and what human needs are appealed to most.

3. Do a check on yourself to learn if you may have been obeying the "buy" messages more than you realized. Walk through your house, with special attention to the kitchen and pantry, and look at recent purchases. Why did you buy them? Did you really need them? Why did you buy a particular brand name? Were food products the most nutritious and health supporting? Is the same item available in a less expensive, less packaged form?

4. Talk to children about their reactions to commercials. Do they believe commercials always tell the truth? What is the purpose of commercials? Since children don't normally buy cereals, why are the cereal commercials in children's programs? When they buy toys advertised on TV, are they ever surprised when they open the box?

Advertising/Worksheet No. 6

## HOMEWORK: ANALYZING COMMERCIALS

| Questions | Commercial No. 1 (Product: ______________ ) | Commercial No. 2 (Product: ______________ ) |
|---|---|---|
| 1. What is the commercial's view of me as a person? | | |
| 2. What is it asking me to do/believe? | | |
| 3. Do I need the product services advertised? | | |
| 4. How will I feel if I buy/use it? | | |
| 5. Will it be life-supporting, life-giving? | | |
| 6. If it intrigues me or tempts me to buy, how did it do that? | | |

Advertising/Worksheet No. 7

## HOMEWORK—ANALYZING PRODUCTS

How essential are the things we see advertised on television? This exercise will help you and your family discover the answer for yourselves.

1. Each person in the family should make a list of three things that are absolutely essential for survival for one week (besides air and water).

2. Out of those lists, the family as a group, must develop a list of six things that the family would need in order to survive for a week.

3. Discuss the list. How many of the things on it are advertised on Television? How many of them existed as products before Television? How would you know about them if it weren't for advertising?

Chapter 5

# TELEVISION AND CHILDREN

## Regulating TV at Home

*Kate Moody*

*KATE MOODY is a writer/producer with MARC. A former teacher, she is a contributor to* The New York Times Encyclopedia of Television *and author of* Growing up on Television: Effects on Children *to be published soon by The New York Times Book Company.*

Ever since I became involved with studies of television and education, I've received lots of letters from people who ask about the effects of TV on children's development and how to use TV in the growing up experience. They inquire—as I too inquire—because they love their children deeply and want to be good parents.

Today's parents—like elders of the past—join in the early learning of the young. They want to help teach the tiny human how to learn. Even the most serious grown up faces respond to baby's smile, which then brings delight to young and old. And when the child first utters a sound that vaguely resembles some other sound, they rejoice that the child is "talking"—and the news goes out. . .

Soon parents begin to wonder what part television plays in the child's learning and perception. They sense (rightly so) that habitual TV viewing affects kids' attitudes about the world, family relationships, violence, sexuality, their appetite for junk foods, and more. They want to know: Will TV viewing make my children aggressive? Can it teach them to read? To think? To imagine? To make judgments? Can TV teach kids to *care*?

One parent wrote: "Even though my eldest child is only three, I am concerned about the television shows she watches. She enjoys *Sesame Street, Mr. Rogers* and *The Electric Company*—but she also expresses her desires to watch *Happy Days, LaVerne & Shirley* and other sitcoms. I would like to know how to decide just how much and what is advisable for her to watch."

As the mother of two boys, I have thought about these questions too. As a writer and teacher, I have absorbed the research, but there is nothing in the research that can tell a parent what to do.

People can make judgments about how to live with television after getting in touch with their own values, intuition and common sense.

Today's parents have no guides or role models for providing information about how to raise children with television. Our elders can't tell us or show us how to raise children with TV or deal with other aspects of rapid change. Our parents didn't have to deal with all of this; they didn't have to develop child-rearing practices which take into account the sexual revolution, drug culture and millions of flickering TV sets—so they can't pass this on to us. Then who are our guides?

Coming to terms with television is a personal matter. I can't give "answers" about other people's children's TV viewing, but I can share what I do with my own children. As parents (peers) we can help each other by sharing experience about how to use TV. Since we have no elders to help in this, we must take cues from each other. I would like to hear of your experiences, and would learn by knowing what *you do*.

For my part, I believe that children have a basic need for protection from the influences of powerful television technology which can cause them to become immobile and passive (itself a dangerous effect) while teaching values which may be exactly the opposite from those we would want them to believe when they are older. Further, there are factors in the perceptual development of young children which limit their ability to take on TV viewing as a major experience while they are so young. (See Dorothy Cohen's article.)

In the U.S. the situation is more serious than in other countries because children here watch more hours than children elsewhere, and as a society we have an increasing tolerance for "violence as entertainment" and for cheap uses of sexuality on the screen. These facts are colored by another truth: in their formative years children need to experience the world through their senses, to learn by doing, and to have lots of human interaction which TV cannot provide.

With such beliefs it is inevitable that I will limit the *time* my children spend with TV, and sometimes the *content*. My husband concurs in this thinking. However, because I was the one who was home for many more hours each week when the children were very young, it was up to me to implement the viewing policies we made, to be the "keeper of the media gates," at least in our home. I wasn't exactly thrilled about that—but I asked myself, "If I don't protect my children from all that is available at the spin of the dial, who will?"

### No TV?

I remember the times before television—during the first two years of my son's life when we lived in Brussels, Belgium in a small flat.

There were many entertainments for us as we did our morning marketing on foot, stopping in parks along the way. Each day after "naps" we'd go to the church yard and play with a large red ball until "Daddy" came home from the University on his bicycle.

Parks, play, walks, talks, naps and nursery rhymes gave structure to our days and provided basic elements of Scott's early education. We spent lots of time repeating and repeating nursery rhymes: One, two, buckle my shoe, three, four, shut the door. . . . Three little kittens, they lost their mittens and they began to cry. . . . Wee Willie Winkie runs through the town, upstairs, downstairs. . . . We did this for entertainment, but I was fully aware that Scott was developing listening skills and an ear for rhyme which are important building blocks of reading ability.

To be without TV, even though by accident of living in a foreign country, was a blessing by making it necessary for us to develop daily rituals based on real-life experience, not TV habits.

When we moved back to Grand Rapids we acquired a TV set. I was a regular viewer of one program—The *Today Show*. I liked to see the news (and Barbara Walters) when I worked in the kitchen at breakfast time. It was in the kitchen with a toddler and a new baby, John, that I received my introduction to "television awareness training."

One morning Scott dropped his toast and started waving his arms wildly in circles making buzzing sounds like ZZZZZZZ, and then he smacked both hands down hard on the table and shouted, "Kills 'em dead!" I was stunned. I hadn't taught my child these words or behaviors. Where was this coming from? The next day, watching the *Today Show,* I saw a commercial for Raid, the insecticide, in which someone sprays it in large circles on the screen accompanied by the sound ZZZZZZ—then a flyswatter comes down hard (smack) on the little bug. "Kills 'em dead," says the voice-over.

Who ever thought that a two year-old would watch the *Today Show* and learn from it? "OK, that's it," I said to myself. What concerned me was that TV was teaching him lessons which he was learning efficiently even though the televised material was designed not for him—but for adults. (What else was he learning while watching?) I also knew that if the *Today Show* weren't capturing the whole family's attention at breakfast we would be talking with each other more. I realized that TV can run stiff competition for any other real-life activity. It was time for values clarification.

Soon we gave up watching the *Today Show*—not because we dislike television, but because we like what happened in our home when it was turned off. It let us preserve "breakfast time" as a family ritual in which people *pay attention to each other*.

When this TV set broke down, we decided not to replace it—at least temporarily—as one way of maintaining an environment which was natural and personal. At this time (1968), I knew of nothing on TV that was really necessary for preschoolers to see. My goal in children's learning is to provide sensory and social experiences, primarily through play, reading aloud and talking together. However I was seeing that my son's peer group, mostly children of three or four or five years-old, had developed "The TV habit," now watching several hours a day of before-breakfast cartoons, late afternoon re-runs and some evening shows. They had a disturbing tendency to sit mesmerized before the TV set for extended periods; later they would nag incessantly for TV products, leap off tables and "fly" wildly like Batman and show little interest in having stories read aloud to them.

**"There is nothing in the research that can tell a parent what do do."**

### Play Group

In a suburban park I met Allison, a British woman. We became a resource to each other as we shared similar views about children and early learning, chiefly that children should develop their senses through play and the deeds of growing up. They should learn how to use their hands, imaginations and social skills. Together we developed a play group for our own children and five others which was successful partly because it utilized simple material which stimulated the children to use their own creativity in making things—no pre-cut Halloween pumpkins or machine-made valentines. Almost all of the materials were found or donated; we used "junk scrap" such as cereal boxes, egg cartons, milk containers, wood bits . . . also wet clay, powdered paint, washtubs filled with water, boats, blocks, paper, crayons, and flour-water paste that the children measured and stirred up. We employed body movement with songs and rhymes: number play, finger plays, marching, jumping, clapping, dance. We played outdoors everyday, even in winter. Through such kinds of play children learn far more than any kind of TV or "workbooks" can offer.

### "Hour a Day" Rule

*Sesame Street* changed our lives. We acquired a TV set again, established the "hour a day" rule, and as the *Sesame* logo first flashed on the screen I watched with my four-year old son at my side and infant on my lap—all of us enchanted. The four-year-old—a member of the intended audience—soaked up songs and watched those frenetic racecars zoom into place as the voice-over counted 1-2-3-4-5-6. The baby's eyes were mysteriously charmed by movement, light and sound (as all babies' eyes are) and with notebook in hand, I felt like a researcher just discovering a social revolution.

While it is generally recognized that *Sesame* can teach kids letters and numbers, what is not generally realized is the ways in which it has affected parents' attitudes, expectations and parenting behaviors. First, *Sesame* made most parents feel that "TV is good for children" and increased viewing levels generally. Studies show that millions of mothers want their children to watch *Sesame* and start them out on the habit. Another side-effect of the program has been that it set up unfortunate expectations in many mothers that children ought to be learning numbers and letters at age three or four. This is a dangerous notion. A young mother wrote, "My youngest son, who is not quite three, doesn't know a number from a letter, although he has the general idea that a symbol on a printed page is one or the other. So I was surprised recently when I was twice asked by friends if he could recite his alphabet and identify some letters. They informed me proudly that their children—who were both a bit younger than mine—could identify five or six letters. They attributed this accomplishment to the influence of *Sesame Street*. I was struck by the fact that in almost three years it had never crossed my mind that he should be learning letters. I must admit, for a moment I wondered if I were being a negligent parent, but the feeling soon passed."

**"Our goal in creating ways to limit TV, isn't so much to protect the children from specific images but to protect a lifestyle which we know is good for us."**

*Sesame Street* promotes "the viewing habit" at an extremely young age. In the midst of this writing, I called Joan Cooney, the developer of *Sesame* and president of Children's Television Workshop (CTW), and asked her: "If you were the mother of some pre-schoolers today, how would you use *Sesame Street* in your own home?" She said, "I would, of course, limit TV viewing in general. Given the fact that the TV is turned on a great part of the day in most households it is likely that the child will take an interest in *Sesame* at some point. If I had a pre-schooler at home and that child were interested in watching I would permit up to one hour a day of viewing of *Sesame*. I would also make sure that I read to the child and would discuss the TV images that the child watched."

A number of questions have been raised about *Sesame Street,* often criticizing its fast pace and style. Many reading teachers flatly reject the goal of trying to teach letters and reading skills to pre-schoolers—saying that it interferes with later teaching in school.

What is often not considered when the toddler is led to the screen is the rapidity with which children outgrow *Sesame*. By age four, the habitual *Sesame* watchers move on to other channels, taking their addiction to television with them. The fare they find on the commercial stations in the form of cartoons or re-runs of situation comedies will be filled with violence, stereotypes and a barrage of commercial messages (the typical child sees 400 commercials a week).

For all my good intentions in permitting my older son to watch *Sesame* for an hour a day, the habit of "TV and Snacks" at 4 o'clock comes easily. When Scott invited friends in after school (especially in the winter when it gets dark early) they would frequently want to watch TV—*Gigantor, Batman, Spiderman.* Under the "hour a day" plan they would tend to select one of these rather than, say, *Mr. Rogers* or *Sesame*. This is typical of boys in the four-to-seven age range who tend to be the most tenacious viewers. My intuition told me that this wasn't a good idea—but I maintained that only one hour a day wasn't going to damage anybody.

I soon found myself spending an unreasonable amount of energy as "manager" of TV policy. Often I'd hear, "I should get two hours today because I missed yesterday—but John can't watch with me because he watched yesterday at Bobby's house and I didn't. . . ." Or, John wants to watch *Sesame* for "his" hour and Scott wants to watch *Flintstones* later. How shall we monitor the situation so that one boy doesn't watch both programs, thereby creating some unfairness under the present policy? As firmly as I believed that children of this age should be limited to an hour a day of TV viewing, I was often weary from the effort to be consistent in enforcing it. Each day, television became a negotiable issue. The negotiations often had a negative tone and assigned to me (again) a negative function. Frequently the parent who is home during the day has repeated responsibilities to "say no" while the parent who may be away working can arrive home in the evening for dinner, fun and games. More often than not (in the U.S.) it is the woman who is at home, and therefore "Mom" feels the strain in parent-child relationships when limiting or "saying no" is a part of child-rearing. When the broadcaster preaches, "If you don't like it, turn it off," he or she delivers a highly sexist message: one more job for Mother. The practice of limiting TV is a good option, but it is hard work. Parents need as much support as they can get from the broadcasting industry, from their spouses, and from babysitters and grandparents.

**"Each day television became a negotiable issue."**

In addition to my own uneasiness with daily TV viewing, my older son was needing a quiet time in the evening to do homework and the younger son needed to be read to, and we all needed some time for talking with each other.

## Weekend Rule

One September, after a summer of outdoor activity had drawn us all from the house and the TV, my husband and I faced the fact that shorter days and daylight savings time would soon make afternoon TV viewing a question. We discussed what kind of policy might be reasonable to consider and by what kind of process it should be developed. We agreed that the adults in the house should make these kinds of judgments about our children's education. We were tending toward a "weekends only" plan.

Later we asked each child, "What programs do you most like to see on TV?" Scott said, "tennis," and John said, "the World Series." Soon the "weekend" viewing plan came into being, with the exception of the World Series which was then in progress. As a matter of routine, our portable TV goes into the attic on Sunday evening (out of sight, out of mind) and isn't in view again until Friday after dinner. We do not permit viewing on Saturday or Sunday morning because there are chores and family matters to attend to then, and even on Saturday night the set must be turned off by 9 o'clock. In the course of a weekend the children might watch a lot of sports—and inevitably some other stuff which I don't care for: *Mission Impossible, Eye Witness News,* etc. Real life activities frequently interrupt this potential viewing time; John's soccer team plays on Saturdays, likewise Scott's tennis matches and various street games, while I suggest diversionary events, "Anybody want to go jogging with me?"

In our neighborhood all the kids play at our house—where there's no TV during the week! They play stick ball, street hockey, catch, ping pong, and the specialty of the house which is known locally as "garbage can tennis." If children are encouraged to be inventive, they will learn how to entertain themselves without TV.

During the school week the absence of "electronic wallpaper" contributes to an atmosphere of study, conversation and calm. We settle down with homework or projects such as an on-going game of monopoly. Later there are bedtime stories or "bedtalk." These are rituals in our family—activities which are carried out in a ceremonial way and which, inevitably, have a binding quality for those who share in them. But these things don't usually happen if kids run from dinner to their favorite TV shows—worse yet, eat their meals in front of the set. Our goal in creating ways to limit TV isn't so much to protect the children from specific images but to protect a lifestyle which we know is good for us.

Occasionally, I've heard people say that children should watch a lot of TV because, if they don't, they won't have anything to talk about in school. This is nonsense! TV is a popular topic, but unless the child has rather serious personality problems, he or she can find plenty to talk about. Anyway, how many times do you have to see *Batman* to know what he does with a cape? The other "myth" I hear passed around is that if you limit TV viewing in your home, your children are likely to run away to the neighbors to watch. That's not true. Of course, they will enjoy watching sometimes, but "running away to the neighbors" won't preoccupy them if their home life is an active one.

I know that our children will always live in a world where there is lots of TV available. They will choose to use TV in different ways as adolescents or adults. I am confident that they will bring to the screen *critical viewing skills* because they will have developed *critical thinking skills* via a childhood based in real-life learning. This is the ultimate defense—and rationale for limiting TV in early childhood.

Coming to terms with television involves examining real life experience to see how TV fits in. I believe that television can teach many important facts and concepts and even expand the imagination—but this can happen only if television is integrated into real life instead of having real life squeezed into the spaces between TV shows.

# Television and the Young Viewer

*Eli A. Rubinstein*

*ELI A. RUBINSTEIN is a currently Adjunct Research Professor of Psychology at the University of North Carolina at Chapel Hill. From 1966 to 1971 he was assistant director at the National Institute of Mental Health, where his last major responsibilities were as vice-chairman of the Surgeon General's Scientific Advisory Committee and co-editor of the five volumes of research reports which resulted from that program.*

For *some* children, under *some* conditions, *some* television is harmful. For *other* children under the same conditions, or for the same children under *other* conditions, it may be beneficial. For *most* children, under *most* conditions, *most* television is probably neither particularly harmful nor particularly beneficial. [*Schramm, Lyle, and Parker 1961, p. 1*]

That assessment, made in 1961 by three leading Stanford researchers, was the general conclusion of one of the first major studies of television and children. Almost two decades and about two thousand studies later, their conclusion remains a reasonably accurate evaluation of the complex and differential impact of television on its millions of young viewers. The published research since 1961 has further confirmed the harmful effects to some children of some television under some conditions. At the same time, stronger evidence for the corollary has been found: for some children, under some conditions, some television is beneficial.

It is in the differentiation between *some* children and *most* children and in the distinction between *some* television and *most* television that the scientific findings are still not clear. As in so many other instances of exposure to persuasive messages, it is the cumulative impact over extended periods of time that should be the crucial test of consequences. It seems reasonable to assume that when millions of young viewers each spend on average about a thousand hours per year watching hundreds of television programs, such time spent must have some significant effect on their social development. It is equally reasonable to assume that if the effect is so tangible there should be little difficulty in identifying its characteristics or assessing its strength. Here, however, the evidence is less than definitive—thus the continued applicability of the Schramm, Lyle, and Parker generalization.

While the total impact of television on the young viewer is still unclear, the pursuit of evidence has attracted increasing interest on the part of social sceintists. In the appendix to the 1961 report by Schramm et al., 52 earlier publications dealing with television and children are annotated, and they make up a fairly complete list of relevant prior research. By 1970 (Atkin, Murray, and Nayman 1971) that list of publications had grown to a total of almost 500 citations. By 1974 a total of almost 2,400 publications were cited under the category of television and human behavior in a major review of the field by Comstock and Fisher (1975). Since 1974, research interest has continued unabated as new topics begin to be explored. The concern about televised violence has been augmented by a concern about sex on television, the persuasive power of television advertising on children, and, indeed, the effect of television on the entire socialization process of children.

Two events in the early 1970s provided the most compelling influence toward that growth of interest. One was the development of the program *"Sesame Street,"* in which formative research was pursued in partnership with the production of the program itself. The other was the completion in 1972 of a major federally funded research program, now known as the Surgeon General's program, to evaluate the relationship between TV violence and aggressive behavior in children.

## Sesame Street

*Sesame Street* provides the most extensive example of how television can be made beneficial for some children under some conditions. The story of its growth and development has been effectively told by its educational director (Lesser 1974). From the research perspective, *Sesame Street* marks a major innovation: it is the first intensive and continuing partnership between education specialists from academia and the creative and technical specialists responsible for putting television programs on the air. How that partnership evolved into not just constructive interchange but productive results is itself a lesson in formative research.

The educational goals of *Sesame Street* were developed over a series of working seminars in 1968, with participation by experts from all relevant specialties. The instructional goals were precisely formulated in a series of specific state-

This article first appeared in *American Scientist,* Dec. 1978.
It is reprinted with permission.

ments on five major topics: (1) social, moral, and affective development; (2) language and reading; (3) mathematical and numerical skills; (4) reasoning and problem-solving; (5) perception. The target audience was disadvantaged inner-city children.

Overall, the development of the total program toward the achievement of those goals was structured by the continuing interplay of production people and academic researchers in a feedback system involving observation of children viewing programs as they were produced. Happily, all the major participants in this innovative enterprise were completely dedicated to the larger task. Of equal good fortune was the initial allocation, from private and public sources, of $8 million for the start-up and production costs for the first eighteen months of *Sesame Street.* While those three ingredients—dedicated talent, adequate start-up time, and ample funding—do not always produce success, they certainly represent a good beginning and were put to constructive use in the development of the program.

Some ten years after its introduction, *Sesame Street* has become not just a national but an international phenomenon. In addition to being the most widely viewed children's program in the United States, *Sesame Street,* in both the original version and in foreign-language versions, is broadcast in more than 50 countries around the world. The Children's Television Workshop, the parent body for *Sesame Street* and all the other educational programs, cites more than 120 articles and books (Children's Television Workshop 1977) primarily on *Sesame Street* and *The Electric Company.* Additional publications through early 1978 bring the total above 150 research reports. These range from studies of attention during the actual program viewing to major evaluations of both *Sesame Street* and *The Electric Company.* The total literature provides an unprecedented body of research on how the entertainment appeal of television can be put to educational use.

Most, but not all, of the evaluations are positive. One major study suggests that *Sesame Street* has been less useful to disadvantaged children than advantaged children (Cook et al. 1975) because of differences in viewing interest. And, in England, for example, grave concern was initially raised that the very format of *Sesame Street* is inimical to the learning process because the program over-emphasizes sheer attention-getting devices and because it links learning too closely to a commercial entertainment format. (Subsequently, *Sesame Street* was aired in Great Britain and achieved much viewer success.)

**"Television is now a dominant voice in American life. It is a formidable teacher of children."**

But these are issues for the educators. From the standpoint of research on television and the young viewer, the history of "Sesame Street" has been of great significance. Formative research has achieved a new status through the ongoing efforts to evaluate the progress and achievements of what Joan Ganz Cooney, the director of the entire enterprise, has called "a perpetual television experiment." The program has clearly demonstrated that television can teach children while still holding their voluntary attention as well, if not better, than conventional television programming. It is the clearest example of the positive potential of television translated into performance.

## Televised Violence

At about the same time that *Sesame Street* was being prepared for broadcast in 1969, a major federal research program was initiated to assess the effects of televised violence on children. The history of that research enterprise has been thoroughly described and evaluated from a variety of perspectives (Bogart, 1972; Cater and Strickland, 1975; Rubinstein, 1975). The belief is fairly widespread that the body of research, published in five volumes of technical reports, provided a major new set of findings.

There is much less agreement about the report and conclusions of the advisory committee itself, because of the cautious language used. Even now, years after the report was published, the conclusions are debated. The debate was sparked initially because of a misleading headline in a front-page story in *The New York Times* ("TV Violence Held Unharmful to Youth") when the report was first released. A careful analysis of the subsequent press coverage revealed how influential that headline was in further confusing the interpretation of the findings (Tankard and Showalter, 1977). The committee had unanimously agreed that there was some evidence of a causal relationship between televised violence and later aggressive behavior. However, the conclusion was so moderated by qualifiers that it was, and still is, criticized and misinterpreted by industry spokesmen as being too strong and by researchers as being too weak.

The effect of televised violence has been an issue of public concern almost from the inception of television in the early 1950s. Periodically, over the past 25 years, a variety of congressional inquiries as well as commissioned reports have drawn attention to the problem of televised violence. In almost every instance concern was raised about harmful effects. In all these reports, however, relatively little new research on the problem was produced. Even the prestigious Eisenhower Commission, which was asked by President Johnson in 1968 to explore the question within its total inquiry into violence in America, devoted its attention primarily to a synthesis of existing knowledge rather than to collecting new scientific information. Its conclusion—that violence on television encourages real violence—was seen as less than persuasive and were largely ignored, especially since attention was then focused on the new Surgeon General's Committee.

The Surgeon-General's program provided the first major infusion of new monies into research on television violence, which in turn has stimulated a "second harvest," as Schramm (1976) calls it, of new work on television and social behavior. The debate on the evidence from the five volumes of research reports produced in 1972 by the Surgeon General's program is also still lively. The essence of the debate emerges from two contrasting approaches to assessing the evidence. The Surgeon General's advisory committee, while acknowledging flaws in many of the individual studies, held that the *convergence* of evidence was sufficient to permit a qualified conclusion indicating a causal relationship between extensive viewing of violence and later aggressive behavior. This conclusion, without the qualifications, is endorsed by a number of highly respected researchers, some of whom participated in the Surgeon General's program and some who were not directly involved.

A different and seemingly more rigorous approach to the evidence is adopted by some other experts in the field. This contrasting view is epitomized by Kaplan and Singer (1976), who conclude that the total evidence does no more than support the null hypothesis.

A brief explanation is appropriate here about the theoretical formulations most common in the research underlying the whole question of television and aggressive behavior. Quite simply, three possibilities exist—and all have their proponents: (1) Television has no significant relationship to aggressive behavior. (2) Television reduces aggressive behavior. (3) Television causes aggressive behavior; a variant on this possibility is that both television viewing and aggressive behavior are related not to each other but to a "third variable" which mediates between the first two variables.

The Kaplan and Singer position, to give one recent example, is that the research so far has demonstrated no relationship between televised violence and later aggressive behavior. They characterize this as the "conservative" assessment of the evidence and come to that conclusion by finding no study persuasive enough in its own right to bear the burden of significant correlation, let alone causal relationship. They are not the only ones to conclude "no-effect": Singer (1971) came to the same conclusion, as did Howitt and Cumberbatch (1975).

The second point of view has been espoused primarily by Feshbach (1961). The catharsis hypothesis holds that vicarious experience of aggressive behavior, as occurs in viewing TV violence, may actually serve as a release of aggressive tensions and thereby reduce direct expression of aggressive behavior. This view has not been supported by research evidence, although it emerges time and again as a "common-sense" assessment of the relationship between vicarious viewing of violence and later behavior. Indeed, the thesis itself goes back to Aristotle, who considered dramatic presentations a vehicle for discharge of feelings by the audience. The Surgeon General's committee, in considering this thesis, made one of their few unequivocal assessments—that there was "no evidence that would support a catharsis interpretation."

The third general conclusion, and the one now prevailing, is that there is a positive relationship between TV violence and later aggressive behavior. This facilitation of later aggression, explained primarily by social learning theory, is endorsed by a number of investigators. The basic theoretical formulation is generally credited to Bandura (Bandura and Walters, 1963), who began his social learning studies in the early 1960s, when he and his students clearly demonstrated that children will imitate aggressive acts they witness in film presentations. These were the so-called "bobo doll" studies, in which children watching a bobo doll being attacked, either by a live model or a cartoon character, were more likely to imitate such behavior. These early studies were criticized both because bobo dolls, made for rough and tumble play, tend to provoke aggressive hitting and because the hostile play was only against inanimate play objects. Later studies have demonstrated that such aggression will also take place against people (Hanratty et al., 1972).

Variations on the basic social learning theory are rather numerous. Kaplan and Singer make a useful schema by incorporating three related theoretical branches under the general label of "activation hypotheses." Bandura and his students are included in the category of social learning and imitation. A second branch is represented by Berkowitz and his students, who follow a classical conditioning hypothesis, in which repeated viewing of aggressive behavior is presented to build up the probability of aggressive behavior as a conditioned response to the cues produced in the portrayal. A number of experiments by Berkowitz and his colleagues have shown that subjects viewing a violent film after being angered were more likely to show aggressive behavior than subjects similarly angered beforehand, who saw nonaggressive films (Berkowitz, 1965; Berkowitz and Geen 1966).

In still a third variation on the activation approach, Tannenbaum (1972) holds that a generalized emotional arousal is instigated by emotionally charged viewing material and that this level of arousal itself is the precursor of the subsequent behavior. Any exciting content, including erotic content, can induce this heightened arousal. The nature of response is then a function of the conditions that exist at the time the activation of behavior takes place. Thus, according to this theory, it is not so much the violent content per se that induces later aggressive behavior as it is the level of arousal evoked. Subsequent circumstances may channel the heightened arousal in the direction of aggressive behavior.

In an important examination of the utility of these various formulations, Watt and Krull (1977) reanalyzed data on 597 adolescents from three prior studies, involving both programming attributes (such as perceived violent content) and viewer attributes (such as viewing exposure and aggressive behavior) through a series of correlations. They contrasted three models, which they labeled catharsis model, arousal model, and facilitation model. The first two models are essentially as described above. The facilitation model is identified as a general social-learning model without regard to whether the process is primarily imitation, cueing, or legitimization of aggressive behavior. Thus, the Bandura and Berkowitz studies both fall into the facilitation model.

Through a series of partial correlations, Watt and Krull found (1) no support for the catharsis model; (2) support for a combination of the facilitation and arousal models; and (3) some differences due to age and/or sex, with the arousal model a somewhat better explanation for female adolescents, whereas the facilitation model better described the data for males. (Sex differences in results in many studies of television and behavior are quite common. One of the major studies in the field, by Lefkowitz et al., 1972, found significant correlations betwen TV violence and later aggressive behavior with boys but not with girls.)

What are the implications in this continuing controversy about the effects of television violence on aggressive behavior? As in so many other social science issues it depends on what you are looking for. The dilemma is neatly characterized in a legal case in Florida in October 1977, in which the defense argued that an adolescent boy, who admittedly killed an elderly woman, was suffering from "involuntary subliminal television intoxication." (This term, which appears nowhere in the scientific literature, was introduced by the defense attorney.) In trying to show that the scientific evidence on television's effects on behavior was not directly pertinent to this murder trial, the prosecuting attorney asked an expert witness if any scientific studies indicated that a viewer had ever been induced to commit a serious crime following the viewing of TV violence. The (correct) answer to that question was "no." The judge thereby ruled that expert testimony on the effects of televised violence would be inadmissible and brought back the jury, which had been sequestered during the interrogation of the expert witness. On the basis of the evidence presented

**". . . television should be considered a major agent of socialization in the lives of children."**

to it, the jury found the defendant guilty of murder and rejected the plea of temporary insanity by virtue of "involuntary subliminal television intoxication."

While there is indeed no *scientific* evidence that excessive viewing of televised violence can or does provoke violent crime in any one individual, it is clear that the bulk of the studies show that if large groups of children watch a great deal of televised violence they will be more prone to behave aggressively than similar groups of children who do not watch such TV violence. The argument simply follows from the basic premise that children learn from all aspects of their environment. To the extent that one or another environmental agent occupies a significant proportion of a child's daily activity, that agent becomes a component of influence on his or her behavior. In a recent comprehensive review of all the evidence on the effects of television on children, Comstock et al. (1978) conclude that television should be considered a major agent of socialization in the lives of children.

An important confirmation of the more general influence of television on the young viewer derives from research on the so-called "prosocial" effects of television. Stimulated by the findings of the Surgeon-General's program, a number of researchers began in 1972 to explore the corollary question: If TV violence can induce aggressive behavior, can TV prosocial programming stimulate positive behavior? By 1975, this question was of highest interest to active researchers in the field, according to a national survey (Comstock and Lindsey 1975).

A significant body of literature has now been generated to confirm these prosocial effects (Rubinstein et al. 1974; Stein and Friedrich 1975). Research by network scientists (CBS Broadcast Group 1977) has confirmed that children learn from the prosocial messages included in programs designed to impart such messages. Because the effect of prosocial program content is so clearly similar in process to the effect of TV violence, confirmation of the former effect adds strength to the validity of the latter effect.

In all the intensive analysis of the effects of TV violence, perhaps the one scientific issue most strongly argued against by the network officials has been the definition and assessment of levels of violent content. The single continuing source of such definition and assessment has been the work of Gerbner and his associates (Gerbner 1972). Beginning in 1969 and continuing annually, Gerbner has been publishing a violence index which has charted the levels of violence among the three networks on prime time. The decline in violence over the entire decade had been negligible until the 1977-78 season (Gerbner et al. 1978), following an intensive public campaign against TV violence by both the American Medical Association and the Parent-Teacher Association.

Gerbner's definition of violence is specific and yet inclusive—"the overt expression of physical force against others or self, or compelling action against one's will on pain of being hurt or killed or actually hurting or killing." Despite criticism by the industry, the Gerbner index has been widely accepted by other researchers. An extensive effort by a Committee on Television and Social Behavior, organized by the Social Science Research Council to develop a more comprehensive violence index, ended up essentially endorsing Gerbner's approach (Social Science Research Council 1975).

Perhaps of more theoretical interest than his violence index is Gerbner's present thesis that television is a "cultural indicator." He argues that television content reinforces beliefs about various cultural themes—the social realities of life are modified in the mind of the viewer by the images portrayed on the television screen. To the extent that the television world differs from the real world, some portion of that difference influences the perception of the viewer about the world in which he or she lives. Thus, Gerbner has found that heavy viewers see the world in a much more sinister light than individuals who do not watch as much television. Gerbner argues that excessive portrayals of violence on television inculcate feelings of fear among heavy viewers, which may be as important an effect as the findings of increased aggressive behavior. Some confirmation of this feeling of fear was found in a national survey of children (Zill 1977): children who were heavy viewers were reported significantly more likely to be more fearful in general than children who watched less television.

## TV Advertising

An area of research that has been increasing in importance since the work of the Surgeon General's program has been concerned with the effects of advertising on children. One of the technical reports in the Surgeon General's program described a series of studies on this topic by Ward (1972), which was among the first major published studies in which children's reactions to television advertising were examined in their relationship to cognitive development. That report provided preliminary findings on (1) how children's responses to television advertising become increasingly differentiated and complex with age; (2) the development of cynicism and suspicion about television messages by the fourth grade; (3) mothers' perceptions about how television influences their children; and (4) how television advertising influences consumer socialization among adolescents.

The entire field of research on effects of television advertising—at least academic published research—has only begun to develop in the 1970s. A major review of the published literature in the field was sponsored and published by the National Science Foundation in 1977 (Adler 1977). It is noteworthy that only 21 studies, all published between 1971 and 1976, were considered significant enough to be singled out for inclusion in the reveiw's annotated appendix. The total body of evidence is still so small that no major theoretical formulations have yet emerged. Instead, the research follows the general social-learning model inherent in the earlier research on televised violence.

Despite the lack of extensive research findings on the effects of television advertising on children, formal concern about possible effects began to emerge in the early 1960s. Self-regulatory guidelines were adopted by the National Association of Broadcasters in 1961 to define acceptable toy advertising practices to children. Subsequently, published NAB guidelines were expanded to include all advertising directed primarily to children. An entire mechanism has been established within the industry, under the responsibility of the NAB Television Code Authority, though which guidelines on children's advertising—as well as other broadcast standards—are enforced. In addition, in 1974 the national Advertising Division of the Council of Better Business Bureaus established a Children's Advertising Review Unit to help in the self-regulation of advertising directed to children aged eleven and under. That organization, with the assistance of a panel of social science advisors, developed and issued its own set of guidelines for children's advertising.

The role of research in helping to make those guidelines on children's advertising more meaningful is only now receiving some attention, thanks in part to the NSF review cited above. Two recent events have highlighted both the paucity and the relevance of research in this field. In 1975, the Attorney General of Massachusetts, in collaboration with Attorneys General of other states, petitioned the Federal Communications Commission to ban all drug advertising between 9 A.M. and 9 P.M., on the ground that such advertising was harmful to children. After a series of hearings in May 1976, at which researchers and scientists testified, the petition was denied for lack of scientific evidence to support the claim.

In 1978, the Federal Trade Commission formally considered petitions requesting "the promulgation of a trade rule regulating television advertising of candy and other sugared products to children." A comprehensive staff report on television advertising to children (Ratner et al. 1978) made recommendations to the FTC, citing much of the relevant scientific literature on advertising to children as evidence supporting the need for such a trade rule. At the time of this writing, the entire matter was still under active consideration.

What does the existing research in this area demonstrate? It is clear that children are exposed to a large number of television commercials. The statistics themselves are significant. Annually, on average, children between two and eleven years of age are now exposed to more than 20,000 television commercials. Children in this age group watch an average of about 25 hours of television per week all through the year. The most clichéd statistic quoted is that, by the time a child graduates from high school today, he or she will have spent more time in front of a television set (17,000 hours) than in a formal classroom (11,000 hours). Indeed, all the statistics on television viewing from earliest childhood through age eighteen show that no other daily activity, with the exception of sleeping, is so clearly dominant.

Just as was shown in the earlier research examining the effects of programming content, even the limited research now available on television commercials documents that children learn from watching these commercials. Whether it is the sheer recall of products and product attributes (Atkin 1975) or the singing of commercial jingles (Lyle and Hoffman 1972), the evidence is positive that children learn. More important, children and their parents are influenced by the intent of these commercials. One study (Lyle and Hoffman 1972) showed that nine out of ten preschool children asked for food items and toys they saw advertised on television.

A number of studies have also revealed various unintended effects of television advertising. While a vast majority of the advertisements adhere to the guidelines that attempt to protect children against exploitative practices, a number of studies have shown that, over time, children begin to distrust the accuracy of the commercial message. By the sixth grade, children are generally cynical about the truthfulness of the ads. A recent educational film by Consumers Union, on some of the excessive claims in TV advertising, highlights the problem of disbelief (Consumers Union 1976). There have also been a number of surveys in which parents have indicated negative reactions to children's commercials. In one study (Ward, Wackman, and Wartella 1975) 75% of the parents had such negative reactions.

In the survey of the literature evaluated in the 1977 NSF review, the evidence is examined against some of the major policy concerns that have emerged in the development of appropriate guidelines on children's advertising. These concerns can be grouped into four categories: (1) modes of advertising; (2) content of advertising; (3) products advertised; and (4) general effects of advertising.

Studies of "modes of advertising" show, for example, that separation of program and commercial is not well understood by children under eight years of age. While these younger children receive and retain the commercial messages, they are less able to discriminate the persuasive intent of the commerical and are more likely to perceive the message as truthful and to want to buy the product (Robertson and Rossiter 1974).

The format and the use of various audio-visual techniques also influence the children's perceptions of the message. This influence is clearly acknowledged by the advertisers and the broadcasters, who have included explicit instructions in the guidelines, especially for toy products, to ensure that audio and video production techniques do not misrepresent the product. What little research there is on this entire aspect of format is still far from definitive. What is clear is that attention, especially among young children, is increased by active movement, animation, lively music, and visual changes. (All of this, and more, is well understood by the advertising agencies and those who develop the ads, and they keep such knowledge confidential, much as a trade secret.)

**". . . there is a positive relationship between TV violence and later aggressive behavior."**

One other relatively clear finding on audio-visual technique relates to the understanding of "disclaimers"—special statements about the product that may not be clear from the commercial itself, such as "batteries not included." A study of disclaimer wording and comprehension (Liebert et al. 1977) revealed that a standard disclaimer ("some assembly required") was less well understood by 6- and 8-year-olds than a modification ("You have to put it together"). The obvious conclusion—that wording should be appropriate to the child's ability to understand—is just one of the many ways in which this research can play a role in refining guidelines.

Studies on the content of advertising have shown that the appearance of a particular character with the product can modify the child's evaluation of the product, either positively or negatively, depending on the child's evaluation of the character. It is also clear that children are affected in a positive way by presenters of their own sex and race (Adler 1977). On a more general level, sexual stereotypes in advertising probably influence children in the same way they do in the program content.

Although there is relatively little significant research on the effects of classes of products, two such classes have been under intense public scrutiny in recent years: proprietary drug advertising and certain categories of food advertising. Governmental regulatory agencies are currently considering what kind of controls should be placed on such advertising to children.

Concerning the more general effects of advertising targeted to children, surveys suggest that parents have predominantly negative attitudes about such advertising because they believe it causes stress in the parent-child relationship. Studies on questions such as this, and on the larger issue of how such advertising leads to consumer socialization, are now being pursued. Ward and his associates (Ward, Wackman, and Wartella 1975) have been examining the entire question of how children learn to buy. The highly sophisticated techniques used by advertisers to give a 30-second or 60-second commercial strong impact on the child viewer make these commercials excellent study material for examining the entire process of consumer socialization. Much important research still remains to be done on this topic.

## Sex on TV

Of the many public concerns about television and its potentially harmful effects on children, the issue of sex on television is at present among the most visible and the least understood. If research on the effects of advertising is still in its early development, research on sex on television has hardly begun.

It has been found that children who watch large amounts of television (25 hours or more per week) are more likely to reveal stereotypic sex role attitudes than children who watch 10 hours or less per week (Frueh and McGhee 1975). Research has documented the stereotyping on television of women as passive and rule-abiding, while men are shown as aggressive, powerful, and smarter than women. Also, youth and attractiveness are stressed more for females than males. This evidence of stereotyping was included as one part of an argument by the U.S. Commission on Civil Rights that the Federal Communications Commission should conduct an inquiry into the portrayal of minorities and women in commerical and public television drama and should propose rules to correct the problem (U.S. Commission on Civil Rights 1977). Program content in 1977 and 1978 has given increased emphasis to so-called "sex on television," at the same time that violence on television is decreasing (Gerbner et al. 1978).

Despite all the public concern and attention, including cover stories in major newsweeklies, relatively little academic research has been done on sex on TV. Two studies reported in 1977 provided information on the level of physical intimacy portrayed on television (Franzblau et al. 1977 and Fernandez-Collado and Greenberg 1977). Franzblau, Sprafkin, and Rubinstein analyzed 61 prime-time programs shown on all three networks during a full week in early October 1975. Results showed that, while there was considerable casual intimacy such as kissing and embracing and much verbal innuendo on sexual activity, actual physical intimacy such as intercourse, rape, and homosexual behavior was absent in explicit form.

Fernandez-Collado and Greenberg examined 77 programs aired in prime time during the 1976-77 season and concluded that intimate sexual acts did occur on commercial television, with "the predominant act being sexual intercourse between heterosexuals unmarried to each other." An examination of the data, however, reveals that in this study, verbal statements—identified as verbal innuendo in the study by Franzblau et al.—served as the basis for the conclusion reached. In fact, explicit sexual acts such as identify an R- or an X-rated movie do not occur on prime-time network television.

Even though few published studies have so far examined the question of sex on television, at least two important issues have been highlighted by the two studies mentioned above. The most obvious point is the difference in interpretation of the data by the two reports—unfortunately not an uncommon occurrence in social science research. Labeling and defining the phenomena under examination, let alone drawing conclusions from results, show variations from investigator to investigator. While this kind of difference is not unique to the social sciences, the more complex the data and the less standard the measurement—qualities often inherent in social science studies—the more likely these individual differences of interpretation.

The second point illustrated by these two studies of sex on TV is more intrinsic to the subject matter itself. The public concern about sex on TV suggests that the general reaction is much in keeping with the substitution of behavior for verbal statements, as is found in the study by Fernandez-Collado and Greenberg. And, in fact, there are no scientific data to indicate that verbal innuendo may not affect the young viewer as much—or as little—as explicitly revealed behavior. What is important here is that we do not know the effects of either the verbal description or the explicit depiction on the young viewer. Research findings of the Commission on Obscenity and Pornography in 1970 suggest that exposure to explicit sexuality seems harmless. Nevertheless, public sensitivities are clearly high; whether those sensitivities are justified by the facts still remains an open question. At the very least, studies should be undertaken to give some objective answers to these questions.

One such effort is a recent content analysis by Rubinstein and his colleagues (Silverman et al. 1978), which confirms the absence of explicit sex in network programs aired in the 1977-78 season but documents a continued increase in sexual innuendos. Furthermore, sexual intercourse, which was never even contextually implied in the 1975-76 analysis by Franzblau et al. (1977), was so implied fifteen times during the week of programs analyzed in the 1977-78 report. Clearly the current decrease in violent content is partially offset by added emphasis on sexual content.

## Issues of Policy

What are the policy implications of research on television and social behavior? Perhaps the most fundamental point to be made is that even with fairly clear research findings, the policy to be followed rarely emerges as a direct result of the research.

The history of the Surgeon General's program provides a useful case study of the complexities of this issue. When the Surgeon General's program of research was initiated, the advisory committee was charged by the Secretary of HEW, Robert Finch, with the responsibility for answering the question originally raised by Senator John Pastore, Chairman of the Senate Subcommittee on Communications: Does the viewing of TV violence stimulate aggressive behavior on the part of young children? The committee was specifically enjoined from making policy recommendations, since the HEW has no regulatory responsibility in this area. Thus, when the committee report was issued in 1972, there was no discussion of direct policy implications, nor were there any specific policy recommendations in that final document.

Senator Pastore, on receiving the report in January 1972, was sufficiently concerned, both about the cautious wording of the conclusion and the absence of policy recommendations, to call another set of hearings in March 1972 to clarify the interpretation of the results and to ask the committee members for their policy recommendations now that they were no longer under official constraints. What Senator Pastore learned at those hearings is now a familiar characteristic of scientists speaking out on public policy: their scientific expertise affords them little advantage in the public policy arena. There were relatively few workable and concrete policy recommendations forthcoming. Indeed, the most specific recommendation came from the Senator himself: a request to the Secretary of HEW and the FCC that an annual violence index be published that would measure the amount of televised violence entering American homes. No such official index has ever emerged, although Gerbner has annually produced such a measure, as a continuation of his ongoing research program.

What is clear from an examination of the Surgeon General's program in retrospect is that the advisory committee was correctly confined to the examination of the research question. But the next step was not taken—to set up a different committee, to develop policy recommendations on the basis of that research and in keeping with legal constraints and operational feasibility. Indeed, it might well have taken more time and care to examine the complexities of social policy in order to come to realistic and useful conclusions about a social course of action than it took to evaluate the research findings.

Attempts at policy formation concerning sexual content on TV will bring the complexities of social science research to public attention. For example, as Dienstbier (1977) has pointed out, the conclusions of the Surgeon General's committee affirm the social learning model. The Commission on Obscenity and Pornography, on the other hand, concluded in 1970 that exposure to explicit sexuality seemed harmless. Aside from the fact that the differences in these two sets of conclusions may be partly a reflection of liberal versus conservative value judgments relating to aggression and sex (Berkowitz 1971), there are some intriguing implications for social policy in other differences between the portrayal on television of violence and physical intimacy. Dienstbier suggests that increased portrayal of sex on television may become an important substitute for extensive sex education programs. While such an assertion may provoke considerable debate among social scientists, let alone the public at large, it is worthy of further consideration, as still anther pertinent research question.

Difficulties in arriving at policy guidelines for advertising to children are equally apparent. In connection with the current FTC examination of the merits of a trade rule to regulate television advertising of candy and other sugared products to children, the scientific evidence, primarily derived from the NSF report (Adler 1977) and from the interpretation of that evidence by the FTC (Ratner et al. 1978), is the source of much debate. Some of the scientists who contributed to the literature are publicly complaining that their findings are misinterpreted in the FTC staff report (Schaar 1978).

It is a minor irony that researchers are just as quick to take issue with interpretations—which they say go beyond the data—designed to support some change in policy on television advertising as their colleagues were in 1972 to take issue with interpretations by the Surgeon General's committee which they felt did not go as far as their data indicated. The correct generalization may well be that social scientists find it difficult to accept someone else's interpretation of their findings regardless of the direction of the policy implications.

In all the present examination of research on television and social behavior and its implications for social policy there are a number of important issues to consider. One point that bears repeating is that the research does not by itself identify the policy direction. Nor, for that matter, does the research to date satisfactorily deal with the many research questions that are relevant to the policy directions. At a major conference on priorities for research on television and children held in Reston, Virginia, in 1975 (Ford Foundation 1976), an entire agenda for future research was developed. Topics and methodology recommended ranged from simple experiments to identify effects of disclaimers and warnings in television advertising to long-range studies, including cross-national studies, to study of the effects of television on political and social beliefs.

However, except for the Surgeon General's program of research, plus a new program supported by the NSF in 1976 following the 1975 Reston conference, no major federal program of research exists. Time and again over the past twenty years, following various congressional hearings dealing with the effects of television, recommendations have been made for a "television research center." In early 1978, Senator Wendell Anderson of Minnesota began exploring the feasibility of legislation to develop a "Television Impact Assessment Act," but to date no final draft bill has materialized.

The three major networks, primarily responding to the pressures from the Surgeon General's program, have expended since 1970 approximately $3 million, primarily on the issue of televised violence. The American television industry seems much less willing to examine the need for a major program of research than does the British Broadcasting Corporation, which, in 1976, commissioned an eminent sociologist, Elihu Katz, to develop a comprehensive set of recommendations for a program of social research (Katz 1977). With the American television industry operating at about a $10 billion annual budget, even one-tenth of one percent devoted to social research would amount to $10 million a year. What Katz has recommended to the BBC would serve well both the American television industry and the public: a comprehensive program of research under the auspices of a new foundation funded by a variety of sources, including the broadcasting industry.

What is critical in such an endeavor is that it be seen as a long-term program. In an earlier paper (Rubinstein 1976) I proposed such a long-term instrumentality that would include studying ways of enhancing the value of television to the child viewer. It is likely that important findings still to be uncovered may provide guidelines for making television a more useful agent of socialization than it is at present.

A whole series of new populations of television viewers await the benefits of a constructive examination of the way television influences our lives. The evidence is already clear that older people watch increasing amounts of television. Organizations of older individuals have begun to criticize the televised stereotypes of the helpless and infirm elderly. Recent public broadcast programming such as "Over Easy," directed on an older audience, has shown how television can be of specific interest and benefit to this population.

Another group worthy of special attention includes the institutionalized mental patients, who, in public mental hospitals, watch a large amount of commercial television in their day rooms (Rubinstein et al. 1977). Careful study may provide insights into how this leisure-time activity can be converted into a more meaningful part of the total therapeutic program of the institution. Rubinstein and his colleagues have been studying the effects of TV on institutionalized children (Kochnower et al. 1978).

## A Bridge Between Research and Policy

What was initially a narrow focus on the presumed harmful effects of televised violence has begun to broaden into other areas that may have even more extensive and important policy implications. Social scientists can make important contributions to policy determinations, but there are important constraints that must be understood and accepted. In a persuasive argument, Bevan (1977) makes a case for the role of the scientist in contributing to the policy process. He stresses the need for scientists to "seek active roles in policy-making both in the public and in the private sectors." Fundamental to taking such a role is the need to recognize the difference between the world of the scientist and the world of the public official. There is a basic dichotomy between an emphasis on scientific inquiry and an emphasis on action and decision-making. That dichotomy is just as real between the social scientist looking at television and its effects on the viewer and the television officials who have the daily responsibility for deciding what does or does not go on the air.

All too often the social scientist venturing into television policy considerations makes naively sweeping recommendations with no understanding of the enormous complexity of responding to all the pressures and necessities of production. At the same time, some responsible members of the television industry take refuge in a defensive posture about the implications of the research findings. In this context, a variation on Bevan's recommendation that scientists engage in the policy process would be that the social scientists and the television industry officials engage in a continuing dialogue on how the research on television and children can be more effectively utilized.

Fortunately, some efforts in this direction are already under way. All three networks have a variety of activities in which outside research consultants meet with television personnel on programming for children. Special conferences and workshops on research have been sponsored in recent years by foundations, by the industry, by citizen action groups, and by professional organizations.

Perhaps the most compelling reason for more collaboration among all sectors—industry, researchers, the viewing public, foundations, and government agencies—is the common objectives held. Television is now a dominant voice in American life. It is a formidable teacher of children. Its healthy future should be the interest and responsibility of all of us.

### *REFERENCES*


Adler, R. 1977. *The Effects of Television Advertising on Children.* NSF.

Atkin, C. 1975. *Effects of Television Advertising on Children: First Year Experimental Evidence,* Report #1, Mich. State Univ.

Atkin, C. K., J. P. Murray, and O. B. Nayman, ed. 1971. *Television and Social Behavior. The Effects of Television on Children and Youth: An Annotated Bibliography of Research.* U.S. Government Printing Office.

Bandura, A., and R. H. Walters. 1963. *Social Learning and Personality Development.* Holt, Rinehart and Winston.

Berkowitz, L. 1965. Some aspects of observed aggression. *J. Pers. and Soc. Psych.* 2:359-69.

______. 1971. Sex and violence: We can't have it both ways. *Psych. Today,* Dec. 1971.

Berkowitz, L., and R. Green, 1966. Film violence and the cue properties of available targets. *J. Pers. and Soc. Psych.* 3:525-30.

Bevan, W. 1977. Science in the penultimate age. *Am. Sci.* 65:538-46.

Bogart, L. 1972. Warning, the Surgeon General has determined that TV violence is moderately dangerous to your child's mental health. *Pub. Opin. Quart.* 36:491-521.

Cater, D., and S. Strickland. 1975. *TV Violence and the Child: The Evolution and Fate of the Surgeon General's Report.* New York: Russell Sage Foundation.

CBS Broadcast Group. 1977. *Learning While They Laugh: Studies of Five Children's Programs on the CBS Television Network.* New York: CBS Broadcast Group.

Children's Television Workshop. 1977. *CTW Research Bibliography.* New York.

Comstock, G., S. Chaffee, N. Katzman, M. McCombs, and D. Roberts. 1978. *Television and Human Behavior.* Columbia University Press.

Comstock, G., and M. Fisher. 1975. *Television and Human Behavior: A Guide to the Pertinent Scientific Literature.* Santa Monica, CA: Rand Corporation.

Comstock, G., and G. Lindsey. 1975. *Television and Human Behavior: The Research Horizon, Future and Present.* Santa Monica, CA: Rand Corporation.

Consumers Union. 1976. *The Six Billion Dollar Sell.* Mount Vernon, NY.

Cook, T. D., H. Appleton, R. F. Conner, A. Shaffer, G. Tomkin, and S. J. Weber. 1975. *Sesame Street Revisited.* New York: Russell Sage Foundation.

Dienstbier, R. A. 1977. Sex and violence: Can research have it both ways? *J. Communication* 27:176-88.

Fernandez-Collado, C., and B. S. Greenberg. 1977. *Substance Use and Sexual Intimacy on Commercial Television.* Report #5, Mich. State Univ.

Feshback, S. 1961. The stimulating versus cathartic effects of a vicarious aggressive activity. *J. Abnormal and Soc. Psych.* 63:381-85.

Ford Foundation. 1976. *Television and Children: Priorities for Research.* New York.

Franzblau, S., J. N. Sprafkin, and E. A. Rubinstein. 1977. Sex on TV: A content analysis. *J. Communication* 27:164-70.

Frueh, T., and P. E. McGhee. 1975. Traditional sex-role development and amount of time spent watching television. *Devel. Psych.* 11:109.

Gerbner, G. 1972. Violence in television drama: Trends and symbolic functions. In *Television and Social Behavior,* vol. 1. *Media Content and Control,* ed. G. A. Comstock and E. A. Rubinstein. U. S. Government Printing Office.

Gerbner, G. L. Cross, M. Jackson-Beeck, A. Jeffries-Fox, and N. Signorielli. 1978. *Violence Profile No. 9*. Univ. of Pennsylvania Press.

Hanratty, M. A., E. O'Neal, and J. L. Sulzer. 1972. Effect of frustration upon imitation of aggression. *J. Pers. and Soc. Psych.* 21:30-34.

Howitt, D., and G. Cumberbatch. 1975. *Mass Media Violence and Society.* Wiley.

Kaplan, R. M., and R. D. Singer. 1976. Television violence and viewer aggression: A reexamination of the evidence. *J. Soc. Issues* 32:35-70.

Katz, E. 1977. *Social Research on Broadcasting: Proposals for Further Development. A Report to the British Broadcasting Corporation.* British Broadcasting Corporation.

Kochnower, J. M., J. F. Fracchia, E. A. Rubinstein, and J. N. Sprafkin. 1978. *Television Viewing Behaviors of Emotionally Disturbed Children: An Interview Study.* New York: Brookdale International Institute.

Lefkowitz, M. M., L. D. Eron, L. O. Walder, and L. R. Huesmann. 1972. Television violence and child aggression: A follow-up study. In *Television and Social Behavior,* vol. 3. *Television and Adolescent Aggressiveness,* eds. G. A. Comstock and E. A. Rubinstein. U.S. Gov. Printing Office.

Lesser, G. S. 1974. *Children and Television: Lessons from "Sesame Street."* Random House.

Liebert, D. E., R. M. Liebert, J. N. Sprafkin, and E. A. Rubinstein. 1977. Effects of television commercial disclaimers on the product expectations of children. *J. Communication* 27:118-24.

Lyle, J., and H. Hoffman. 1972. Children's use of television and other media. In *Television and Social Behavior,* vol. 5. *Television in Day-to-Day Life: Patterns of Use,* eds. E. A. Rubinstein, G. A. Comstock, and J. P. Murray. U.S. Government Printing Office.

Ratner, E. M., et al. 1978. *FTC Staff Report on Television Advertising to Children.* Washington. DC: Federal Trade Commission.

Robertson, T. S., and J. R. Rossiter. 1974. Children and commercial persuasion: An attribution theory analysis. *J. Consumer Research* 1:13-20.

Rubinstein, E. A. 1975. Social sciénce and media policy. *J. Communication* 25:194-200.

______. 1976. Warning: The Surgeon General's research program may be dangerous to preconceived notions. *J. Soc. Issues* 32:18-34.

Rubinstein, E. A., J. F. Fracchia. J. M. Kochnower, and J. N. Sprafkin. 1977. *Television Viewing Behaviors of Mental Patients: A Survey of Psychiatric Centers in New York State.* New York: Brookdale International Institute.

Rubinstein, E. A., R. M. Liebert, J. M. Neale, and R. W. Poulos. 1974. *Assessing Television's Influence on Children's Prosocial Behavior.* New York: Brookdale International Institute.

Schaar, K. 1978. TV ad probe snags on Hill, researcher ire. *APA Monitor* vol. 9, no. 6, 1.

Schramm, W. 1976. The second harvest of two research-producing events: The Surgeon General's inquiry and "Sesame Street." *Proc. Nat. Acad. Educ.,* vol. 3.

Schramm, W., J. Lyle, and E. B. Parker. 1961. *Television in the Lives of our Children.* Stanford Univ. Press.

Silverman, L. T., J. N. Sprafkin, and E. A. Rubinstein. 1978. *Sex on Television: A Content Analysis of the 1977-78 Prime-Time Programs.* New York: Brookdale International Institute.

Singer, J. L., ed. 1971. *The Control of Aggression and Violence.* Academic Press.

Social Science Research Council. 1975. *A Profile of Televised Violence.* New York.

Stein, A. H., and L. K. Friedrich. 1975. *Impact of Television on Children and Youth.* Univ. of Chicago Press.

Surgeon-General's Scientific Advisory Committee on Television and Social Behavior. 1972. *Television and Growing Up: The Impact of Televised Violence.* U.S. Government Printing Office.

Tankard, J. W., and S. W. Showalter. 1977. Press coverage of the 1972 report on television and social behavior. *Journalism Quart.* 54:293-306.

Tannenbaum, P. H. 1972. Studies in film- and television-mediated arousal and aggression: A program report. In *Television and Social Behavior,* vol. 5. *Television Effects: Further Explorations,* eds. G. A. Comstock, E. A. Rubinstein, and J. P. Murray. U.S. Government Printing Office.

U. S. Commission on Civil Rights. 1977. *Window Dressing on the Set: Women and Minorities in Television.* Washington, DC.

Ward, S. 1972. Effects of television advertising on children and adolescents. In *Television and Social Behavior,* vol. 5. *Television in Day-to-Day Life: Patterns of Use,* eds. E. A. Rubinstein, G. A. Comstock, and J. P. Murray. U. S. Government Printing Office.

Ward, S., D. B. Wackman, and E. Wartella. 1975. *Children Learning to Buy: The Development of Consumer Information Processing Skills.* Cambridge, MA: Marketing Science Institute.

Watt, J. H., and R. Krull. 1977. An examination of three models of television viewing and aggression. *Human Communications Research* 3:99-112.

Zill, N. 1977. *National Survey of Children: Preliminary Results.* New York: Foundation for Child Development.

Children/Worksheet No. 1

## LOG OF PROGRAM EXCERPTS

### Positive Children's Programming

1. PBS has been known for its children's shows:

   Sesame Street
   —Counting to Ten
   —Naming Animals
   Mr. Rogers—exposing kids to new music
   Zoom—how to make a tree loom

2. Some attempts by the commerical networks to better children's programming:
   Shazam/ISIS—scene of interracial friendship
   Shazam/ISIS—moral regarding responsibility
   CBS's In The News
   Captain Kangaroo
   ABC's School House Rock—spot on interjections

3. Euell Gibbons spot—Secondary message—it's OK to eat wild foods.

4. The Swing
   RaceCar —2 pro-social PSA's promoting alternatives to violence

## Children/Worksheet No. 2

**(for use by groups of threes)**

1. In the film "The Anonymous Teacher" you saw TV program excerpts which many people would consider to present stereotyped roles of people. Imagine your child was watching the programs. Discuss in your group:

   —How do you feel about your child viewing such programs?

   —What kind of information would you like your child to have about the role of women in our society? The role of minority persons?

2. Write down a list of examples your group can think of which demonstrate television's tendency to present stereotyped images of persons.

## Children/Worksheet No. 3

**(exercise for small groups)**

Discuss and react to these two situations:

1. Your child asks permission to watch a show such as where there is sure to be at least one murder.

2. Your child says there is going to be a real-life murder in the neighborhood and asks permission to go and watch.

Questions: What would you do in each case? How different are the two situations? What effect do you think each situation would have on the child? If you say yes to either request, what does this say to the child about your attitude about violence?

## Children/Worksheet No. 4

**(for use in small groups)**

1. **Write five instructions for teachers or child care persons on how you want them to relate to children.**

2. **Based on your experience of viewing television (and film excerpts), write some of the instructions that seem to be given to producers of programs for children.**

3. **Create a list of five instructions you would like to hand to children's program producers.**

# Children/Worksheet No. 5

## (for small group use)

Discuss the creation of guidelines for the use of TV in homes with children. Remember that most children view a great deal of adult programming. As a group, write down the guidelines you feel are the most important:

## Children/Worksheet No. 6

Drawing on your work in your small group and in the sharing of guidelines, now create you own list of guidelines for the use of television in your home:

## Children/Worksheet No. 7

### HOMEWORK

1. Review the guidelines for TV viewing developed in this session and earlier sessions.

2. Re-read the Aimee Dorr article, pp. 217.

3. Begin to use the guidelines for your family. Making changes in how a family uses television can be very difficult and will almost surely cause tension and conflict. It is suggested you follow a routine such as the following:

   a. Before attempting any changes, observe how your family is using TV until you are thoroughly familiar with the patterns of viewing.

   b. Begin slowly. Make a point of viewing with children and asking casual questions about such things as:

      —How people work out problems
      —How different kinds of people are presented
      —How real the children believe programs are
      —What children like and dislike about certain characters
      —The truthfulness of commercials
      —How they tell when things are real or not
      —Do they ever wish they could be like some TV characters
      —Do they worry about things they see on TV

4. Plan a family meeting to talk about television and how your family uses it.

   a. Ask if children know why there are commercials. Clarify any confusion, pointing out that the purpose of commercial TV is to sell products.

   b. Talk about how much your children view and what programs are watched. Talk about some of the things you have observed about their use of TV.

   c. Try to find out if there are other constructive ways children would prefer to be using some of the time now given to television. Discuss possibilities.

   d. Ask children what they think they've learned from TV. Do they think children ever learn bad things.

   e. Ask children if they are ever bothered or worried by what they see on TV.

   f. Explain the workshop session you were in and tell children some of the things you've changed about your own use of television.

5. Discuss with your family a plan for using television differently in your home. You will probably get some protests and some "why" response at this point. Explain why you think it is important. Try to get children to be participants in an experiment of using TV differently. Suggest the following process:

   a. For one week everyone goes on watching the same programs as usual. But one of the following processes is used:

      (1) Adults and children reverse roles as they watch. Children think of themselves as parents who are responsible for young children and evaluate each program and commercial from that viewpoint. Adults watch as though they were young children, evaluating what they see from a child's viewpoint.

      (2) Ask the children to join the adults in a game. You are Martians who have been asked to report back to your leaders with information about what Americans are like. Your only source of information is what you are seeing on the TV screen. Watch TV with that in mind.

*(Worksheet No. 7, Continued)*

b. At the end of the week, meet again as a family group and discuss reactions. Try to get agreement on which programs seem most positive and which seem most negative. Talk about your own growing awareness of how some programs are manipulating you in ways you don't like, making you think of people and life in ways that violate your values. Try to get agreement for the following:

(1) Dropping at least one program that seems the most negative.

(2) Check TV Guide to find a positive sounding program that could be added to replace the one dropped.

c. Continue this weekly process as long as it seems to be working.

d. Other approaches you may want to try:

(1) Everyone agrees to pretend the TV set is broken for one week. At the end of the week, meet and discuss what changes occurred. How did each person feel? Was there any effect on conversation, game playing, going to other persons' houses to view? What other feeling level changes were there?

(2) Decide not to watch TV as such. Instead decide to **watch a specific program.** When you have viewed that program, turn the television set off. If possible, discuss the program in terms of values, violence, stereotyping, and the commericals.

(3) Radically adjust your television viewing for one, two, four weeks or more and view programs on the public television channel. As time progresses, observe whether there are any differences in relationships in the family. You may find family members experience a great deal of discomfort and uneasiness at first as radical changes are made in viewing habits. Arguments and conflict may go up. Does it go down again as new appetities and new program interests are developed? Does the different program fare affect conversation? Does TV viewing stay about the same, increase, or decrease?

(4) Radically regulate your television viewing to one or two hours per day. Allow the family to review program possibilities and to select the programs for the evening. This discipline forces each family member to become selective. Each can discover what programs are most important to him/her. Surely there will be conflict and disagreement. This gives the family a chance to work out in non-violent ways solutions to situations in which persons want different things.

(5) Ask the children to write a report on one program or a commercial each day which they view. Ask them to report on the following questions:

Name of the program
What was the most exciting part? Why did they think so?
Was there violence? Why was the violence necessary?
Can the child think of a way to handle the violent situation without violence?
What commercials were in the programs?
Write about one of them.
What product was advertised?
Was it truthfully described?
Why should someone buy the product?
Would the product be healthy and good for the person? Why?
Can families get along without the product? Why?

Discuss the reports in the family. Apply your own insights and values to the TV messages. Affirm pro-social positive behavior and values; discuss why you are concerned about anti-social, violent or negative behavior and values.

## Children/Worksheet No. 8

Select an hour, perferably from 9 a.m. to noon on Saturday to view children's television. Watch the commercials; make notes during the program. You are looking for the values messages being communicated and the techniques being used by the advertiser. You will want to make notes on:

a. the name of the advertiser
b. the time of play
c. the program it appears in
d. the station or channel
e. any material you find objectionable
f. any material you find commendable

The following questions will help to increase your awareness:

a. Is the product useful, health giving?
b. Can the product be dangerous if improperly used?
c. Does it attempt to deceive or overstate?
d. Does it use language children can understand?
e. Is it clear whether all elements shown are available for the price quoted?
f. Is assembly information clear to children?
g. What assumptions does the advertiser make about
    1. the child as purchaser
    2. boys and men
    3. girls and women
    4. minorities (absence, presence, minor/major roles, smart, dumb)?
h. What values does it communicate? How do these values differ from yours?
i. Are there other secondary messages which are communicated?

## Children/Worksheet No. 9

## HOMEWORK

Interview your children or neighborhood children. Be non-directive. You are trying to find out what the child's perceptions are. Ask questions such as:

a. What is your favorite program?

b. What do you like about it?

c. Who is your favorite TV person?

d. What makes this person special to you?

e. Is the person good most of the time or bad?

f. What does he/she do that's good/bad?

g. Do your favorite TV people tell the truth?

h. Do they ever lie or steal or cheat? What do you think about that?

i. Do you think commericals tell the truth?

j. Can you think of any commercials that don't tell the truth?

k. Tell me about violence on television. Do you like to watch shows that are violent?

l. What do you think about guns and shooting people?

m. Is that the way it is in real life?

n. Do you think your parents should let you watch violent programs? Why?

o. Do you ever get to stay up late to watch TV? How late? How often?

p. Do you ever get scared watching TV? What happens?

q. Do you think you learn things from TV? What do you learn? Tell me more about that.

After the interview, write down information which you think is important. Exchange experiences with others.

Chapter 6

# TELEVISION AND HUMAN SEXUALITY

## There's More to Sexuality Than Sex

*Elizabeth J. Roberts*
*Steven A. Holt*

*ELIZABETH J. ROBERTS is Executive Director of the Project on Human Sexual Development, based at Harvard University, and president of Population Education, Inc.*

*STEVEN A. HOLT is Assistant Director of the Project on Human Sexual Development.*

One's sexuality involves the total sense of self as male or female, man or woman, as well as perceptions of what it is for others to be female or male.[1] Human sexuality is part of one's basic identity—it is expressed through our lifestyles, and our social roles, in the ways in which we express affection and intimacy, as well as through our erotic behaviors and our attitudes towards the social and economic consequences of our sexual conduct. Human sexuality is expressed in the full range of interactions with others.

To understand what sexuality means, we must consider what it means to be masculine or feminine in our culture, how we feel about our bodies, what it means to get married, divorced or remain single, to live alone or with friends. Throughout our lives we are sexual beings and understanding sexuality is essential to our establishment of self image, self-understanding, personal identity and to the formation of human relationships.

Such an expanded notion of sexuality necessitates a new look at the way in which individuals learn about sexuality. It is a lifelong process, beginning in earliest childhood and continuing into old age. It involves not only learning new information, but also the development of attitudes, values, beliefs, fears and fantasies. It involves considerably more than a sixth grade lecture on menstruation or venereal disease or a-once-in-a-lifetime parent-child talk about the "facts of life." Sociologist John Gagnon has said of sexual learning:

> In any given society, at any given moment in its history, people become sexual in the same way they become everything else. Without much reflection, they pick up directions from their social environment. They acquire and assemble meanings, skills and values from the people around them. Their critical choices are often made by going along and drifting. People learn, when they are quite young, a few of the things they are expected to be and continue, slowly, to accumulate a belief in who they are and ought to be throughout the rest of childhood, adolescence and adulthood.[2]

**". . . seldom are erotic relationships between people seen or discussed in the context of a warm, loving, stable relationship."**

Seen in this light, very little of the information and few of the attitudes relevant to understanding sexuality are learned either by children or adults in a formal manner or through formalized sex education. Sexual learning takes place in informal, incidental ways. It occurs through interaction with our children, our spouses, our friends and colleagues; it takes place in the home, the school, the church and on the job. We learn and develop attitudes from reading advice columns in the newspapers and from watching the facial expressions, body postures, actions and reactions of our mass media heroes and heroines. What is learned, how it is learned and who the teachers are may change as we grow older, but we may be sure that sexual learning is going on all the time.

### Learning About Sexuality

Given the importance of television in the lives of American children and youth, the growing concern about the impact of television on the sexual learning of its viewers should come as no surprise. We are all well aware of the statistics that report that over 98% of American households have at least one television set (more than have indoor plumbing!) and that many of today's young people spend from one fourth to one half of their waking hours watching television, with only sleep surpassing television viewing as a time consumer.

Not only has television changed our sleeping and eating patterns, but it has affected our personal interests, our family and social interactions, our private moments alone and our perspective on and expectations of the world around us.

A growing body of literature suggests that television has become one of the chief storytellers of our society, supplementing and in many instances taking the place of the home, the church and the school.[3] Through its entertainment and dramatic programming, television makes our culture audible and visible to its members. It encourages people to perceive as normal and real that which fits the established fantasies of our society. Today young and old use what they learn on television as the cultural standard by which to judge the larger society. This role of television as a common socializer and storyteller has become particularly conspicuous when it comes to sexual messages.

In a recent study of family life and sexual learning mothers and fathers from all educational and economic backgrounds, ranked television as the major source of their children's information about sexuality (aside from what they, as parents, teach). Many of these same parents, however, voiced deep concern about the accuracy of the information and the appropriateness of the values and attitudes conveyed.

But what does television programming communicate about sexuality? Clearly, if sexuality is more than "sex" (and there is virtually no "sex" actually portrayed on commercial television), then many aspects of television programming need to be explored to understand the full meaning of television's sexual curriculum.

Just as it is difficult to separate sexuality from the rest of our lives, it is difficult to isolate television's sexual content from the rest of programming. Certainly sexuality on television encompasses more than a special on adolescent pregnancy, the double entendres of a variety show or the VD theme of a situation comedy. Sexuality is an ever-present and often subtle part of each and every human interaction portrayed. The themes, settings, storylines, characterizations and interpersonal dynamics portrayed on television provide considerable insight into what it means to be a man or woman in our society; how affection and intimacy are expressed; how erotic conduct fits into daily life; how men and women arrange their personal and professional relationships; who is treated with respect, who with disdain; what values are trivial or important.

To gain insight into our culture's current sexual norms and myths, as well as to appreciate more fully the role television plays in reinforcing and/or creating these attitudes, we consider five content areas that we believe are essential to understanding the meaning of human sexuality in our lives and television's role in the process of sexual learning.

Where research findings are mentioned, keep in mind that figures and statistics are drawn from an almost five-year time span. It is certainly clear to us that the medium has changed over time and continues each season to change; it is a very different medium now than it was five years ago. Also, the very nature of research on television tends to isolate specific pieces of information. The situation is far more complex than "black and white" statistics might lead one to believe. Much of the research doesn't even touch on areas that are of keen interest to us in exploring issues related to television and sexuality. Subtleties of intentions and actions have a strong impact on the viewer and yet are difficult to count, graph and chart. Nevertheless, this data presented in each section does provide a basis of common understanding that may facilitate future consideration and discussion.

**". . . the emphasis for males is on strength, performance and skill development . . . the emphasis for females is on attractiveness and desirability . . ."**

## Gender Roles

Learning about "appropriate" masculine and feminine attributes and roles is one of the most important aspects of learning that affects the development of sexual attitudes, values and behaviors. The first question asked about a newborn child is "Is it a boy or a girl?" Based on this information, parents—indeed all of society—treat and respond to the infant differently depending on the child's biological sex. Infant girls and boys are handled differently by their parents; and as they grow and develop they are encouraged to behave in different ways and to develop different skills and talents.

Perhaps the single most important element in this process of gender role learning is the way in which males and females are segregated—physically and psychologically. Physically, they are directed to play in different ways with different toys and later are encouraged to participate in different school curricula (shop or home economics) and to consider different work roles. Most important, however, boys and girls, men and women are segregated psychologically, by continuously having it reinforced that how one gender behaves, the other should not. Even the term "opposite sex" implies this segregation. Thus if girls can cry, boys do not; if men are competitive, women are not.

Substantial research indicates that, for the most part, television programs perpetuate the notion of men and women as opposites. On television, men are stereotypically ambitious, competitive, smart, dominant and violent. They think logically, are seldom beset by strong emotions (unless it is anger) and solve their own problems usually without the help of others.[5]

On the other hand, women on television tend to be sensitive, romantic, warm, submissive, timid. They are often over-emotional, unable to solve problems (theirs or anyone else's) and usually depend upon their father, husband, colleague or boyfriend to come up with the solutions.[6]

On television, similarity between men and women seems to breed conflict. A man and woman who share similar personality traits or perform similar tasks are more likely to have a relationship characterized by conflict and violence than are men and women with different personalities and roles.[7] For example, if a woman on television is nurturant and man independent, their relationship is more likely to be peaceful than if both are independent.

This segregation of the sexes on television is not a matter of "separate but equal." In a variety of ways, males dominate the prime-time television screen. There has been an increase in the number of female characters in general, and, in a number of television programs they do difficult and daring jobs. However, many television critics have pointed out that frequently such shows simply serve to illustrate how the traditional roles are "dished up in new guises." Often in these programs, the leading female character loses her cool more readily than male colleagues, becomes more emotionally involved and evokes more concern for her safety. "In subtle ways, such programs often manage to repeat the sexual stereotypes prevelant in most other television programming."[8]

Clearly, television communicates many messages about gender—about what it means to be a man or woman in our society. And most research concludes that gender stereotyping and physical and psychological segregation of the sexes on television is frequent and pervasive. While the majority of studies have concentrated on the portrayals of women on television, role stereotyping is evident in the portrayal of male characters as well. Always in control, coolly planning, emotionally uninvolved, the man is rarely seen exhibiting other human traits such as vulnerability, nurturance or emotional expressiveness. For the viewer, the pervasiveness of these images on television may imply that not only are these gender roles the prevailing ones, but that there are no alternatives.

## Body Image

Learning to view our bodies as a source of pride, pleasure and satisfaction or a source of embarrassment, shame or guilt also contributes importantly to our feelings about ourselves as sexual women or men. On television (and in most of society), body learning is closely related to gender role learning. In general, the emphasis for males is on strength, performance and skill development—"what can my body do?" The emphasis for females is on attractiveness and desirability—"how do I look?" Ninety percent of women on television are under 40.[9] In addition to being young, most women on television are attractive, well groomed and fashionably dressed.[10] And while male roles seldom are limited by an actor's wrinkles, baldness or other signs of aging, few women have significant leads on television once they no longer fit the conventional, youthful romantic roles.

**". . . mothers and fathers from all educational and economic backgrounds, ranked television as the major source of their children's information about sexuality . . ."**

In recent years, there has been particular concern about the use of women on television as "sex objects." Many critics point to story-lines and settings that require revealing or erotically enticing female costumes, or to camera angles that dwell on certain aspects of a woman's body. Such treatment not only casts women as sexual objects, but communicates that men are only interested in women for their erotic potential. And while male characters are allowed wider latitude in terms of physical appearance, they too are frequently typecast, particularly on action-adventure programs. In these shows, the prize (either fame, money, or a woman) usually goes to the strongest, swiftest and toughest man around. And a predominant way of demonstrating such attitudes on television is through the use of male bodies in physical violence and combat. Police officers and detectives on television are often latter-day gladiators required to prove their physical prowess (and their importance) over and over again.

Postures, gestures, the range of body movements, what's often called "body language,"[11] as well as how clothing is used and who "commands" space all communicate messages about the body. Watching some television with the sound turned off will let you see just how much is communicated without a word being spoken.

## Affection, Love and Intimacy

As social beings, we communicate and receive many messages about our needs for affection and affiliation. Through verbal and nonverbal communication, we learn how, when and with whom it is appropriate to share our intimate thoughts and feelings. Once again, this learning is different for males and females. Girls, more than boys, are encouraged to express affection through hugging, touching, nurturing and caretaking behavior. Indeed, males are often systematically discouraged from kissing, hugging, being gentle or nurturant, or asking for comfort and help. By the time most individuals are adults, they have learned whom to kiss and whom to hug, when we turn a cheek and when to extend a hand or settle for a slap on the back.

Research has concluded that the majority of "close relationships" on television are between partners who work together.[12] However, in these televised relationships, one is struck by the lack of genuine intimacy portrayed—particularly on dramatic or action-adventure programs. Although the heroes and heroines of these shows are portrayed as leading exciting and rewarding professional lives, they appear to endure austere private lives, lacking in physical or verbal expression of tenderness.[13]

As mentioned earlier, because television programs tend to portray men and women as opposites and because women are most frequently cast as the affectionate, romantic and vulnerable ones, it is difficult for males on television to demonstrate those aspects of their humanity at all, especially in television drama. In order for a male to ask for help, cry in frustration or unabashedly hug a friend, spouse or even a child, he is likely to be cast in a comedy role. A recent study[14] which looked at the expressions of intimacy that are observed in different kinds of programs revealed that displays of affection such as kissing, hugging or embracing appear more often in situation comedies than on crime, adventure, or dramatic shows.

**". . . one is struck by the lack of genuine intimacy portrayed . . ."**

The image of intimate and affectionate relationships portrayed on television is indeed a limited one. Women, primarily concerned with attracting men, are often cast in competition with other women, making close, affectionate, and warm relationships between them unlikely. Men, cast as all knowing, competitive and aggressive, seldom share their need for affection or their intimate feelings with anyone. Affectionate interchanges generally limited to situation comedies are deemed irrelevant and inappropriate to the "real world" of action and drama.

## Marriage and Family Life

The pattern and meaning of our close relationships and our expressions of intimacy are often communicated through the ways in which we arrange, integrate and manage our life style. Our family patterns and life styles—decisions to get married, to remain single, to live alone or with friends, to balance work and family responsibilites—are related to personal and/or social views of acceptable sexual roles and relationships. For many Americans today, this aspect of sexuality—especially the balance of work and family responsibilities—is being seriously examined. With increasing numbers of women entering the paid labor force, men, women, and children alike are struggling to find new ways to structure family roles and responsibilities.

Television programming seldom portrays the joys and difficulties inherent in these life style changes. In fact, most television does not reflect these changes at all. On television, marriage and family life are of concern primarily to females.[15] Most women on television are married, and if single, widowed or divorced, they are usually preoccupied with "getting a man." About 90% of television characters have no children,[16] but if a woman is a mother, she is even less likely to have a life outside her family (fewer than one out of every ten mothers on television holds a job).[17] While family life is clearly portrayed as a woman's domain, she seldom enjoys significant authority within her home. In charge primarily of all cooking and cleaning, she relegates most authority and responsibility to male characters.

The exception to this rule—i.e., successful working woman character—usually must pay the price by having problems in her personal relationships with lover, husband or child. A recent study[18] showed that women are significantly more often portrayed as either very rich or very poor—with wealth generally achieved by marriage or family background, rather than by work. On those few occasions where wealth was achieved by work, it was usually gained at the expense of happiness.

Such a picture is in striking contrast to the male lifestyles portrayed. For men, the important world is outside the home. Male characters rarely have much of a personal life, and family must take a second place to the more rewarding demands of a job. Television programs usually don't even tell us the marital status of most male characters and only 20% of male interactions have been found to focus on marital or family relationships.[19] As noted earlier, the vast majority of males on television are portrayed as strong, adventuresome and independent. An exception to this stereotype is the television husband and father, often portrayed in the situation comedy format. Unlike his male peers in the action-adventure shows, he is frequently characterized as inept and bungling.

One message, then, is that for women, marriage is an all consuming life style. Women who deviate from this life style risk their happiness and jeopardize the well being of their loved ones. For men, marriage and family may be seen as largely irrelevant or a lifestyle to which the less able and the ineffectual are relegated.

## Erotic Conduct and Its Social Consequences

In an effort to expand the notion of sexuality on television, we do not want to eliminate that topic which is most often thought of when discussing "sex on TV"—erotic conduct and its social consequences. How and when young children learn to identify certain modes of conduct as "sexy" or erotic is still a process not clearly understood. However, what each of us learns to define as arousing or erotic certainly influences the way in which we interpret the multitude of signals in our society related to sexual behavior.

There are few (if any) visual portrayals of explicit erotic activity on commercial television programing. There are, however, numerous cues that television provides to communicate to the viewer about the events that presumably will happen or have happened off screen. Unfortunately, little research exists at the moment to shed light on TV's portrayals of attitudes toward erotic activity. The limited research that has been done, however, suggests that frequently on television, erotic activity is linked with violence.

When sexual behavior is verbally

**". . . frequently on television, erotic activity is linked with violence."**

addressed in an action-adventure or dramatic program, it usually results from the discussion of rape or other sex crimes.[20] When not victims of sex crimes, women use their eroticism to entrap men. Whether prostitute or police woman, sex is a major vehicle for women in achieving their goals. In scripts, seldom are erotic relationships between people seen or discussed in the context of a warm, loving, stable relationship.

Contraceptive responsibility seems to be a "nonissue" on most television programs, and pregnancy or venereal disease are often used as the punishments for sexual activity on the part of young people.

Research is needed before we can understand the full impact television has on our development of erotic attitudes and behavior patterns. Even without such conclusive evidence, however, we might ask ourselves: Does television today suggest that erotic activity is an uncontrollable drive for men? Something that just happens to women? That sexual activity is a source of personal pain and social problems? That it is pleasurable? Or primarily a duty? That it occurs only for beautiful people? That being single means you swing, while marriage robs you of your "sex appeal" (appeal that can only be replenished by whiter than white teeth, bigger automobiles and multiple vitamins)? We might ask ourselves also if we see the values, attitudes, roles and relationships on TV today that we want our young people to emulate tomorrow.

As suggested earlier sexuality is more than "sex" and televised sexuality is not simply a program on pregnancy, venereal disease, homosexuality or rape. Continued investigation of these various aspects of sexual learning may help clarify the breadth of issues that should be considered in discussion about sexuality and television.

The research findings mentioned here by no means present a definitive picture of television programming. For most of us, the effect of television on our attitudes and behaviors comes as a result of a repetitive stream of nearly identical messages. Certainly a particular program may have a significant impact on our learning because of its excellence, its portrayal of moving incident, or the close fit between the program theme and our own individual needs. However, television principally affects our attitudes and behavior through its repetition of a limited range of human relationships, meanings and feelings communicated daily by its regular fare. Television's everyday portrayals of the roles and relationships between people may be far more important and have far more impact on a viewer in the long run than any one program.

Give this perspective, there is no one right way for sexuality to be portrayed on television. No one right way to handle a bedroom scene, a dual career family, or a young boy's first love affair. It is true that a list of do's and don'ts might reassure public interest groups and seem to make life easier for the industry executive. However, that kind of specificity would only perpetuate misunderstanding about televised sexuality by suggestion that it can be conveniently compartmentalized, monitored, and regulated.

Responsible television does and must deal with issues of sexuality. These issues are central to our humanity and they are at the core of human comedy and drama. Through discussion and exploration of such issues as masculinity/femininity, intimacy, love, relatedness, vulnerability and affection, perhaps the public and the television industry can come to understand the fullness of human sexuality and television's responsibility in presenting the diversity of this human experience with honesty, compassion and accuracy.

## REFERENCES


1. *Human Sexuality: A Preliminary Study*. The United Church of Christ, United Church Press, New York, 1972.
2. John Gagnon, *Human Sexualities,* Glenview, Ill., Scott Foresman and Company, 1977. Quoted in Janet Kahn and David Kline, *Towards an Understanding of Sexual Learning and Communication: An Examination of Social Learning Theory and Nonschool Learning Environments* (unpublished), 1978.
3. "Television as Teacher," a series of papers commissioned by the National Institute of Mental Health, 1978.
4. Elizabeth J. Roberts, David K. Kline, and John H. Gagnon, *Family Life and Sexual Learning,* Population Education, Inc., Cambridge, Mass., 1978.

5&6. See: Linda J. Busby, "Sex Roles Presented in Commercial Network Television Programs Directed Toward Children: Rationale and Analysis," Ph.D. dissertation, The University of Michigan and "Sex-role Research on the Mass Media," *Journal of Communication,* Vol. 25, No. 4, Autumn 1975. And: Nancy S. Tedesco, "Patterns in Prime Time," *Journal of Communication,* Vol. 24, No. 2., Spring, 1974. And: Jean C. McNeil, "Feminism, Femininity, and the Television Series: A Content Analysis," *Journal of Broadcasting,* Vol. 19, No. 3, Summer, 1975.

7&8. Hilde T. Himmelweit and Norma Bell, "Television as a Sphere of Influence on the Child's Learning About Sexuality" (unpublished), 1978.

9. Craig E. Aronoff, "Sex Roles and Aging on Television," unpublished study cited in *Journal of Communication,* Spring, 1974.
10. Michele L. Long and Rita J. Simon, "Roles and Status of Women on Children and Family TV Programs," *Journalism Quarterly,* Vol. 51, No. 1, Spring, 1974.
11. See Nancy M. Henley, *Body Politics,* Prentice-Hall, Englewood Cliffs, N.J., 1977, for an interesting look at nonverbal communication.
12. George Gerbner, "A Preliminary Summary of the Special Analysis of Television Content, undertaken for the Project on Human Sexual Development (unpublished), 1976.
13. See: Susan Franzblau, Joyce N.

Sprafkin, and Eli Rubinstein, *A Content Analysis of Physical Intimacy on Television,* Brookdale International Institute, Stonybrook. N.Y., 1976.

14. L. Theresa Silverman, Joyce N. Sprafkin, Eli A. Rubinstein, *Sex on Television: A Content Analysis of the 1977–78 Prime Time Programs,* Brookdale International Institute for Applied Studies in the Mental Health Sciences, Stony Brook, N.Y., September, 1978.
15. Long and Simon; and McNeil.
16. Judith Lemon, "A Content Analysis of Male and Female Dominance Patterns on Prime Time Television," Harvard Graduate School of Education Qualifying Paper, August, 1975.
17. N. Kaniuga, T. Scott and E. Gade, "Working Women Portrayed on Evening Television Programs," *Vocational Guidance Quarterly,* Vol. 23, December, 1974.
18. J. F. Seegar, *Journal of Broadcasting,* 19, 1975.
19. Long and Simon; and McNeil.
20. Franzblau, Sprafkin, and Rubinstein, 1976.

## Sexuality on the Screen

*Susan H. Franzblau*

*SUSAN H. FRANZBLAU is a doctoral candidate in developmental psychology at the State University of New York At Stony Brook. She has written and published and taught in the area of sexuality, sex-roles, and effects of televised aggression on girls.*

Sex is probably the most private and misconstrued of any social behavior we think and talk about, and engage in. Although public and private disagreement about the nature and allowability of most sexual activities has always existed, the capacity in today's world of various media to reach large segments of the population has created a feeling of immediacy in dealing with this issue.

It is clear that people differ widely in their opinions concerning the effects of sexuality displayed in books, movies, and TV. Television, in particular, has caused much concern inasmuch as it has become, since the late 1940s, a major socializing agent in this country. Research indicates that children and adults learn from models portrayed on TV and use this information to establish norms and guidelines for how they will act in a variety of situations. Growing awareness of television's pervasive power to teach has led to much concern about the sexual messages communicated via the television screen. As a result of this concern some action has been taken. For example, the term "edited for TV," indicates that when movies not specifically made for television are shown on TV, the more explicit sexual scenes have been removed. This act of censorship is indicative of the knowledge that TV affects a vast number of people and should not be allowed to convey the sexual information that movies now generally permit.

Even with such action taken, recent reports note that many parents believe that their children still receive large amounts of sexual information from television but that parents are uncertain about the accuracy of the information about sexual activities and values and quite uncertain about the necessity of broadcasting any programs which involve sexuality. Some groups have asked that sex be completely removed from prime-time TV.

In the early 1970's, as a response to public pressure about sex (and violence), the National Association of Broadcasters (NAB) set aside a period of time during which both sex and violence would potentially appear less frequently. This period of time was known officially as Family Viewing Time (FVT). Although the Family Viewing Code was declared unconstitutional by a California Federal judge in November of 1976, the NAB declared its intention to informally abide by a sex and violence-free period of prime-time programming. However, despite public pressure to remove sex from TV and the NAB's response to this pressure, no one had investigated, in any systematic way, the nature and amount of sex that appeared on prime-time television. In other words, no one had characterized televised sexuality to determine what was being censored and if censorship was, in fact, needed.

**"The viewer sees a picture of sex as harsh, hurtful and manipulative."**

Sexuality, despite disagreements as to its value, is a part of life and culture. Clearly, our opinions about sex and each individual's femaleness and maleness are derived, in part, from the information our society gives us. Inasmuch as television is a powerful socializer, it has the capacity to provide this information through its presentation of sex-role relationships and physical intimacy. As one researcher puts it:

> "It is hard to imagine what a society tolerates in its mass media as a portrayal of sexual reality will not come to be the kind of sexual reality that a society's next generation lives."

Today's social reality, however, reflects confusion about sex. By the time children reach puberty they are laden with prejudices and misconceptions about sexuality. Although research efforts in this area do much to dispel unfounded myths about sex, body image, and reproduction, these efforts do not always reach the population for whom sex education can do the most good—parents and their children. Some researchers have criticized the light in which TV has presented sexuality; that is, they view television as artificially separating sex from the larger context of social behavior within which its value has shifted historically and situationally. They claim that these biased and limited portrayals adversely affect the developing sexual attitudes of its young viewers.

There is obviously an urgent need for research that would help clear up the conflicting views regarding televised sexuality and its effects. Before making generalizations concerning what should or should not be portrayed on prime-time television and how sex might influence the lives of TV viewers, we need to explore the nature and extent to which sex is portrayed on television today.

## Physical Intimacy

The research presented here is an attempt to approach the issue of sex on TV from the broader perspective of physical intimacy. In this way we are able to examine a wide range of behaviors and verbal communications that involve intimacy (e.g., hugging a friend) but may or may not be defined as explicitly sexual. As such we are able to look at a continuum of sexual behaviors taking into account the concept of intimacy/sexuality as behaviors appearing in social contexts and judged in relation to these contexts. We are asking how often, under what conditions, and with whom particular physically intimate behaviors and references to such behaviors occur. Two studies, one selecting from programs aired during 1975 and the other selecting from programs aired during 1977, coded and analyzed the physically intimate content of programs shown between 8-11 p.m. by the three major networks, CBS, NBC, and ABC. The 1977 study added the category movies/specials to the four program types analyzed in 1975:

1. situation comedies
2. variety shows
3. dramas
4. crime-adventures

The general list of behaviors examined in both studies included:

1. kissing
2. hugging
3. flirting/suggestiveness
4. touching—nonaggressive
5. touching—aggressive
6. homosexual behavior
7. prostitution
8. rape
9. masturbation
10. intercourse

All categories took into account both implied and direct physical behaviors and implied and direct verbal references to behaviors. An average of 62.5 programs were recorded and analyzed in the 1975/1977 studies.

## What Is Sex on TV?

In both the 1975/1977 samples we found that the behaviors that appeared most often were kissing, embracing, aggressive touching, and interpersonal casual touching (e.g., handshakes). Casual touching was by far the most frequently coded behavior. As in 1975, the 1977 study found that controversial sexual behaviors rarely appeared. For example, in 1975 there were some verbal references to rape and prostitution, but these tended to appear on dramas and crime-adventure shows as characters talked about solving crimes:

> *e.g., On the crime-adventure show,* Baretta: *Baretta talks to another cop about a criminal they know. "He's been into prostitution."*

The differences between the 1975 and 1977 samples are reflected in two major changes. First, flirting behavior quadrupled in frequency from 1975 to 1977. Less than one behavior an hour occurred in 1975 while in 1977 there were more than three/hour. Innuendoes, the verbal analog for flirting, increased from one to about seven an hour. Second, although in 1975 virtually no direct behavioral or verbal reference to intercourse was coded, the 1977 sample included fifteen verbal references to sexual intercourse. None of these references, however, were explicit.

## Where Is Sex On TV?

In both samples the results showed that kissing appeared most often on situation comedies and least often on crime-adventure shows. Variety shows and dramas had approximately the same number of acts of kissing, falling midway between those appearing on situation comedies and crime-adventure shows. Also, kissing was spoken or sung about more often on situation comedies and variety shows than other, more serious programs. Crime-adventure programs showed less embracing than any other program type. But there was far more aggressive touching (e.g., punching, karate, general street brawling) on these shows than on lighter comedy shows. And the reverse was true for casual touching. Situation comedies and variety shows contained far more instances of casual touching than did crime-adventure programs. It is interesting to note, however, that even though variety shows are light fare, they contained significantly more acts of aggression than we would have expected considering the type of program.

Since 1975, significant program type changes on two of the three networks have influenced the frequency with which each network portrayed particular physically intimate behaviors. For example, post FVT crime-adventure shows (9-11 p.m.) were reduced by one-half since 1975 on ABC and NBC. NBC, however, replaced this show type by an equally aggressive movie/special. ABC moved to the situation comedy, which meant that sexuality was exchanged for violence on ABC. Whereas the 1975 study found no significant difference by network, the 1977 sample points to a heavily violent NBC and a heavily humorous/sexual ABC. We must, however, use this information with caution. Although some sexual behaviors are on the increase, both analyses suggest that from 1975 through 1977 the most frequent physical act involving another human being on network TV was non-sexual and casual.

We note an increase in the tendency to entice the viewing audience with flirting and subtle, indirect verbal references to sexuality (i.e., the innuendo). In both samples, situation comedies and variety shows presented more innuendos than did any other program type. More verbal references to flirtation and seductive behavior appeared on variety shows than any other type. The following innuendo gives a good idea of the kind of suggestiveness involved:

> *During the Lear comedy,* Maude, *Maude tells Walter, "You can't teach an old dog new tricks, which I pointed out to you at 2:30 this morning."*

Looking at physical and verbal acts of intimacy for all categories, it seems that variety shows and situation comedies have the most physically intimate acts and that these acts have increased, at least during the hiatus between the two studies. I would characterize the program types in this way:

On situation comedies and variety shows, the characters touch casually, kiss, embrace, and indirectly, through innuendo and flirtation, suggest a more sexual kind of physical intimacy. Flirtation and seductive talk appear most often on variety shows. Aggressive touching is almost nonexistent on situation comedies but appears more often on variety shows. Physical intimacy on movies, specials and crime-adventure programs, however, is primarily aggressive with not much of the more tender behaviors such as kissing and embracing. In fact, even indirect statements about intimacy (i.e., the innuendo) are rarely heard on these programs. The dramas stand somewhere in the middle, representing the more moderate examples of physical intimacy: kissing, embracing, and casual touching.

**"Sexual freedom calls for the ability to act responsibly. This is not possible with misinformation."**

Television's message about sexuality is not direct or clear-cut. Its message is cloaked, increasingly, in subtle innuendoes and provocative flirtations. Suggestive messages, however, are often not left to be interpreted by the viewer inasmuch as canned laughter and laugh cards tease out into the open the innuendoes producers want to be emphasized. We are told what the producers want us to think about sex; the viewer is enticed to laugh because the audience is laughing. Clearly, television prescribes and manipulates how we interpret subtle sexual remarks. In fact, it probably helps make us think of those remarks as sexual in the first place. Television also tells us, over and over, that sex is an acceptable subject if it is cloaked in humor or ridicule, or viewed as a despicable crime.

The sexual humor of variety shows, despite the glamour and rowdiness of those programs, follows standard sex-role themes, joking about sex within a very conventional framework. For example, women are more likely than men to act seductively and are often portrayed as "sex objects," and mindless ones at that.

## Sexuality On Situation Comedies

The situation comedies, on the other hand, have potentially given the television viewer a different look at both traditional and non-traditional sexuality. New ideas for television such as interracial marriage (The Jeffersons), love between the elderly (*All in the Family, Doc*), problems of single men and women *(One Day At A Time),* alternative sexual life styles and extramarital relationships (*Three's Company, Soap, Dallas*) appear again and again. The method used, however, is still innuendo for the most part.

The outcome was and remains superficial. These comedies present a casual look at sensitive issues which deserve more serious treatment. We might well conclude that sex is funny because sex appears most often on funny shows. The reality, as we well know, is not slapstick.

This disappointing finding is further punctuated by the *lack* of sexuality on serious shows. The shows which could provide a framework for dealing responsibly with such significant issues as rape, abortion, prostitution, pregnancy and childbirth, or homosexuality seem to be lacking in any but infrequent and lurid glances at these issues. Within these shows, sex tends to often become a tool, used by criminals of one sort or another to wield power over others. Rather than showing sex as a part of a more complete relationship, sex is displayed as the only aspect of a relationship. Children who watch television after nine o'clock, and there are many, get the idea that people who engage in sex are one-dimensional, and often bad. Thus, sex becomes pure exploitation and TV for the most part conveys the message that women are the exploited. The viewer sees a picture of sex as harsh, hurtful, and manipulative.

Unfortunately, instead of offering a contrast to this picture, the heroes and heroines of crime-adventures and drama programs seem to lead austere private lives, although their public lives dealing with criminals are exciting and extremely elaborate. If the hero is married, the viewer gets little or no glimpse of that relationship. The heroine is even more remote from sexuality. Sex is never a direct and integral part of an otherwise picturesque life.

## What Are The Alternatives?

If TV is to be a reflection of society, we might suggest that TV more accurately present a picture of sexuality as it reflects real life. Our relationships are not often ribald jokes. Our concerns about sex are serious and as persistent as our other concerns. In a sense, TV continues to be afraid to give more meaning to an area of deep concern. If we learn from TV, is there not some way we can treat these matters as we treat our other concerns? Is there not some way our children can have the opportunity to learn in an honest, direct way about a most important part of their lives? If TV reflects things to come, do we honestly want TV to teach us, on one hand, to laugh at everything having to do with sex, and on the other hand, to fear it?

It is interesting to note that 72 years ago, in 1907, Sigmund Freud spoke to this idea:

> ". . . everything sexual should be treated like everything else that is worth knowing about . . . the curiousity of children will never become very intense, for at each stage in its inquiries it will find the satisfaction it needs. Explanations about the specific circumstances of human sexuality and

> some indications of its social significance should be provided before the child is eleven years old . . . A gradual and progressive course of instruction in sexual matters such as this, at no period interrupted . . . seems to me to be the only method of giving the necessary information that takes into consideration the development of the child and thus successfully avoids ever-present dangers."

The "ever-present danger" seems to be the danger of receiving misinformation, leading to a distortion of reality. It seems to me quite possible that if we are not more careful, this is what television will continue to do.

I said earlier that there has been a great deal of concern about the presence of sexuality on television. Our research supports the validity of that concern, not because sex is there, but because what is there so inadequately represents the positive nature of human relationships and sexuality.

Often, parents support sex education in schools because of their fear that the television medium and other impersonal modes of communication appeal to intimacy at earlier and earlier ages. However, parents are also often reluctant to discuss sex openly for fear that discussion of sex will eventually lead to the very behavior they view education as supposed to prevent. This paradoxical situation compounds the initial problem of learning about sexuality and learning to act responsibly as a sexual being. But sex education continues in one way or another. I believe the larger issue is that not taking responsibility for its form will eventually lead to misinformation—a significantly more dangerous consequence than the consequence of good education. Concealing sexual activities under a rubric of misinformation or silly innuendoes can only alienate a child from her/his sexuality. This is especially true if no one takes the responsibility for disseminating such information in an intelligent way. Sexual freedom calls for the ability to act responsibly. This is not possible with misinformation. Good sex education must not be equated with the recommendation of indiscriminate indulgence in sex nor must it be equated with the view that sex is either a joking or lurid matter.

On TV it tends to be all right to laugh about sex but not all right to take it seriously as a natural part of a loving relationship. It is all right to show, quite explicitly, a woman being raped (presumably because the rapist is a criminal and the scene does not express approval of sexual behavior), but it is not right to show a positive, loving intimate act.

Is that what we want television to teach us and our children about sexuality and interpersonal relationships? I do not suggest complete freedom from censorship but I suggest that prime-time TV take us seriously, as total persons. And I suggest that we, as citizens, have a right and obligation to demand change from ridicule to reality, from sillyness to sensitivity, from perversion to a positive education toward a moral and informed sexuality.

> Human sexuality is not confined to the bedroom, to the nighttime or to any single area of the body. It is involved in what we do, but it is also what we are. It is an identification, an activity, a biological and emotional process, an outlook and an expression of self . . . It is an important factor in every personal relationship and in every human endeavor from business to politics.
>
> *Human Sexuality*
> The American Medical Association

## Sexuality/Worksheet No. 1

# LOG OF PROGRAM EXCERPTS

**1. Innuendos**

WKRP Cincinnati—Woman suggestively tells man that she has been out with other guys.
Welcome Back Kotter—Sweat Hogs joke about Vinny spending his time in the backseat of a car.
Movie—Two guys discuss women "school teachers make you do it over until you get it right."
LaVerne and Shirley—LaVerne says, "this fur coat is the second best thing I ever felt."
Three's Company—Mrs. Roper, "If it were any good, I wouldn't fight it".
Mork and Mindy—Gal says, "Whatever turns you on."

**2. Casual Sex—An encounter not sanctioned by long-term commitment**

Dallas—Embracing couple confronted by man's wife
Movie—Couple in bed discuss the man's lovemaking style
Movie—Couple in bed making love
Movie—Man invites woman to have an affair

**3. Sexual activity not commonly sanctioned by society**

Movie—Man in woman's bathhouse
Alice—Man discusses his sex life and preference for catsup
Alice—Man impersonates a waitress
Love Boat—Man dressed up as a policewoman
Soap—Jody (homosexual character) proposes to a woman

**4. Romantic Sex—Episode between long-term or committed partners**

Centennial—Man and woman in bedroom
Family—Reunited husband and wife cuddle
Movie—Husband is too tired to respond to wife's advances

# Sexuality/Worksheet No. 2

Feeling responses to sexual images on TV.

| | 1 | 2 | 3 | 4 | 5 |
|---|---|---|---|---|---|
| **Titilated** | | | | | |
| **Amused** | | | | | |
| **Disgusted** | | | | | |
| **Disturbed** | | | | | |
| **Interested** | | | | | |
| **Surprised** | | | | | |
| **Shocked** | | | | | |
| **Scared** | | | | | |
| **Angry** | | | | | |
| **Happy** | | | | | |
| **Self-Conscious** | | | | | |
| **Inadequate** | | | | | |
| **Excited** | | | | | |
| **Aroused** | | | | | |
| **Put Down** | | | | | |
| **Manipulated** | | | | | |
| **Sad** | | | | | |

# Sexuality/Worksheet No. 3

The following situations are present in society and, to some extent, on TV. Select one or more of these situations (or one you create yourself) and write a brief description of how you would develop the situation(s) for TV in a way that provides helpful accurate information and is in keeping with your own values. You may use any TV format and any broadcast time. Be aware that people do need information about sexuality issues and are presently getting information about sexuality from TV. You may want to refer to WORKSHEET #4 to determine what you believe needs to be communicated to viewers. Consider using existing shows and dramatic series with present performers.

A young woman is living with a man she cares a lot about in an open, warm, sexual relationship.

A couple live together in a homosexual relationship.

Middle aged wife wants a separate bedroom so that she can be more in control of her life.

Young married couple experiences first sexual intercourse in a tender lovemaking scene.

Female police officer encounters a rape; how does she handle it?

Celebrity couple discusses their sexual and personal incompatibility, and their decision to secure a divorce.

Police officer is returning a minor to her home after her arrest for prostitution.

In a short comedy sketch a mother deals with her 13-year-old about masturbation.

Create a television commercial for prophylactics.

Create your own situation and human sexuality issue.

# Sexuality/Worksheet No. 4

This worksheet is for your own information and for sharing as you desire with the group. Check space when you agree with statement.

| | This is a common human experience | People (I, my family, others) need info about | It could be treated sensitively on TV | I have seen it treated on TV Good | I have seen it treated on TV Bad | Not for TV |
|---|---|---|---|---|---|---|
| Kissing | ______ | ______ | ______ | ____ | ____ | __ |
| Caressing | ______ | ______ | ______ | ____ | ____ | __ |
| Petting | ______ | ______ | ______ | ____ | ____ | __ |
| Intercourse | | | | | | |
| Extra-martial | ______ | ______ | ______ | ____ | ____ | __ |
| Pre-martial | ______ | ______ | ______ | ____ | ____ | __ |
| Non-married | ______ | ______ | ______ | ____ | ____ | __ |
| Marital | ______ | ______ | ______ | ____ | ____ | __ |
| Masturbation | ______ | ______ | ______ | ____ | ____ | __ |
| Wet dreams | ______ | ______ | ______ | ____ | ____ | __ |
| Menstruation | ______ | ______ | ______ | ____ | ____ | __ |
| Involuntary Erection | ______ | ______ | ______ | ____ | ____ | __ |
| Impotence | ______ | ______ | ______ | ____ | ____ | __ |
| Inability to have orgasms | ______ | ______ | ______ | ____ | ____ | __ |
| Absence of lubrication | ______ | ______ | ______ | ____ | ____ | __ |
| Premature ejaculation | ______ | ______ | ______ | ____ | ____ | __ |
| Feelings related to sex | | | | | | |
| Joy and pleasure | ______ | ______ | ______ | ____ | ____ | __ |
| Frustration and sadness | ______ | ______ | ______ | ____ | ____ | __ |
| Inadequacy | ______ | ______ | ______ | ____ | ____ | __ |
| Jealousy | ______ | ______ | ______ | ____ | ____ | __ |
| Envy | ______ | ______ | ______ | ____ | ____ | __ |
| Possessiveness | ______ | ______ | ______ | ____ | ____ | __ |
| Anger | ______ | ______ | ______ | ____ | ____ | __ |
| Anxiety | ______ | ______ | ______ | ____ | ____ | __ |

(Worksheet No. 4 Continued)

| | This is a common human experience | People (I, my family, others) need info about | It could be treated sensitively on TV | I have seen it treated on TV — Good | I have seen it treated on TV — Bad | Not for TV |
|---|---|---|---|---|---|---|
| Body image concerns | | | | | | |
| Fear of aging | ______ | ______ | ______ | ____ | ____ | __ |
| Breasts | ______ | ______ | ______ | ____ | ____ | __ |
| Illness, mutilation | ______ | ______ | ______ | ____ | ____ | __ |
| Disability | ______ | ______ | ______ | ____ | ____ | __ |
| Lifestyles | | | | | | |
| Homosexual | ______ | ______ | ______ | ____ | ____ | __ |
| Communal (Sexually active) | ______ | ______ | ______ | ____ | ____ | __ |
| Communal (Sexually inactive) | ______ | ______ | ______ | ____ | ____ | __ |
| Extended family | ______ | ______ | ______ | ____ | ____ | __ |
| Non-married (Sexually active) | ______ | ______ | ______ | ____ | ____ | __ |
| Celibate | ______ | ______ | ______ | ____ | ____ | __ |
| Pregnancy | ______ | ______ | ______ | ____ | ____ | __ |
| Childbirth | ______ | ______ | ______ | ____ | ____ | __ |
| Menopause | ______ | ______ | ______ | ____ | ____ | __ |
| Abortion | ______ | ______ | ______ | ____ | ____ | __ |
| Sterilization | ______ | ______ | ______ | ____ | ____ | __ |
| Contraception | ______ | ______ | ______ | ____ | ____ | __ |
| Divorce | ______ | ______ | ______ | ____ | ____ | __ |
| Incest | ______ | ______ | ______ | ____ | ____ | __ |
| Group sex | ______ | ______ | ______ | ____ | ____ | __ |
| Rape | ______ | ______ | ______ | ____ | ____ | __ |
| Pornography | ______ | ______ | ______ | ____ | ____ | __ |
| Prostitution | ______ | ______ | ______ | ____ | ____ | __ |
| Transvestism | ______ | ______ | ______ | ____ | ____ | __ |
| Transsexualism | ______ | ______ | ______ | ____ | ____ | __ |
| Voyeurism | ______ | ______ | ______ | ____ | ____ | __ |

## Sexuality/Worksheet No. 5

### HOMEWORK

Select a variety of program trypes to view. Note presence or absence of sexual incidents (refer to Worksheet #4) and how they are treated. Be aware of positive or negative, types of persons involved, (age, gender, race, marital status, gay or straight, attractive or unattractive character) visual or verbal, straight or innuendo.

| **Program** | **Incident/Characteristics** | **Comment** | **Overall Rating Positive—Negative** + | − |
|---|---|---|---|---|
| | | | | |
| | | | | |
| | | | | |
| | | | | |
| | | | | |
| | | | | |
| | | | | |
| | | | | |
| | | | | |
| | | | | |
| | | | | |
| | | | | |
| | | | | |
| | | | | |

Chapter 7

# TELEVISION NEWS

## News and Values

*George C. Conklin*

*GEORGE C. CONKLIN is Director, Center for Media Studies at the Pacific School of Religion, Berkeley, California. He is a national T-A-T trainer.*

Most of what we've been taught to expect from television news has been learned from watching television news. What we don't learn about are the tensions in values which are part of the news system. Those tensions, which hide behind the friendly open faces of TV news, profoundly affect the kind of news we get.

The primary tension is between the nature, function and tasks of the journalist in society and the growing role, particularly at the local station level, of television news programs as a profit center for the broadcasting industry. It is a tension between social needs and show business, between journalistic excellence and winsomeness on camera, between accurate perceptive reporting and that magic image (if you are seeking employment as an anchorperson) of being authoratative, warm and friendly. These qualities are not mutually exclusive, but again and again the choice is made on the side of audience appeal.

Charles Kuralt, who had been "On the Road" for CBS-TV for eight years, spoke to the 1975 annual meeting of the Radio and Television News Directors Association: "I have spent a lot of time watching the local news on television (out there). My overwhelming impression of all those hours in all those years is of hair. Anchormen's hair . . . Hair carefully styled and sprayed, hair neatly parted, hair abundant and every hair in place . . . I remember the style, but not the substance. And I fear the reason may be there wasn't much substance there . . . I recognize that an anchorman can be brilliant as well as beautiful, and that some are . . . I don't care what station managers say, I don't care what the outside professional news advisers say. I don't care what the ratings say, I say this is the continuing disgrace of this profession. The plain truth is that in a society which depends for its life on an informed citizenry, and in which most citizens receive most of their information from television, millions are getting that life-giving information from a man or a woman whose colleagues wouldn't trust to accurately report on his or her afternoon round of golf . . . I think we all ought to be ashamed, that 25 years into the television age, so many of our anchormen haven't any basis on which to make a news judgment, can't edit, can't write and can't cover a story."

### The Nature of Communication

Mass communication came into being as the tribes of humanity grew larger and larger throughout history. The quiet voice sharing new information around the camp fire became the loudest speaker of the tribe, talking now to a larger group. Early picture and symbol language developed into written languages, and humanity began the slow climb to literacy, speeded by the invention of printing devices in China, Korea, and Germany. Within a very short span of history (the last one hundred years) electronic technology has radically changed the flow of information in society. Awesomely large audiences see and hear such events as landing on the moon, Olympic Games and the U.S. presidential activities of travel, debates, and resignation. Because we view TV programs in the privacy of our homes, we tend to forget that television gathers communities—persons sharing a common experience in numbers undreamed of fifty years ago. For example, an estimated 90 million people watched the first televised debate between presidential candidates Carter and Ford.

The early tribes of humanity had needs which communication helped meet. While their histories are lost in time, we may guess they needed to know where food could be found, where they would be safe from harm, where they could prosper and grow in numbers. Students of communication over the years have sought to identify the functions of communications in society. The most commonly used theories cluster human information-sharing events under four headings:

—First, the need to know what is going on around us, to look about the environment (or hear, or smell) and know what is happening, to be aware of opportunity and danger.

—Second, to help the group figure out how to respond, or perhaps not respond, to what's happening out there.

—Third, to share with each new generation of children what the tribe has learned about living in the world.

—Fourth, to entertain, amuse and play as a group.

These four functions may be described simply as news, editorial (in the sense of the elders saying this is what is happening, this is our response), education and entertainment. The functions can overlap and often do, but the distinctions are a helpful frame of reference.

### Primitive News Coverage

In the primitive tribe the news

coverage of the surroundings was provided by individuals, who were scouts, watchers and observers. They ranged afield and reported back on the presence of friends or enemies, the availability of food and shelter. The decision makers of the tribe (it could have been a single leader or the entire group) made decisions for their future on the basis of the reports. The skills of the tribe in hunting, fishing, farming, shelter and clothing making, doing battle and governing themselves (or being ruled) would be taught to each generation. After the hunt, battle or feast, the story tellers and jesters would recount the story for their listener's pleasure.

The functions of television news today are the same. But there is a major problem in identifying whether the communication role of television news is that of providing news or entertainment. The tribal storyteller could enlarge upon the tales told; the tribal scout could not. The safety of the tribe depended on accurate reporting, informing wise decision making.

## What is News?

The definitions of news today are many. One broadcast journalist suggests that it is knowing (and telling) about an event that is of interest to at least one other person. The more other persons that are interested, the bigger the story. Another defines news as information people receive second hand about worlds, events, and processes which are not available to their immediate personal experience. A good working definition says it is the responsibility of the journalist to provide us with a picture of the world upon which we can act.

This last definition expands to include the accuracy and relevance of the picture of the world that is reported. It also includes reporting on our reactions—individual responses such as conserving or not conserving energy—and community, state, national and world responses through our many organizations.

We need that picture, that information, that *news* about what is happening and how we are responding. We need it just as the primitive society needed to know where the game was and what the plan was to hunt it. We need the information quickly and accurately. And because people live in different ways, one person may need the information in greater detail than others.

All this means that the task of the journalist is not an easy one. Selection has to be made from the almost infinite happenings of each day. The selection may need to be set in the perspective of prior events. Research is often needed to find related information, which may be hidden, or not immediately apparent.

Other criteria for evaluating television news can be found in these excerpts from the ethics codes of newspaper and broadcasting associations.

"The primary function of newspapers is to communicate to the human race what its members do, feel and think. Journalism, therefore, demands of its practitioners the widest range of intelligence, or knowledge, and of experience, as well as natural and trained powers of observation and reasoning. To its opportunities as a chronicle are indissolubly linked its obligations as teacher and interpreter."

CODE OF ETHICS
THE AMERICAN SOCIETY
OF NEWSPAPER EDITORS

"Television is seen and heard in nearly every American home . . . Television broadcasters . . . are obligated to bring their positive responsibility for professionalism and seasoned judgment to bear on all those involved in the development, production and selection of programs . . . television programs should not only reflect the influence of culture, but also expose the dynamics of social change which bear upon our lives . . . Television broadcasters and their staffs occupy positions of unique responsibility in their community's needs and characteristics in order to better serve the welfare of its citizens . . . a television station's news schedule should be adequate and well balanced. News reporting should be factual, fair and without bias . . . Good taste should prevail in the selection and handling of news: Morbid, sensational or alarming details not essential to the factual report, especially in connection with stories of crime or sex, should be avoided. News should be telecast in such a manner as to avoid panic and unnecessary alarm.

TELEVISION CODE
NATIONAL ASSOCIATION
OF BROADCASTERS

## Definition vs. Performance

In San Francisco several years ago, the television audience heard the top-rated local newsman on an ABC-owned television station say at 10 p.m., "A male sex organ has been found beside a railroad track in Oakland. Details at 11."

When quizzed about the incident during a CBS network documentary on tabloid news broadcasting, the newsman said there had been violent crimes in the area and there must have been a victim, concluding with the statement, "We didn't cut it off and put it there."

Between the measured words of news definitions and codes, and the real world of television journalism there is a gap. It comes in large measure from the conflict between the role of a journalist and the economics of television.

When word was announced of network television's first million-dollar-a-year contract (for a five-year period), Walter Cronkite is reported as saying he felt "the sickening sensation that we were all going under, that all of our efforts to hold network news aloof from show business had failed."

Ratings are the key element in television economics. The larger the audience, the higher the ratings. A top-rated news program can charge more for advertisements and more advertisers want to be on the program. The salaries of television news anchor persons are salaries of stars, not journalists. Industry publications report that several local news anchor persons in the largest cities are paid $250,000 per year, while news directors at the stations earn $25,000. Salaries for on-camera people in the top ten markets are often in the $100,000 range and for the top fifty markets, $25,000 to $75,000 is not uncommon.

There is a substantial difference between the economics of network and local television news. Network news is not profitable. The three networks together spend some $150 million on television and radio news. How much they get back from advertising is a corporate secret, but estimates run from 30% to 60%. Local station profits are also closely guarded secrets; there are reports that in major markets, newscasts and documentaries may bring in 50% of a station's highly profitable advertising revenues.

## Characteristics of TV News

The task of the television journalist is also made difficult by a number of characteristics of television news. Some are particular to network or local news. *These characteristics are common to both:*

- Visual values are given priority over real news values, the question is often asked, "Is there good footage?" (of film or tape)
- Polarization is often overemphasized in the effort to provide a balanced treatment of issues and events. The search for equal and opposite opinion spokesperson often yields a report with no acknowledgement of a middle ground or efforts at negotiation or reconciliation.
- The tradition of the "scoop", aided by the instant reporting ability of new portable television cameras, often means that the viewer and journalist discover an event at the same time. There is no time for the journalists to do their work, to sort out news values, set the event in perspective or even decide that the event is not news and not worth reporting.
- Events, particularly those which are part of an ongoing social process or issue, may be represented as fragmented and out of context. The pressures of time both within the format of the news program and for background research, make it difficult to present an event within its historical context.
- Building audience is often in the minds of the news staff. The television ladder of success in employment teaches news people to remember that audience size and salaries are related to each other.
- Events, particularly on-going social processes, are fragmented. There is no time to research for background information which would set the story in perspective.
- The image of human history seems more often to be one of problems than resolutions, and the problems are seen in isolation rather than as part of larger wholes.
- Minor stories with strong visual action share equal time, and thus importance, with larger events which shape the life of community, nation or world.
- There is little time to develop a story in proportion to its importance. When commercials are subtracted from the half hour news program, only 22 minutes remain for news. Newspapers vary the length of stories over a wide range. (Note: as this is written, the networks are discussing expanding the length of news programs to accommodate more news in greater depth.)

*At the network level,*

- Network news staffs and technical facilities, and thus news reporting, are concentrated in New York, Los Angeles and Washington. Reports from other locations may involve the added costs of travel time, obtaining facilities (equipment) from local stations, and line charges to send the picture back to Los Angeles or New York.
- The anchor men and women, who are skilled journalists, are office and studio bound with little time to use their talents in the field.
- The local affiliated stations of the network are reluctant to have network news expanded. This might cut into the length of the highly profitable local news programs.

*And, at the local station,*

- There is a strong pressure for increased or continuing high profitability. The station's sales staff says in effect, "Give us a news program we can get out there and sell."
- In many communities television station news departments are in brisk competition with each other. Each station wants to be number one in news; staff and format are changed in an effort to beat the competition rather than for reasons of journalistic excellence.
- The training and competence of television journalists is uneven. One of the expectations of a news director is that he or she will get better ratings than the other news programs in the market. The tension between winsomeness and ability is very much present when that news director hires staff.
- National consulting firms are hired to analyze newscasts and audiences, and recommend new presentation styles. They tend to stress more stories of shorter length, use of more film, casual banter among the anchorpersons and sports and weather reporters, a formulated news sequence which will include stories on police, fire, sex, violence and human interest (children and animals) topics.
- New equipment exists, such as the ENG (electronic news gathering) cameras which can transmit live pictures from most locations in the station's community—so events must be found which happen during the time of the station's newscasts. The journalist becomes a tour guide.
- The nature of local television news does not lend itself to the development of what newspapers have called the "morgue", a reference library of past stories and other information.

Unlike a newspaper, whose stories can be clipped and filed in appropriate folders or other storage systems, film and video tape are expensive and space consuming to store. With current technology they are also extremely difficult to cross reference and retrieve.

- Behind the increasingly familiar phrase, "We have a live minicam report," lies the possibility of television news orchestrating some aspect of community life. Planned news events, as the speech of a U.S. Senator at a rally, are timed not to the ebb and flow of the actual happening but to the specific time when a station news schedule calls for the live report.

• In some communities broadcast newscasters still function as sales persons during commercials for merchandise. The roles of reporter and product presenter are intertwined to the detriment of professionalism in journalism.

## The Role of the Audience

There is another element in the equation which affects the kind of television news we have—the audience. There is a strong economic pragmatism at work in the television industry. Large audiences of viewers, sometimes the majority of viewers, select the news programs with the stylized hair and format. Fire, crime, sex, and violence do draw an audience. Human interest stories do provide a needed relief, do suggest that all is not tragedy. The preoccupation with tragedy and trivia is partly a lesson taught by years of news watching. It is also a part of our being human to watch for the outcome of a "cliffhanger" story, to interact at a deep level with pain or stress of strangers. Sexuality is vital to the perpetuation of humanity and our interests are deep and strong. The television industry finds responsive chords for their orchestration of the events of the world.

Obviously, some of the needs of viewers are met. The question remains, is this a picture of the world upon which we can act? Is the half-hour or hour of news, viewed by half the nations households, selected on the basis of what is important for us to know or on what will attract us?

It is not easy to look at television, or its news programs, critically. Our pattern is to be passive recipients of a television fare which is primarily entertainment. First steps in breaking that pattern include thinking about standards for good television journalism and watching with clock, pad and pencil in hand. The kinds of stories and their lengths, the appearance of men and women on camera, the visual content of stories as opposed to their news value—all these form patterns which often come as a revelation to the television viewer.

## Regulation and the Courts

The responsibilities and rights of television came increasingly to the fore as the industry entered its third decade in our society. Legislative proposals and court decisions shape both the news function of the industry and its posture in various public and legal pleadings.

During the second decade of its existence, television frequently stated its need for parity with newspapers, particularly with regard to the industry's unique obligations under the Fairness Doctrine and the Equal Time Rule. The former, applying to issues with conflicting points of view, and the latter to candidates for elective office and ballot issues, were seen as an unfair burden out of proportion with the industry's obligation to serve the public interest incurred through the use of the public's air waves. More recently the emphasis seems to be on identifying television, and its rights and obligations, with other industries in society.

In television news there are conflicting, and not easily reconciled, claims for justice. The industry, acutely sensitive to court decisions regarding journalists' research notes and intentions in reporting a story, says they are second class citizens with respect to the first amendment, that they must have freedom of speech to properly represent the right of the public to know. Members of the public, feeling themselves misrepresented or libeled, ask for legislation or court cases to assert their rights. The industry counters with its own approaches to legislation and litigation.

It is difficult to assess the effect of this debate on television journalists. For some, it may be an occasion for increased adherence to professional standards and ethics, for others it may be a cause for greater timidity in investigative reporting.

## A Balancing of Powers

The television news industry is a complex fast-paced endeavor. It is beset by divergent claims ranging from requests for publicity to complaints about fairness, accuracy, balance and libel. The industry has its own concerns relating to freedom of the press issues and uses of responsibility. A model for a considered approach to the complaints and needs of industry and viewer is offered by the National News Council. Composed of people from newspaper and broadcast news, and citizens from various walks of life, the Council works through two committees: the Grievance Committee, which considers complaints from individuals and organizations concerning inaccuracy or unfairness in news reporting, and the Freedom of the Press Committee which deals with complaints from news organizations concerning the restriction of access to information of public interest, the preservation of freedom of communication and advancement of accurate and fair reporting.

The National News Council welcomes requests for information and reports of citizen's problems with the news media. The Council is located at 1 Lincoln Plaza, New York, N. Y. 10023.

### Quality of Life

There is growing emphasis in our society on the quality of life. The phrase catches up the priorities of a wide range of broad and narrow interest groups which suggest that the environment, public education, availability of art forms and cultural events, the nature and setting of work, housing, food supply, national economy, international relations and other factors shape the reality of our lives and our perceptions of its quality or lack thereof.

Television news has a dual role in this quest for quality of life. It is our extended scouting system which can inform us both of the quality of life and the factors which influence it, factors which we might help shape. Television news itself is also an important part of the quality of our collective lives. There is, then, a correlation between the quality of news and the quality of life.

With polls showing that nearly 65% of U.S. adults rely on TV as their principal source of information, the medium's importance in shaping public opinion has given TV broadcasters extraordinary power in their communities and has fostered concern over the integrity of the news product almost equal to the concern over integrity in Government."

*Les Brown*
The New York Times Encyclopedia of Television

---

There are three primordial forces in the world today—atomic power, genetic engineering and television.

*Daniel Schorr*

---

Jay Epstein, in his book *News from Nowhere,* points out that the closed-circuit lines connecting networks with their affiliates in various cities normally can transmit programs in only one direction, from a network to a Denver affiliate, for example. This means that the affiliate cannot usually feed a story to the network in New York on the connecting line. A decision by the network to use the story means paying additional line charges, a fact that can influence the decision.

---

What are implications of the fact that news (which gives us the violence disaster-oriented messages) is often the last thing people see before they go to bed?

In our hurried world, right after we turn out the light, we turn to what may be our only close relationship of day.

*Ben Logan*

---

Richard Salant:

Would you give us a general critique of network news?

We don't spend enough time on important stories. We don't get the nuances and the complexities of stories on network hard news. Because we only have 23 minutes. We're better than a headline service but if you put a headline service over on this side and a 70-page newspaper over on this side, we are somewhere closer to the headline service than we are to the 70-page newspaper. And that I think is one of the great sins of today's reporting and today's society. It's a very, very complicated world; it's a very, very complicated society. And one of the worst things you can do under your responsibilities to a working democracy is to oversimplify. To make heroes and villains, black hats and white hats. We all have a tendency to do that. By making our stories as bang-bang-bang as we do, we reinforce that very dangerous tendency. We leave out the (qualifications) that should be left in.

*Richard Salant*
Broadcasting, *Feb. 26, 1979*

---

Television's first mission is not to inform. It is not even to entertain. It is to move goods, to round up viewers for the main event—the commercial.

Ron Powers



---

## News In An Entertainment Medium

*Walter Lister*

*WALTER LISTER is a Staff Producer with CBS News. A newspaperman for seventeen years at the New York Herald Tribune he joined CBS News in 1963. Experience includes on-air reporting, field producer for special reports, producer-writer of "What's It All About?" informational specials for children, and producer of "In the News" briefs for children.*

In the early years of television, news was an afterthought. Entertainment was the name of the TV game, and programmers naturally picked entertainers who had proved themselves in television's electronic antecedent, radio. Similarly, TV news at first was so much like radio that if you closed your eyes you could easily think it *was* radio. News items came from wire service reports which someone would simply read to a TV camera.

Even when film crews began going out to cover news events for television—chiefly fires and news conferences—the TV crews were seldom expected to report news. Their primary assignment was to bring back interesting film. Print press reporters (John Chancellor called them the "Gutenberg boys") often resented it when TV news would use a statement that had been made only after a pertinent question was asked by a newspaperman—who never got any credit.

But as TV sets became more commonplace in American homes, TV news began to get more respectable. News documentaries—notably those pioneered at CBS News by Edward R. Murrow and Fred W. Friendly, each of whom came from radio news—did much to improve the image of news on television. And for regular TV news broadcasts managers began hiring trained newspaper reporters to send out with camera crews.

Television news also attracted attention in the 1950s and '60s with live coverage of major events: the national political conventions, for example, and those rocket launchings that took American astronauts—and TV cameras—out into space and eventually onto the surface of the moon. Such coverage was expensive, but events like these were scheduled in advance, so that TV news teams had time to prepare.

A crucial test came on November 22, 1963. The assassination of President John F. Kennedy forced network news to get on the air and stay on. Entertainment programs were canceled. The network news departments suddenly had to keep a shocked nation fully informed through a time of tragedy. Those four days of blanket coverage mark the time that network news came of age.

**"Nearly two-thirds of American adults get most of their news from television."**

Coincidentally there was a major turning point for TV news in 1963. It was the first year, according to the Roper Organization, that people depended more on television than on newspapers to tell them what was going on in the world. Now, Roper's polls indicate that nearly two-thirds of American adults get most of their news from television. Some say they depend on both newspapers *and* television for news, but fewer than half give newspapers any mention.

Meanwhile, as awareness of TV news has shot up, radio news has slipped. Twenty years ago more than a third of the population sample mentioned radio news; now less than one-fifth count radio as a source of their news. Trailing further back are magazines and word of mouth.

Widespread dependence on television for information has worrisome aspects, because two innate characteristics weigh against news on television.

First, television is a *time* medium. The time available for news broadcasts has expanded but it is always limited. If all the words spoken on a half-hour network news broadcast were printed in newspaper format, they would not fill even one page of *The New York Times*. Doubling that air time to a full hour—which the networks would gladly do, if local stations would only let them—would stretch the wordage out to barely one and a half newspaper pages.

A network half hour, by the way, after commercials, titles, and credits, holds only twenty-two minutes of news time. Obviously a lot of what gets printed in newspapers does not get on television.

TV time is not only limited; it keeps moving. There is no turning back for a second look. The viewer can never skip around, as every newspaper reader does. All a viewer can do is switch channels or turn off the set—which raises the other inbred characteristic of American television that creates problems for newsmen.

Television is a professional entertainment medium, and it is very profitable. Programs that attract more viewers charge more money for the commercials that go with them. This basic fact of TV life means that news broadcasters have conflicting interests. Their journalistic mission is to inform the public, but their best efforts are lost if the public is not watching. So TV news directors do their damnedest to keep viewers watching and to attract as many more as possible.

**" . . . when news documentaries compete head-on with entertainment, they usually land far down in the ratings."**

There will always be controversy over what is news and what is not. But the men and women in charge of TV news still define news in much the same way that most newspaper editors define it. News is what interests people, or what people ought to know—preferably both. It may be an unusual event that could have happened to almost anyone, or did happen to many people, or is going to affect many people. It may be something that happened to someone well known, or it may simply be a fascinating—or shocking—real life story.

Beyond that, a TV news editor inevitably has to ask: What pictures go with this story?

Right here it should be noted that network news and local news are not the same.

Putting news into an entertainment medium naturally invites show business techniques. Almost any story with good action pictures has a good chance of being broadcast, while a complicated economic report is hard to illustrate on television and has to be very important to rate more than passing mention.

When a picture story is coming from Europe, the Middle East, Asia, Africa, South America or any domestic location outside of New York or Washington, there is an additional cost factor. Because television viewers have come to expect today's news today—and because of competition among networks—many news stories require video transmission by satellite or telephone lines, or both. Satellite charges (which vary in quixotic patterns) have actually become less expensive, but the TV networks are spending more because their use of satellites has greatly increased.

Although the cost of remote video transmission can run anywhere from a few hundred to a few thousand dollars, network news producers rarely consider costs—unless the story's newsworthiness is marginal. Transmission costs are almost always secondary to judgments on picture quality and news value.

A major underlying difference is that the commercials on network news broadcasts do not come close to paying the high costs of worldwide news-gathering operations, while local news broadcasts have become sources of high profits for many individual stations.

When local stations began to learn that news could be very profitable, they set aside more time for it and they sought outside advice on how news might be packaged to attract and keep viewers. That advice often called for good-looking, mellifluous anchorpersons and on-air camaradarie among news team members.

Although there has been some backlash against "happy" news and sensationalism in local TV news, the way a newsperson looks and performs on television definitely counts as much as good news sense.

Network news takes itself more seriously, but it is also anxious to win viewers. Nielsen ratings are not sought for direct profits (virtually impossible), but not just out of pride, either. Big national audiences for network evening news are important because these broadcasts lead into what is called "local access" time, which is another area of high potential profits for network affiliates, and then into the prime-time hours where the networks' ratings battles are won or lost.

The decision makers who are trying to attract bigger shares of the available TV audience to their network like to see an audience build early, in those evening news broadcasts. The theory is that once a TV set is tuned to one channel, it tends to stay on that channel.

Leadership in network news is no guaranty of leadership in prime time ratings, however, perhaps because many viewers who do not watch any news do watch the prime time entertainments. Both CBS and NBC score several rating points ahead of ABC in regular evening news, but ABC has been winning the prime time ratings war.

Nevertheless, all three networks fully believe that attracting more viewers to news broadcasts helps to hold viewers for prime time. And ABC, as part of its campaign to stay on top, is trying very hard to increase the audience for its evening news.

Today, news may be found all over the TV schedules. There are news broadcasts in the morning, at midday, in the early evening, late at night, and on weekends. There are news documentaries, news magazines, news interviews, news for children, news headlines, and news specials. Most TV news coverage of daily events, however, is planned as a story for the evening news, network or local. Those are the *prime* news broadcasts.

When some network executives recently wanted to expand their evening news to a full hour, they were flatly rebuffed by network-affiliated local stations. The reason was simple: money. Any expansion of network news takes air time away from either local news or local access time. The local stations saw no way of making up the profits they expected to lose to an hour-long network news broadcast.

**"Putting news into an entertainment medium naturally invites show business techniques."**

Despite its inbred limitations, television does offer two big advantages for news: picture power and immediacy.

Television, like radio, always had the ability to give the public a news bulletin much faster than the story could be published and distributed in a newspaper. But film, if available at all, takes time to process and edit. And just getting film into a TV newsroom used to take hours—twenty-four hours or more in the case of some foreign stories.

Microwave relays and satellite transmissions shortened that time greatly, and now portable electronic cameras make it possible to get news pictures on the air in living color almost instantly—sometimes while the story is happening.

Even on a day with no war breaking out, nor any major disaster, a network newsroom will schedule several satellite feeds from overseas and across the country. It has become almost as routine as getting wire service reports from around the world, except that satellite transmission of pictures—whether they are used or not—costs a great deal more than sending words by wire.

Electronic news stories brought in by a network are often available for local stations to use—either after they have been aired by the network, or rejected by the network and made available through a syndication service.

But local stations normally emphasize local news and their own correspondents. Local assignment editors decide where to concentrate their manpower, and the decisions they make produce the main ingredients for local news broadcasts. If a local story is covered at all, it has a good chance of getting on the air. (If it turns out to be a *major* story, it may be offered to or be requested by the network.)

Both local and network news editors try to cover stories they think will be worth using on air, but there is a large quantitative difference. While local uses most of the stories it covers, at least briefly, network always covers far more than could ever be squeezed into the available news time. That might seem wasteful, from a cost accounting point of view, but it is the only way a network can hope to illustrate the day's top news events in a medium where pictures may be worth more than thousands of words.

Television's unique advantage is its power to show the visual drama in news events.

America's military involvement in Indochina coincided with explosive growth in television viewing, so that the fighting in Southeast Asia became our "living room" war. Without any editorializing, scenes on the nightly news—like the one in which an American G.I. ignited the grass roofs of a Viet village with his cigarette lighter—did much to spread doubts about our role out there.

Later, after *The Washington Post* persisted in calling attention to the Watergate affair, television enabled the American public to watch while the Senate and House held hearings that led to impeachment charges. *The Washington Post's* circulation is about 600,000. The Watergate hearings played to audiences in the tens of millions.

Audience size alone does not fully explain TV's power. Coupled with the large number of viewers (even for a low-rated broadcast) is the dramatic impact of real events being shown on a private screen.

**"If all the words spoken on a half-hour network news broadcast were printed in newspaper format, they would not fill even one page of *The New York Times*."**

People who try to turn television's power to their advantage are not necessarily successful. The pitfall is that TV's camera eye penetrates mercilessly. A performance may be improved by editing—as is done in thirty-second commercials—but TV news coverage, particularly live coverage, is likely to catch and may even exaggerate any weaknesses.

In the early years of television news, demonstrators for or against good or bad causes quickly learned that it was easy to get on the air, especially if they told TV news desks where and when the action would take place. TV assignment editors have become more aware that they can be used, but it still happens.

In covering racial disturbances at night in the '60s, TV crews found that the lights they had to use to record anything on film often stimulated activists and could make a violent situation worse. Responsible TV news executives eventually ordered a general policy of lights out, at the risk of missing some action, rather than risk being the cause of violence. Later, with faster film and then with electronic minicameras, TV news has become able to take pictures under very low-light conditions, even at night.

While television and newspapers compete for some advertising dollars, there is less competition in covering news. Television and newspapers actually complement each other. TV attracts attention, gives headlines, and shows dramatic pictures. Newspapers give fuller explanation, carry texts and tables, offer opinions, and provide reader services totally impossible for television—such as the opportunity to clip out and keep any item of particular interest.

An analysis of data from a recent survey by W. R. Simmons (mainly on magazine readership) suggests that the people who watch network news also tend to read newspapers.

In TV news the major competitions are among the three networks and among stations in the same viewing area. Because of competitive pressure, TV news generally cannot afford to cater to special interest groups. Air time, no matter how squandered it may seem to some viewers, is always precious to those who control it. Any news treatment that fails to interest a majority of the audience runs the risk of losing viewers.

On the other hand, some situations may invite special services for a particular audience. If many potential viewers speak Spanish, for example, giving the news in Spanish could make good economic sense. And in pursuit of viewers who are hard of hearing, some stations have tried captioned news or sign language.

News-oriented programs for children are cropping up more frequently, but these are generally restricted to time periods when the potential audience has a large percentage of children.

There is increased interest in news programming on television, even though most news broadcasts are not ratings winners. The regularly scheduled network evening news broadcasts, which compete mostly with each other, have ratings that are at best barely half the Nielsen rating for some prime time entertainment programs. Half the rating means roughly half the viewing audience. And when news documentaries compete head-on with entertainment they usually land far down in the ratings.

Documentaries and news specials are aired more for prestige than for ratings. A newsworthy special that is well received, even though its rating may be relatively small, gives a luster to the network that carried it—and it may help to bring in a few more viewers another time. News documentaries are usually aired on a preemptive basis, and they usually appear in time periods when broadcast executives expect low ratings will do a minimum of damage.

There is one big exception: *Sixty Minutes.* This magazine-style, investigative news broadcast has built a strong following and has climbed right up among programs with the highest ratings. One week it was the most watched program of all.

*Sixty Minutes* is spared direct competition from entertainment programs, because the Federal Communications Commission has reserved its time period, 7 to 8 p.m. Sundays, for either children's programs, or news, or public affairs. Adult-oriented entertainments are not allowed. Nonetheless, *Sixty Minutes* has earned its success, as proved by a host of imitators.

Imitation is the path program planners pursue nine times out of ten; it seems safer to copy any formula that works rather than risk trying something different. Among the few different things that are tried, however, news seems to be sneaking into some successful entertainment formulas.

News has a key element that gives it an edge over fiction. By definition, news is factual and fresh. There are no plots to wear thin, no comic situations to grow stale. News is what's new. It may seem implausible, but it is believable because it actually happened. And repetition serves mainly to make news more interesting—like lightning striking twice in the same place.

Producers of dramatic programs, frustrated by television's voracious appetite for scripts, frequently turn to the news of yesterday. So viewers have been treated to dramatic recreations of historical events. Successful entertainments have been built on such realistic foundations as slavery, Nazi genocide, Pearl Harbor and Watergate. The practice suggests that news in the guise of history is infiltrating prime time entertainment.

One large uncertainty in the future, is the development of cable TV. All-news channels are among the services being planned for cable subscribers. But with the multitude of channels that cable is capable of offering, no one really knows what effect, if any, this may eventually have on the news broadcasts carried by the networks and stations that earn their income by selling air time to advertisers.

Meanwhile, news keeps happening and television news keeps trying to cover it. And more and more Americans are looking to television, one way or another, to give them their daily news.

News/Worksheet No. 1

# LOG OF PROGRAM EXCERPTS

## I. SENSATIONALISM

Teaser, Patty Hearst—group sex
Intro to news—fire, murder, Nixon love letters
Gossip on Nixon love letters
Girl forced into prostitution, camera shot of her in underpants
Decapitation murder, camera shot of head under box
Dart sniper still on the loose
Camera shot of blood on street

## II. JUXTAPOSITION, LACK OF SENSITIVITY TO SERIOUSNESS, INTRUSION OF COMMERCIAL INTERESTS

Story on earthquake victims blends into garden spot
Story on women killed by train blends into ad for laundry detergent

## III. TRIVIA—NON-NEWS

Doc Severensen's Divorce
Carter's teeth
Jaw's machine goes bananas
Ali won't spit cherries
Ford doesn't know how to eat tamales
Howard K. Smith on little green men—similarity to "war of worlds" situation

## IV. CHIT-CHAT FORMAT, HAPPY TALK

Chuck and Berry chat
Tom Snyder jokes around
Goofy weatherwoman

## V. POSSIBILITIES FOR IN-DEPTH NEWS

60 minutes
Robert McNeil report

## VI. CRONKITE—"THAT'S THE WAY IT IS"

## News/Worksheet No. 2

**Write what attracted your attention in the program excerpts or watch a half-hour news program and record your negative or positive reaction:**

# News/Worksheet No. 3

1. Pretend you are part of a tribe which has just selected scouts to go out and seek information for you about dangers, enemies, resources, opportunities and living conditions on the trail or in the days ahead. Use the space below to write down a list of instructions you will give to those scouts.

2. Based on your experience with television and your analysis of news programs, use the space below to write down the guidelines that TV news directors, reporters and producers seem to follow.

# News/Worksheet No. 4

**Write down some guidelines you would like to use for your viewing of TV news.**

**List ways you can supplement the information your TV "scouts" obtained for you.**

News/Worksheet No. 5

## HOMEWORK: ANALYZING TV NEWS

The purposes of this exercise are to analyze:
1) Kinds of stories reported
2) Where they are placed in the newscast
3) Amount of time given to each story category

Number each story as it appears. Place the number in the appropriate category. Using a watch with a second hand, time each story.
At the end of the newscast, total the time used for each category.

STATION ________ TIME ________ DAY ________

| | local | state | national | international |
|---|---|---|---|---|
| **Political/economic/affairs of state**<br>Political/government | | | | |
| Economic | | | | |
| Social Issues | | | | |
| Military & War | | | | |
| **Disaster**<br>Police and crime | | | | |
| Fire | | | | |
| Natural disaster | | | | |
| Tragedy | | | | |
| **Sports** | | | | |
| **Weather** | | | | |
| **Business and Finance** | | | | |
| **Human Interest** | | | | |

"Because of the structure of entertainment television, it should not be surprising that the business works on three simple principles: keep the audience up, the cost down, and the regulators out. It is these principles that have shaped the kind of television entertainment we see.

*Robert Liebert, Ph.D.*
Television Awareness Training
*First Edition*

---

"The communications business, that is, the business of transferring information from one place to another, now accounts for half of the gross national product of the United States—and that percentage is rising . . . That figure, from a Commerce Department study, includes the cost of everything from the entire operation of a newspaper to the inter-office communications system used by a private company like the Union Pacific Railroad. But the message is clear: Information is power and everyone is doing everything he or she can to acquire the information needed to do a job as quickly and efficiently as possible."

*Larry Kramer*
Washington Post
*Mar. 4, 1979*

---

"We must come to know ourselves as we really are, not as advertising would have us be."

*The Public Trust*
The Carnegie Corp.

---

Commercials have worked—with success—toward revision of many traditional tenets of our society. As we have seen, reverence for nature has been replaced by a determination to process it. Thrift has been replaced by the duty to buy. The work ethic has been replaced by the consumption ethic . . . . . . . . . . . .
Self-love is consecrated ritual. The woman caressing her body in shower or bathtub has become a standard feature of commercials. A woman applying perfume says: "It's expensive, but I think I'm worth it!"

Erik Barnouw
*The Sponsor*

---

Better-educated adults now watch more television than uneducated adults watched ten years ago; and the two groups of adults watch the same programs.

George Comstock
*Television and Human Behavior*

---

Chapter 8

# TELEVISION: STRATEGIES FOR CHANGE

*Stewart M. Hoover*

*STEWART M. HOOVER is Consultant for Media Education and Advocacy for the Church of the Brethren. He is co-founder of T-A-T, holds an M.A. in Ethics from the Pacific School of Religion and is currently a graduate student at the Annenberg School of Communications in Philadelphia.*

It is now almost impossible to conceive of American life without broadcasting. The TV screen lights up the average home for more than six hours a day. It is a soother, an irritant, a seller of goods, a baby sitter, a companion, an entertainer, a supplier of information, a teacher—a kind of all-inclusive window on the world. And that window has massive influence on what we know and think, on how we act.

Yet, how much do we know about broadcasting—about how it operates—about what our rights are? Or does that matter? Do we really need to know anything about TV, for instance, except how to turn it on and off?

I think we do. First of all, we, the public, have rights and responsibilities.

Secondly, broadcasting can be improved, can better serve our needs. To be a participant in that change process, we have to know something about the broadcasting system.

Television station owners do not own the airwaves. They hold the airwaves they use in trust for you and me and every other citizen.

The *Communications Act* of 1934 established the American system of broadcasting. The Act held that this system should be primarily concerned with serving the needs of the public at large. Built into the Act was the idea that broadcasting frequencies and channels are a *scarce resource* and held in trust for the public. To insure that the public is served by those who use its frequencies, an agency (now called the Federal Communications Commission) was established. The FCC has authority to determine whether a given radio or television station is fulfilling its responsibility to best serve the public interest or whether another individual or group might be able to do a better job of serving the public.

It is now accepted by the FCC that the best way to ensure such service is through active involvement by the public at large. This suggests that broadcasting needs you and your active involvement if it is to be the helpful resource it has the potential to be. The system is designed to give you specific points to "plug-in" if you have comments and criticisms. As a member of the public, you have a *responsibility* to the system which stems from your *right* to have input into it.

The Communications Act has been seen for many years as a document sorely in need of revision. It was written before the ages of television and cable dawned. Recent years have seen an increase in interest in rewriting and updating the Act. You should keep on top of these efforts to see how any changes will affect the situation in your community. It is my purpose here to explain how you fit into the present system of broadcasting.

It is important to realize that you are different things to different parts of the system. To the FCC and the stations, you are a member of the public that is to be served by those who are licensed to use the airwaves.

To the networks, stations, and sponsors, you are an audience, one of many people who watch a given program and thus affect how valuable that air time is. And you are a potential buyer of products.

To both broadcasters and sponsors you also are part of a wider public, one who has an "image" of what broadcasting is doing. No station owner, network, or advertiser, wants to have a bad "public image." They learn what the public thinks of them from individuals. This gives you power.

To members of Congress, which oversees the FCC and often considers legislation that affects broadcasting, you are a constituent. You have the right to be in touch with your Senator or Congressperson about broadcasting just as you do about any other legislation or concerns.

## Changes at Home

There are ways to begin working on change individually—at home. It means developing your skill as a television consumer, reviewer and critic. Be aware of what you are watching, and of the subtle and overt messages you are receiving. Are the programs gratuitously violent? Do they show methods of problem-solving you consider unrealistic? Are women and minority groups treated in abusive or biased ways? Are the commercials too loud, unrealistic, or exploitative?

Learn to be honestly critical of what you see, keeping in mind that this broadcast service is intended to serve your needs and interests. Begin to watch television systematically, with an eye to the range of programming presented and how that programming can be improved. Keep a log of what you watch with comments on the programs. Note, if you can, the source of the program, whether local or network, and any or all commercials that appear on it.

A good way to begin studying television's service in your home is to look at *TV Guide* or your local paper's listings of programs for a given week. Categorize the programs according to the list of program types considered by the FCC ". . . to be necessary for a station to serve the public interests, needs, and desires." (The list is printed at the end of this chapter.) Analyzing the TV service can give you a new perspective on your television experience.

## Changes in the Community

Once you have developed concerns about the broadcast service in your area, it is possible to "feedback" to the broadcasters with letters and phone calls. Your local stations are the most logical place to begin. They are required to be receptive to your comments and criticisms. (Many broadcasters are quite eager to hear from viewers.) Send carbon copies of your letter to the FCC as well (the address is at the end of this chapter) where they will be filed and may be used when your station applies for license renewal. Your letter also will be kept on file at the local station, and will be seen there by other people who are interested in the station's performance.

Phone calls do not usually get the same consideration as letters. If your comment is anything more than ". . . I liked (or didn't like) such and such . . ." it is probably better to write. However, a volume of spontaneous phone calls (jamming the switchboard) has been known to have quite an impact. Stations do log phone comments.

Another strategy that has proven to be quite successful in recent years has been direct contact with program *sponsors.* A number of people have found that sort of action useful on the local level, where a local business is very sensitive to the "public's" reaction to their sponsorship of specific programs. National organizations have found national sponsors to be responsive to their concerns as well. A number of church groups, for instance, were successful in getting major advertisers to institute policies against sponsorship of violent programs when they found that they held stock in those companies and approached them as shareholders. Sponsors are necessarily sensitive to their public image, so letters to them regarding the kinds of programs their ads appear in can have an impact.

**"There are ways to begin working on change individually—at home."**

Remember that most sponsors do *not* "sponsor" a given program. They usually merely "buy time" with a specific kind of *audience* in mind, paying little attention to what specific programs are aired.

The *networks* tend not to be very helpful or responsive to questions and criticism from individuals. The networks are not licensed or regulated by the FCC. They are not legally required to be concerned with our specific needs and interests. Also, they attract a much larger and more diverse audience than local stations, and one viewer is a very small part of that audience. Still, one viewer who bothers to write is seen as somewhat significant to the networks, so don't be surprised if you get a personal reply. Your impact could be minimal, or it could make quite a difference.

Very often the operators of local stations are not the owners. Owners will also be interested in your comments. Broadcasting is very profitable and an owner wants the station to operate in a way that will insure its being re-licensed by the FCC. Each broadcast station is licensed to operate because the owners convinced the FCC they would be best able to satisfy the ". . . public interest, convenience, and necessity . . . " through their service to your community. Stations reapply for license every three years. *Each time* they must survey the needs and interests. While the stations themselves decide how they go about their service, they rely on input from the public to establish those "needs and interests" that must be served.

Broadcasters on the whole have done a poor job of letting people know how the system works and what role the public plays. The "ascertainment" (active researching of the public's needs and interests) has often taken place without the knowledge of the general public, with stations relying on contacts with "community leaders" for input.

The FCC has made several recent rulings which correct some of the deficiencies in station "ascertainment." One ruling requires ascertainment to include a systematic survey of the public at large using a process set down by the FCC, called the "ascertainment primer." Another recent ruling requires each station to make announcements on the first and sixteenth day of each month telling the public about its rights, and the existence of the "public file."

Local stations are required to be interested in your needs and your comments about their performance. Because your personal exposure to their performance is usually limited, you, as an individual viewer, are often ill-prepared to question or challenge station performance. The stations are required to provide a source that can help you evaluate their service. Called the *Public Inspection File,* it includes materials concerning the station's performance of its mandate to serve your needs and interests.

This "Public File" is open for your inspection during regular business hours all year long. You are free to copy any or all of it. You are free to spend as much time as you wish pouring over the file. Among other things this file contains:

- The station's application for its license.
- A yearly report of its ascertainment including its listing of community needs and problems.
- Its annual programming report, which indicates how it went about meeting those needs and problems.

- A series of representative "logs," showing minute-by-minute what was broadcast, including commercials.
- An FCC booklet, "The Public and Broadcasting Procedural Manual," which explains the broadcasting system.
- All letters of comment and criticism.

By going to inspect this file, you will be doing two things. First you will learn how that particular station sees its role in your community and how other members of the public have been responding to the station. Second, you will be one of a very few people who bother to go to the station and investigate it. Because of this, someone from the station will likely wish to speak with you. That will be an opportunity to talk face to face with someone representing the station. Your very presence is a reminder to the station that you have legal rights and that you can influence that station's ability to renew its license.

## Reform Groups

If you decide on changes you feel need to be made in the broadcast service for your community, it is a good idea to form a group of people who share your concerns. You may find there are community organizations already working on the problem. Other organizations have priorities related to broadcast programming. The PTA, Library Association, YMCA, YWCA, and church groups might all be concerned about the effects of television on children, for instance. Women's groups would be interested in stereotyping on television, as would minority organizations. Farm people might be interested in how agricultural issues are dealt with. These kinds of concerns must be given serious consideration by your stations.

The station's public file is also a potential source of contacts since you will find letters there which comment on the station's performance. Look through these carefully. They indicate not only *who* has expressed concern, but also *what* issues have been raised by other people in your community.

It is important that you develop clear and definite issues to pursue. It is also important that you be able to document your concerns. Investigation of the Public File can provide evidence of the station's service (or lack of it), for instance. The organizations listed at the end of this chapter all have done systematic research on TV, and can provide concrete information on how to obtain documentation for your concerns.

As you become an organized group of concerned citizens, it might be useful to be in touch with national or regional organizations which can provide expert consultation. Such help will be particularly important if your group chooses to challenge the renewal of the license of one of your stations.

**"No action strategy should skip the step of direct consultations with the management of local stations."**

No action strategy should skip the step of direct consultation with the management of local stations. Do not go in with a chip on your shoulder, determined to find an enemy of the people. Broadcasters are people, too. They have children and may themselves be concerned about children's programming. They are members of your community and may have their own doubts and concerns about what they broadcast. You need to be prepared to listen to them and to understand how they operate and what pressures they may be under. However, stations are a business and have financial interests to protect. It is natural that they will be self-protective, so don't be put off by easy "public relations" kinds of answers. When you talk to a station manager, friendly but *firm* is a good approach.

It is important to remember that only the local station is licensed to use the airwaves in serving you and your community. That individual station must decide what is good service. They cannot give this responsibility to someone else—a network or sponsor for instance. Any local station has complete freedom not to broadcast something that the network supplies.

Such "discretion" means that your local station's relationship to the network and to sponsors can be significantly affected by the concerns and actions of groups such as yours. While you cannot directly affect network decision-making from your base as a local organization, significant changes could come about if a local station, in response to your interest and/or pressure, refused to broadcast a specific program or advertisement. Stations receive many complaints, many of them irresponsibly made, some very hostile. To avoid being categorized as a crank, have your "homework" done when you go in. Know who is on your side. Mention individuals and organizations working with you or supporting you. This adds credibility to your cause. Also, be clear on what your concern is. Just saying that the programming is bad won't do. (They might even agree and say there's nothing they can do about it.) Have specific concerns or questions. If you've done your research on their public file, you'll be able to point out specific ways you and others feel they've failed to serve the community's needs and interests. Many broadcasters are quite eager to hear from viewers. Send carbons of your letter to the FCC also.

## Fairness Doctrine

Two issues you will undoubtedly confront are the *Fairness Doctrine* and *equal time*. Although they are not the same thing they both deal with fairness of sorts, and you should be aware of them.

The Fairness Doctrine is a rule established by the FCC in 1964 which requires that broadcasters do two things to insure that their coverage of important issues is adequate. First, they are required to seek out and cover (in programming or news) "controversial issues of public importance." They should not shy away from controversy whatever it may be. They are required to make sure that whatever that coverage is, it is balanced either within a specific program or elsewhere in their broadcast service. If you think that a station is failing to cover an issue of public importance or that its coverage of an issue is unbalanced, you should complain to the station (copies to FCC) or directly to the Commission if the station is unresponsive in citing the Fairness Doctrine.

Equal time is entirely different from the Fairness Doctrine. It requires that if a candidate for office receives air time, other candidates for that office can demand equal time. You will probably have little occasion to deal with equal time unless you are running for office.

Don't be intimidated by counter attacks, such as accusations that you are meddling in a private business. You have a right to be there. The station is required to provide service. It is a promise they made to the FCC, *and to you,* when their license was issued. A talk with station management can often get results, particularly if you've prepared well and seem to mean business.

Your cause is always bolstered by results of *research* you or others have done. Research can be something simple, such as examining the composition logs in the public file to find the amount and type of programming for children. As a result, you may decide to show that the station's service to children is inadequate. More ambitious research could include monitoring a station. You, or your group, might count, for example, the number of violent acts on a given station over a period of time.

The organization and publications listed in the *Resources* section of this chapter will provide an idea of the kinds of research which will help you document your specific concerns.

You will find publicity to be an important part of your strategy. Any organized community action attracts attention. Local newspapers, church and community newsletters, and even broadcast stations will be interested in what you are doing. Such publicity helps you put your case before the public, gain allies, and bring additional support.

Think about the publicity you are getting and what kind you want. Make members of your group available as speakers for community meetings. Let the media know what you are doing. A series of press releases (announcements to the press, written by you and illustrating your concerns) can help your cause.

### Petition to Deny

If you still have not achieved your goals through the strategies mentioned so far, it may be necessary for you to proceed to action against a station at license-removal time. Licenses come up for renewal each three years. (See the schedule on page 122 for renewal dates in your state.) At the time of renewal, a formal legal document called a "Petition to Deny Renewal" can be filed. The intent of such a petition is to take the license from the present holder, and give it to another party.

**"The power of the individual is a relatively new idea in broadcast reform, but we have proof that the idea works."**

Usually, the reason for such a petition will be your dissatisfaction with the station's performance in a given area, such as children's programming. Such citizen petitions often do not result in denied license-renewals but result in settlement of issues through *negotiation.* Station management usually will become much more interested in making changes that deal with your concerns once they find you have drawn up a specific petition-to-deny. At that point, you may be able to negotiate a settlement with the station and then withdraw or not file your petition.

Upon receiving your petition, the FCC can return it, or if they feel that it has merit, they will withhold the renewal of the license until they have had a chance to study the petition. They may also move a further step and hold hearings on the petition.

Moving into this area of legal action against the station is a complicated process, and requires the services of an attorney. It is often possible to obtain the volunteer services of local lawyers who are interested in your cause. The "Lawyer's Sourcebook" listed in the *Resources* section of this chapter should be read by any such volunteer legal counsel.

### Two Case Histories

*Action for Children's Television* is a good example of the impact a well-organized national group can have on broadcasting through the use of research, dialogue, publicity and legal action. ACT began when a group of concerned parents, teachers and physicians decided they didn't like what their children were viewing on TV. They particularly objected to children being seen as a market. The group's actions have resulted in significant reforms in children's television. For example, ACT has:

—Actively encouraged television industry decision-makers to tone down violence during hours when children are watching.

—Been instrumental in getting advertising reduced by 40 per cent on children's programs.

—Succeeded in improving the quality of advertising on children's shows. Ads that show violence have been discouraged. Ads telling children to ask their parents to buy a given product have been eliminated. Commercial pitches by hosts, such as Captain Kangaroo, have been stopped.

An organization which has successfully used the "petition to deny" is the *Committee for Children's Television* in San Francisco. CCT began as a group of parents and teachers concerned about children's television. Repeated contacts with San Francisco stations proved fruitless. Children's TV got worse instead of better. CCT then decided to concentrate on the issue of local produced children's shows. Their goal was local shows serving the unique needs of San Francisco children. CCT filed a petition on the local-programming issue and the FCC withheld the license of one of the San Francisco stations for three years, while evaluating the petition. FCC eventually renewed the license in spite of the CCT petition, but even so the story is one of success. CCT and other parent groups began getting remarkable concessions and cooperation from all of the stations, especially the challenged one. In fact, when the license was renewed, the FCC was able to point out that there had been substantial improvement in the local situation. This improvement would not have come about without the concrete action taken by CCT. The support and awareness they gained has continued as CCT goes on working for changes in children's television.

As you develop a group of concerned individuals with specific skills in media reform, you gain credibility for your cause. Such citizens are often asked to testify before legislative groups who are evaluating laws governing broadcasting. The increase in citizen involvement in broadcast license-renewal has, among other things, caused the broadcast industry to seek self-protective changes in the laws which would make it harder for you to be involved. You should be prepared to write letters and testify on this as well as on other specific concerns about broadcasting.

The FCC has become more and more interested recently in public input on changes in their regulations. These changes can often result in more progress nationally than any activity on a purely local level. The "FCC Actions Alert" Newsletter and "FCC Feedback" Newsletter are designed to aid you in having input. Subscribe to them. Write comments when the Commission is dealing with issues that concern you. Your experience and reflection on your local situation are invaluable if the regulatory process is ever to reflect your needs and interests. The Telecommunications Consumer Coalition, Action for Children's Television, Consumer Federation of America, and the FCC Consumer Assistance Office all have guides and helps prepared for you to be involved in this way.

I began by saying that you, as an individual, are different things to different parts of the broadcasting system. Each of these parts or "steps" requires your input and involvement if our television service is ever to reflect the diversity of people in a community, and the kinds of needs that we have. Thus, you can involve yourself at a variety of levels and places.

The goal of each of us needs to be that of becoming an informed, conscious consumer of media messages—and an informed "party of interest" in the operation of stations that are licensed to serve our community. The power of the individual is a relatively new idea but we in broadcast reform have proof that the idea works.

Our rights as individual citizens to have a strong voice in broadcasting have been clearly established. With our rights established, I believe we must assume the responsibility of using those rights and work on constructive changes in a communication medium that has become one of the most powerful influences in our society.

Program categories defined by the Federal Communications Commission (FCC) as "generally required for programming in the public interest."

1. Entertainment Programs
   - *Situation Comedies
   - *Action-adventure
   - *Drama
   - *Variety
   - *Made-for-TV Movies
   - *Theatrical Films *(originally made for theaters)*
2. Opportunity for Local Self-Expression
3. The Development and Use of Local Talent
4. Programs for Children
5. Religious Programs
6. Educational Programs
7. Public Affairs Programs
8. Editorialization by Licensees
9. Political Broadcasts
10. Agricultural Programs
11. News Programs
12. Weather and Market Reports
13. Sports Programs
14. Minority Audience Programs

# RESOURCES

## For Decisions and Action

### I. Organizations

A. C. NIELSEN CO.
1290 Avenue of the Americas
New York, N. Y. 10019
Attention: Bill Behanna,
Director of Press Relations
(for general information and pamphlets on ratings)
212-956-3538

ACTION FOR CHILDREN'S TELEVISION (ACT)
46 Austin Street
Newtonville, MA 02160
617-527-7870

AMERICAN BROADCASTING COMPANY (ABC)
1330 Avenue of the Americas
New York, New York 10019

AMERICAN COUNCIL FOR BETTER BROADCASTS
120 East Wilson Street
Madison, WI 53703
603-257-7712, 255-2009

ANTI-DEFAMATION LEAGUE OF B'NAI B'RITH
315 Lexington Avenue
New York, N. Y. 10016
212-689-7400

CBS TELEVISION NETWORK
51 West 52nd Street
New York, N. Y. 10019

CHILDREN'S ADVERTISING REVIEW UNIT
National Advertising Division,
Council of Better Business Bureaus, Inc.
845 Third Avenue
New York, N. Y. 10022
212-754-1320

COMMITTEE ON CHILDREN'S TELEVISION
245 S. Carmelia
Los Angeles, CA 90049
213-476-5181

CONSUMER FEDERATION OF AMERICA
1012 14th Street, NW
Washington, DC 20005
202-737-3732

COMMUNICATION COMMISSION
National Council of Churches of Christ in the U.S.A.
475 Riverside Drive
New York, N. Y. 10115

CITIZEN'S COMMUNICATION CENTER
1914 Sunderland Place, NW
Washington, D.C., 20036
202-296-4238

FEDERAL COMMUNICATIONS COMMISSION (FCC)
Consumer Assistance Office
Washington, D.C. 20554
202-632-7000

FEDERAL TRADE COMMISSION (FTC)
Office of the Secretary
Washington, D. C. 20580
202-523-3598

GRAY PANTHERS MEDIA WATCH
1841 Broadway, Room 300
New York, N. Y. 10023

MEDIA ACCESS PROJECT
1910 N. Street, NW
Washington, D. C. 20036
202-785-2613

MEDIA ACTION RESEARCH CENTER, INC.
475 Riverside Drive, Room 1370
New York, N. Y. 10115
212-865-6690

MEXICAN AMERICAN LEGAL DEFENSE and EDUCATION FUND
145 Ninth Street
San Francisco, CA 94103
415-626-6196

NATIONAL ASSOCIATION FOR BETTER BROADCASTING
P. O. Box 43640
Los Angeles, CA 90043
213-474-3283

NAACP LEGAL DEFENSE AND EDUCATION FUND
10 Columbus Circle
New York, N. Y. 10019
212-586-8397

NATIONAL BLACK MEDIA COALITION
244 Plymouth Avenue, South
Rochester, N. Y. 14608
715-325-5116

NATIONAL CITIZEN'S COMMITTEE FOR BROADCASTING
1530 "P" Street, N. W.
Washington, D. C. 20005
202-462-2520

NATIONAL NEWS SERVICE
One Lincoln Plaza
New York, N. Y. 10023
212-595-9411

NATIONAL ORGANIZATION FOR WOMEN MEDIA TASK FORCE
425 13th Street, NW
Washington, D. C. 20004
202-632-8622

NATIONAL PTA TV ACTION CENTER
700 North Rush Street
Chicago, IL 60611
Toll-free action line, 800-323-5177
(IL. residents call 800-942-4266)

NATIONAL SISTER'S COMMUNICATION SERV.
1962 South Shenandoah
Los Angeles, CA 90034
213-559-2944

NBC TELEVISION NETWORK
30 Rockefeller Plaza
New York, N. Y. 10020

OFFICE OF COMMUNICATION
UNITED CHURCH OF CHRIST
105 Madison Avenue
New York, N. Y. 10016
212-683-5656

PRIME TIME SCHOOL TELEVISION
120 South LaSalle Street
Chicago, IL 60603
312-368-1088

PUBLIC BROADCASTING SERVICE (PBS)
Education Services
475 L'Enfant Plaza SW
Washington, D. C. 20024
202-488-5046

TELECOMMUNICATIONS CONSUMER COALITION
105 Madison Avenue
New York, N. Y. 10016
212-683-5656

UNITED METHODIST COMMUNICATIONS
475 Riverside Drive, Suite 1370
New York, N. Y. 10015
212-663-8900

UNITED METHODIST WOMEN'S DIVISION
Board of Global Ministries
475 Riverside Drive
New York, N. Y. 10015
212-678-6068

## II. Publications

### A. Available from the Office of Communication, United Church of Christ 105 Madison Avenue, New York, N Y 10016

—*How To Protect Your Rights in Television and Radio,* by Ralph Jennings and Pamela Richard. $5.50
—*A Lawyer's Sourcebook: Representing the Audience in Broadcast Proceedings,* By Robert W. Bennett, $5.50
—*Parties In Interest,* by Robert Lewis Shayon. One copy free. In quantity, $.60 each
—*A Short Course in Cable,* One copy free.
—*Check Your Local Stations* Newsletter, Contribution.

### B. Available from the Superintendent of Documents, U.S. Government Printing, Office, Washington, D. C. 20402

—*The Communications Act of 1934 and Packets* 1-5. $1.25
—*Consumer News.* HEW Department of Consumer Affairs. $4.00 per year.

### C. Available from the Office of Reports & Information, FCC, Washington, D. C. 20554. Free

—*The Public and Broadcasting: A Proceedural Manual*
—*1960 Programming Statement,* FCC 60-970
—*FCC Actions Alert Newsletter.*
—*FCC Feedback Newsletter*
—*FCC Information Bulletin 1-A.* (lists FCC documents available from the U.S. Government Printing Office.)

## III. Where to Write

### A. Government

—Federal Communications Commission
1919 M Street, NW
Washington, D. C. 20554

—Federal Trade Commission
6th & Pennsylvania Avenue
Washington, D. C. 20580

### B. Networks

—American Broadcasting Co., Inc.
1330 Avenue of the Americas
New York, N. Y. 10019

—Columbia Broadcasting System, Inc.
51 West 52nd Street
New York, N. Y. 10019

—National Broadcasting Co.
30 Rockefeller Plaza
New York, N. Y. 10020

—Public Broadcasting Service
485 L'Enfant Plaza West-SW
Washington, D.C. 20024

### C. Trade Associations

—National Association of Broadcasters
1771 N Street, NW
Washington, D. C. 20024

—Television Information Office
745 Fifth Avenue
New York, N. Y. 10022

**D. Ten of the Biggest TV Advertisers**

| *Company* | *Products* |
|---|---|
| PROCTER & GAMBLE<br>301 E. Sixth Street<br>Cincinnati, Ohio<br>45201 | Big Top Peanut Butter, Biz, Bold, Bonus, Bounty Towels, Camay, Cascade, Charmin Paper Products, Cheer, Cinch, Clorox, Comet, Crest, Crisco, Dash, Downy, Duncan Hines, Duz, Folgers, Gain, Gleem, Head & Shoulders, Ivory, Jif, Joy, Lava, Mr. Clean, Oxydol, Pampers, Prell, Puff, Safeguard, Scope, Secret, Spic & Span, Tide, Top Job, Zest. |
| GENERAL FOODS<br>250 North Street<br>White Plains, NY<br>10602 | Alpha-Bits, Awake, Baker's Chocolate, Birds Eye Foods, Burger Chef, Cool 'N Creamy, Cool Whip, D-Zerta, Danka, Dream Whip, Gaines Dog Food, Good Seasons Dressings, Grape Nuts, Gravy Train, Jell-O, Kool-Aid, Log Cabin, Maxim, Maxwell House, Minute Rice, Post Cereals, Prime Dog Food, Raisin Bran, Sanka, Shake 'N Bake, Start, Tang, Swans Down, Thick & Frosty, Toast 'Ems, Top Choice Dog Food, Yuban Coffee. |
| AMERICAN HOME PRODUCTS<br>685 Third Avenue<br>New York, NY 10017 | Beef-a-Roni, Chef Boy-Ar-Dee, Aero Wax, Black Flag, Easy Off, Easy On, Sani-Flush, Wizard, Woolite, Brach Candy, Aero Shave, Anacin, Bisodal, Dristan, Heet, Infrarub, Preparation H, Quiet World. |
| GENERAL MILLS<br>P. O. Box 1113<br>Minneapolis, MN<br>55447 | Buc Wheats, Cheerios, Cocoa Puffs, Count Chokula, Franken Berry, Golden Grahams, Kix, Lucky Charms, Nature Valley Granola, Torta, Trix, Wheaties, Betty Crocker, Bisquick, Gold Medal, Snackin' Cake, Bac-os, B. C. Breakfast Squares, B. C. Pie Shops, Red Lobster Inns, Gorton Seafoods, Red Hots and Slim Jim Snacks, David Crystal, Haymaker. |
| GENERAL MOTORS<br>General Motors Building<br>Detroit, MI 48201 | Buick, Cadillac, Chevrolet, Oldsmobile, Opel, Pontiac; AC Spark & Filters; Delco Products; Fisher Body; Frigidaire. |
| BRISTOL-MYERS<br>630 Fifth Avenue<br>New York, NY<br>10020 | Ban, Bromo Quinine, Bufferin, Drano, Endust, Excedrin, Fitch Shampoo, Mum, No-Doz, Sal Hepatica, Score, Vitalis, Vote, Clairol, Metrecal, Pal Vitamins, Tany A, Shape, Nutrament, Vanish, Windex. |
| MCDONALD'S CORP.<br>McDonald's Plaza<br>Oak Brook, IL 60521 | Hamburgers and Egg McMuffin. |
| FORD MOTOR COMPANY<br>The American Road<br>Dearborn, MI<br>48121 | Fords: Galaxy Mustang, Maverick, Pinto, Thunderbird, Torino; Lincoln-Mercury: Capri, Cougar, Cyclone, Ma Marquis, Montego, Continental; Philco-Ford Appliances; Autolite Spark Plugs. |
| LEVER BROTHERS<br>390 Park Avenue<br>New York, NY<br>10022 | Golden Glow, Good Luck, Imperial and Promise Margarine, Butter-Worth's syrup, Lucky Whip, Spry, Aim, Close-up, Pepsodent, All, Breeze, Caress, Dove, Final Touch, Lifebuoy, Lux, Pears, Phase III, Rinso, Whisk, Lipton's, Wish-Bone, Spray 'n Vac, Good Humor. |
| CHRYSLER<br>P. O. Box 1687<br>Detroit, MI 48231 | Airtemp Division, Boat Corp., Chrysler, Dodge, Plymouth, Simca, Sunbeam. |

## IV. License Renewal Dates For Television Stations by State. *(Petition to deny deadlines are one month earlier.)*

**Alabama** April 1, 1979; 1982; 1985
**Alaska** February 1, 1981; 1984; 1987
**Arizona** October 1, 1980; 1983; 1986
**Arkansas** June 1, 1979; 1982; 1985
**California** December 1, 1980; 1983; 1986
**Colorado** April 1, 1980; 1983; 1986
**Connecticut** April 1, 1981; 1984; 1987
**Delaware** August 1, 1981; 1984; 1987
**District of Columbia** October 1, 1981; 1984; 1987
**Florida** February 1, 1979; 1982; 1985
**Georgia** April 1, 1979; 1982, 1985
**Hawaii** February 1, 1981; 1984; 1987
**Idaho** October 1, 980; 1983; 1986
**Illinois** December 1, 1979; 1982; 1985
**Indiana** August 1, 1979; 1982; 1985
**Iowa** February 1, 1980; 1983; 1986
**Kansas** June 1, 1980; 1983, 1986
**Kentucky** August 1, 1979; 1982; 1985
**Louisiana** June 1, 1979; 1982; 1985
**Maine** April 1, 1981; 1984; 1987
**Maryland** October 1, 1981; 1984; 1987
**Massachusetts** Arpil 1, 1981; 1984; 1987
**Michigan** October 1, 1979; 1982; 1985
**Minnesota** April 1, 1980; 1983; 1986
**Mississippi** June 1, 1979; 1982; 1985
**Missouri** February 1, 1980; 1983; 1986
**Montana** April 1, 1980; 1983; 1986
**Nebraska** June 1, 1980; 1983; 1986
**Nevada** October 1, 1980; 1983; 1986
**New Hampshire** April 1, 1981; 1984; 1987
**New Jersey** June 1, 1981; 1984; 1987
**New Mexico** October 1, 1980 1983; 1986
**New York** June 1, 1981; 1984; 1987
**North Carolina** December 1, 1981; 1984; 1987
**North Dakota** April 1, 1980; 1983; 1986
**Ohio** October 1, 1979; 1982; 1985
**Oklahoma** June 1, 1980; 1983; 1986
**Oregon** February 1, 1981; 1984; 1987
**Pennsylvania** August 1, 1981; 1984; 1987
**Rhode Island** April 1, 1981; 1984; 1987
**South Carolina** December 1, 1981; 1984; 1987
**South Dakota** April 1, 1980; 1983; 1986
**Tennessee** August 1, 1979; 1982; 1985
**Texas** August 1, 1980; 1983; 1986
**Utah** October 1, 1980; 1983; 1986
**Vermont** April 1, 1981; 1984; 1987
**Virginia** October 1, 1981; 1984; 1987
**Washington** February 1, 1981; 1984; 1987
**West Virginia** October 1, 1981; 1984; 1987
**Wisconsin** December 1, 1979; 1982; 1985
**Wyoming** October 1, 1980; 1983; 1986

# Change/Worksheet No. 1

**Update your personal set of guidelines for your use of TV. Check guidelines developed in earlier T-A-T. sessions and reorganize here for easy use.**

## Change/Worksheet No. 2

Evaluate a station's programing day. Using *TV Guide* or a newspaper listing, make a check mark for each half hour of programing type in the appropriate time period. Then evaluate the diversity of the station's programing against the time of day the types of programing are presented. Is the schedule balanced? Is there diversity? Are community needs being met?

STATION ________ WEEK ________ NETWORK ________

| | TIME OF DAY | | | | | | | |
|---|---|---|---|---|---|---|---|---|
| Program Type | 6–9 a.m. | 9–12 a.m. | 12–1 p.m. | 1–4 p.m. | 4–6 p.m. | 6–8 p.m. | 8–11 p.m. | 11— p.m.– |
| **Entertainment:** Situation comedies | | | | | | | | |
| Action adventure | | | | | | | | |
| Drama | | | | | | | | |
| Variety | | | | | | | | |
| Made for TV movies | | | | | | | | |
| Theatrical movies | | | | | | | | |
| Game Shows | | | | | | | | |
| Sports | | | | | | | | |
| Soap Operas | | | | | | | | |
| **News and Public Affairs:** Newscasts | | | | | | | | |
| News documentaries | | | | | | | | |
| Cultural | | | | | | | | |
| Minority programs | | | | | | | | |
| Educational programs | | | | | | | | |
| Religious programs | | | | | | | | |
| Local programs | | | | | | | | |
| Agricultural programs | | | | | | | | |
| Children's programs | | | | | | | | |

# Change/Worksheet No. 3

Evaluate a week of television programing in your community. Using TV guide or a newspaper listing, tally the total number of half-hour programs for each category per week. Use different colored pens for different stations.

| **Entertainment:** | Sun. | Mon. | Tues. | Wed. | Thurs. | Fri. | Sat. | Total half-hours |
|---|---|---|---|---|---|---|---|---|
| Situation comedies | | | | | | | | |
| Action adventure | | | | | | | | |
| Drama | | | | | | | | |
| Variety | | | | | | | | |
| Made for TV movies | | | | | | | | |
| Theatrical movies | | | | | | | | |
| Game Shows | | | | | | | | |
| Sports | | | | | | | | |
| Soap Operas | | | | | | | | |
| **News and Public Affairs** | | | | | | | | |
| Newscasts | | | | | | | | |
| News documentaries | | | | | | | | |
| Cultural | | | | | | | | |
| Minority programs | | | | | | | | |
| Educational Programs | | | | | | | | |
| Religious programs | | | | | | | | |
| Local programs | | | | | | | | |
| Agricultural programs | | | | | | | | |
| Children's programs | | | | | | | | |

## Change/Worksheet No. 4

**List programs or types of programs that would be broadcast on your ideal TV station.**

**Why do you think your ideal shows aren't on the air?**

## Change/Worksheet No. 5

# THE PUBLIC FILE

One of the ways to evaluate the performance of a given station is to inspect the public file. This is a file that the Federal Communications Commission requires be made available to the public during normal business hours. You should make an appointment to see the file, but you do not have to be any more specific than giving your name, your address, and the general purpose of your inspection. Other than that, the file is there at your disposal. You may make copies of any pages you wish to retain, or have the station copy them. (They can charge you for copies.)

Use the following scratch sheet to evaluate the public file of the station

1. Is the booklet "Public and Broadcasting" there? (It is a *Federal Register* publication of September 5, 1974, and looks mimeographed. It is supposed to be there. On the last page, under section 60, you will find the items that are supposed to be in the file. Look at that list.) Ask for the following information by form number if the station representatives ask which forms you want, or you can just ask to browse through.
2. Who makes the decisions at the station? Who owns it and who manages it? *(ownership report, Form 323, and application form, Form 303.)*
3. What are relative percentages of program time devoted to: News; Public Affairs; Entertainment? *(Form 303-A and others.)*
4. How many public service announcements are carried by the station? *(composite week logs, Form 303-A, and elsewhere)*
5. What are percentages of women and minorities at various levels of employment at the station, and what progress has been made on that score? *(employment forms, Form 395.)*
6. What are the top ten problems found by the station in its ascertainment? *(Annual listing of community problems and needs.)*
7. How many letters from other viewers are on file?
8. In major categories, what are most of the letter writers concerned about?
9. Did you get a chance to talk with someone from the station? Who? Note impressions of your conversation.

Items in Public File:

1. Application for Renewal *FCC-303*
2. Ownership Report *FCC-323*
3. Annual Employment Report *FCC-395*
4. Letters from Public
5. Annual Listing of Problems of community
6. Annual programming Report *FCC-303-A*
7. Composite Week Logs

"Some genius determined that just one game on Monday night attracted more viewers than 'Gone with the Wind,' 'The Sound of Music' and 'From Here to Eternity' combined. And not as many people had seen 'Hamlet.'

"Weighing this in my head, I knew that values were twisted somewhere. I am not completely pure of mind and spirit, but I knew this was crazy. I got disappointed in myself and I felt I had to get away. I had lost my taste for the show. . . ."

*Don Meredith*
*of Monday Night Football*
*as reported in* TV Guide *Nov. 9, 1979*

---

The rituals of our time take place not on Sunday mornings, but at football games, prime time and political conventions.

*Wm. Kuhns*

---

. . . if sports reflects or molds a society, then much could be learned by attending to those features which regulate play. A definition of civilization might be how a society curbs violence . . . perhaps the best control of violence is in the effectiveness with which it curbs violence in it's games. Here, when well enforced and properly respected by fans and players, the society is its own best definition of civilization.

C. Pierce, "Super Stress Competitions"
*(unpublished paper)*

---

Nick Johnson, head of National Citizens Committee for Broadcasting (NCCB) in Washington once told a group of broadcasters and advertisers: "What you get for one minute of time in the Super Bowl game is NCCB's annual budget."

---

Chapter 9

# TELEVISION AND SPORTS

## The Interdependence of Sports and Television

*Donald E. Parente*

*DONALD E. PARENTE is an Associate of the Mass Communication Research Center at Southern Illinois University.*

Since World War II, sport has become an integral part of the leisure industry. It employs significant numbers of people, accounts for a significant portion of the Gross National Product, and has been ingested into the practical income-generating sphere of American social life. Sport also provides a major source of drama for its public and generates a rich mythology and iconography as well as abundant statistical and verbal lore. It is perhaps the most stylized and widely participant ritual of contemporary life and therefore a major vehicle through which meanings are developed and communicated.

Today's professional sport is intimately involved with the institutions of mass communications and, in particular, with television. Television and, significantly, the financial support it can offer have become so important to professional sports that they have molded, adapted, and changed their rules to meet the desires and needs of television. It has been suggested that once a sport, league, or team has had its "product" bought by television for use as programming, that entity can seldom exist thereafter, at least in the same style or manner, without the financial support of television. Similarly, television has become dependent upon sports to fulfill many of its programming needs. The result has been an intimate relationship between two essentially dissimilar entities. This relationship may be called a symbiosis.

The association between television and sports dates back almost to the beginning of broadcasting. Sports events were a ready-made, accessible, and inexpensive source of programming for television. Moreover, sports were the only type of programming that would attract sizable audiences on Saturday and Sunday afternoons. Television was relatively unimportant to sports until the end of the 1950s when organized professional team sports began to look at television as a potential major source of revenue. The interdependency between television and sports became very strong during the 1960s as payments for television rights became a significant part of the revenue of professional teams and leagues. As Horowitz (this symposium) points out, the growth in broadcast revenues from 1956 to 1976 increased over 1,000 percent, or from $10 million to over $112 million.

**"In a broad sense sports, like television, have become a part of the advertising industry."**

Television has become an important factor in decisions involving sports for at least four reasons:

1. For many sports, television rights payments represent a substantial portion of gross revenue.
2. Broadcast revenue is normally a stable source of income that is less subject to the changing whims of fan allegiance than is typically the case with attendance.
3. Television is one of the few sources of income for many organizations that has potential for increase. Many teams and events have little room for growth in attendance and little opportunity to raise ticket prices which are about as high as the market might bear.
4. The decision-makers in sports have apparently found it easier to change the nature of their sport to appeal to the desires of television rather than to the wants of the live spectator.

Of the reasons stated above, the fourth is, perhaps, the most interesting. Prior to the sixties, changes in sports generally were made to improve the sport itself either for "sporting" reasons or to make it more interesting to spectators in order to stimulate attendance. Gradually, entrepreneur types of sportsmen saw an opportunity for greater profits by making slight changes within their sports to appeal to the desires of television. These changes seldom affected attendance adversely, although there were some notable exceptions. Eventually, sports unabashedly began "marketing" themselves for television.

Tradition was often abandoned as rules, styles, and playing fields were changed so that organized sports would be more attractive to television. For example, networks like neat time packages to increase a program's saleability to prospective sponsors. So, to help them out, the NFL cut the halftime intermission from 20 minutes to 15 minutes so that the program would fit more comfortably into a two-and-a-half-hour time segment. Golf changed from match to medal or stroke play in order to guarantee prospective television sponsors attractive celebrity golfers for their telecasts in the final stages of an event. Tennis introduced the tie-breaker to end the drawnout deuce games. NHL hockey changed its center line to a broken line so it would show up better on television.

## Sports and Advertising

As changes began to occur within sports—frequently as a result of requests from television—advertisers increasingly came to see sports as a source of programming that was highly attractive to the kinds of audiences they were interested in reaching. For this reason networks too increasingly began to care about the composition of their audience, for they wanted a quality audience to attract advertisers and to help the overall image of the network. Thus,

**"Sports have now entered the mainstream of the entertainment business."**

as exemplified in the late sixties by CBS's decision to cancel several popular prime-time programs such as *Andy Griffith, Gomer Pyle, Red Skelton,* and *Ed Sullivan* primarily because those programs attracted proportionately more elderly, lower-income, rural audiences, the networks began to revamp their schedules to appeal to the young, urban, affluent audiences who not only had more cash to spend, but who were also more likely to try new products.

Among the most difficult demographic segments of the population to reach are the adult males. Sports programming provides advertisers with vehicles to reach them in large numbers without the waste that would normally be realized through other types of prime-time programming. Thus the desirability of football for television programming can be attributed not only to its high audience ratings but also to the composition of its audience, for according to Nielsen ratings, the television audience for football is the only TV sports audience which is composed of at least 50 percent adult males. In contrast, although baseball may be as popular as football, it reaches fewer, less affluent men and therefore is less attractive to advertisers (1).

Spectator sports have always competed with other forms of entertainment for the discretionary income spent on leisure time activities *outside* of the home. In fact, many professional and intercollegiate teams are deemphasizing the prospects of winning their promotion and instead are stressing the fun aspects of coming to the stadium (3). However, as evidenced by the increasing number of sports events being telecast and the most prominent time slots being given over to them, spectator sports have found another area in which to compete in the entertainment industry.

In the 1960s, although sports programs were occasionally televised in prime time, directly opposite expensive programming, for the most part sports programming was relegated to afternoons, a time period that was unattractive to nonsports programming. Typically these were local broadcasts of local professional teams by strong independent stations. However, during the 1970s, the televising of sports in prime time increased, emanating from strong independents, syndicated networks, and, most importantly, the three major networks. As society has changed over to a leisure-oriented, "service economy," sports viewing, along with television viewing, has become an increasingly popular way to spend one's time. Much of this popularity can be directly attributed to the highly conscious efforts that sports have made to shape their "products" so that they become more appealing to television.

Williams and Comisky, Bryant, and Zillman point out that watching a sports event on television is distinctly different from seeing it in person in a number of respects. Williams notes that television imposes its own structures and provides its own ideological viewpoints which mediate the viewers' experience of the event. Comisky, *et al.*, argue that the televised event is *not* an unbiased and reliable, "true" representation of the game seen by the fans in the stadium. Bryant, Comisky, and Zillman (this symposium) report that sportscasters employ a wide variety of dramatic embellishments in their commentary. These embellishments, as well as the lack of commentary, can alter the viewers' perception of the event and thus affect the enjoyment of the telecast.

Watching a sports telecast may also affect a viewer's subsequent enjoyment of participating in sports. Goldstein and Bredemeier observe that "televised coverage of many sport events convey particular social values to observers, and these values help shape the meaning, and hence the effects, which subsequent participation in sports may have." The television and sports relationship

**"Golf changed from match to medal or stroke play in order to guarantee prospective television sponsors attractive celebrity golfers for their telecasts in the final stages of an event. Tennis introduced the tie-breaker to end the drawnout deuce games. NHL hockey changed its center line to a broken line so it would show up better . . ."**

then has its own socializing effects.

By passing laws, issuing rules and guidelines, filing suits, threatening to file suit, and acting in many other influential ways, the government has profoundly influenced the structure of sports. Horowitz holds that certain legislative concessions and regulatory restrictions have worked to the mutual advantage of the three major television networks and the sports franchises (and against the public interest).

Professional sports can be very profitable. The monopoly and quasi-monopoly position that many professional clubs find themselves in are not generally available to ordinary businesses. As a result, the restrictive agreements which prevent new teams in a league from invading the territory of existing teams, and certain exemptions from antitrust laws as well as tax advantages, largely through generous depreciation allowances, all help keep professional sports teams generally a profitable business to own. Morover, the financial position of even the weakest teams is not nearly as critical as is commonly believed. In the past, Federal tax laws have made it possible for an owner in a partnership of "subchapter S" corporation to lose nearly $1 million on the books and $600,000 in cash flow and still break even in his personal finances.

The Federal government will continue to play a major role in the television/sports relationship. A case in point has been legislation prohibiting blackouts in a home team's territory if all tickets have been sold 72 hours or more before a game. Although the NFL has steadfastly opposed this legislation for various reasons Siegfriend and Hinshaw recommend that it be extended indefinitely.

In the future, then, it seems clear that this strong interdependent relationship between television and sports will continue unabated. And as television technology becomes more sophisticated and TV screens become significantly larger, we may begin to see some sports existing as studio sports, with only token live audiences for background and atmosphere, where in fact as well as effect, the new stadium is TV.

***REFERENCES***

1. "A Look at Sports." *Nielsen Television Index*. Chicago, Ill.: Media Research Division, 1970 and 1975.
2. Noll, R. G. and B. A. Okner in U. S. Senate, Committee on the Judiciary, Subcommittee on Antitrust and Monopoly. *Professional Basketball*. Hearing. 92nd Congress, 1st Sess., on S. 2373, a bill to allow the merger of two or more professional basketball leagues, September 22, 1971.
3. Parente, D. E. "Sports Promotion and Public Relations." *In Making Advertising Relevant*. Columbia S.C.: American Academy of Advertising.
4. Parente, D. E. "A History of Television and Sports." Unpublished doctoral dissertation, University of Illinois at Urbana-Champaign, 1974.
5. Parente, D. E. "A History of Television and Sports: 1939-1960." Unpublished manuscript.
6. Parente, D. E. "Television and Sports: a Symbiotic Relationship." In J. H. Goldstein (Ed.) *Sports, Games and Play*. Hillsdale, N .J.: Lawrence Erlbaum, in press.

Journal of Communication, Summer, 1977. Reprinted by permission.

# TV Sports and the Viewer

*Chester M. Pierce, M.D.*

*CHESTER M. PIERCE, M.D., Professor of Education and Psychiatry, Harvard Graduate School of Education, and Harvard Medical School, Cambridge, Massachusetts.*

For thousands of years people have focused on their games and athletes. The Greek Olympic victor was fed for life at public expense. The Roman emperor, Hadrian, for all his good deeds, was best remembered in the centuries following his death by the coins which memorialized the magnificent games he sponsored.

Anyplace in the world, a visitor can engage in a spirited conversation with citizens on the subject of sports. The visitor might be hesitant to discuss subjects such as family, politics, economics or national history. A subject most available for discussion between two strangers, however, is sports.

We accept this fact as both unremarkable and ordinary.

Today it may be true that the best recognized person on earth is an athlete, Mohammed Ali. It may be true too that no statesman, scientist or artist commands such respect as Pelé, the soccar superstar. The civil war in Biafra was suspended during Pelé's visit to that country so that everyone would have the privilege of seeing him perform. It can be argued that humans forever, and perhaps everywhere, have had a special regard for sports. The mass media of modern times, both print and electronic, have made tens of millions of people even more aware and sensitive about athletes, games and sports activities. Sports have meant a great deal to television and television has meant a great deal to sports. Further TV sports have changed the lives of the spectators.

In the United States, where the TV set is typically turned on for 6½ hours each day, virtually everyone watches TV at some time. Since the presentation of sports occupies a regular place on newscasts as well as particular sports events, televised sports consumes a great deal of our time. It is hard to say whether the sportscasts reflect or mold society. For these reasons, viewers should be more aware—more discerning—while watching sports on TV.

**". . . sports are 'considered by many the perfect program form for television, at once topical and entertaining, performed live and suspensefully without a script, peopled with heroes and villains, full of action and human interest and laced with pagaentry and ritual.'"[2]**

## What Sports Mean To Television

To capitalize on the universal interest in sports, television has given us increasing sports coverage. During the 70s in the U.S. each network spent eighty million dollars or more for sports rights and production. By the mid-70s advertising revenue for 1,100 hours of network sports coverage was $315 million. Of course, many millions more were spent in local and syndicated sports coverage.[1] Therefore, whatever else sports has meant to TV, it has meant big money.

Sports has meant something else to television. Sports have influenced creative organization and imagination of the medium. *The New York Times Encyclopedia of Television* begins its analysis of sports saying that sports are "considered by many the perfect program form for television, at once topical and entertaining, performed live and suspensefully without a script, peopled with heroes and villains, full of action and human interest and laced with pagaentry and ritual."[2]

## What Television Means To Sports

Qualities that distinguish humans from other animals are useful to consider in elaborating the meaning that television has to sports. It is important to realize that there are human-peculiar proclivities to violence. Arguably humans are the animals that commit suicide and are the most likely to exploit or eradicate their own and other species.

Perhaps a basic cause of the timeless love of sports is that they provide a direct (or vicarious) satisfaction for humans' particular taste for violence and entertainment. Television provides a means (probably never excelled) in which humans can be entertained according to their appetite for violence. The medium is good at doing this because it offers unparalleled intimacy with real life action without endangering the viewer.

Viewers have had the opportunity to acquaint themselves with more sports and with superlative performances because of TV. Sports are better appreciated by more people as a result of TV spectatorship. This has probably driven up the quality of sports performance as connoisseur audiences demand to see only the very best. Thus, sports have contributed to better technical television production, and that in turn has contributed to better sports performance and wider acceptance of diverse sporting events.

In addition, participation in certain sports has soared. People who had not been acquainted with tennis were introduced to it via TV—and there was a tennis boom. After Olga Korbut's performances on TV, thousands of youngsters took up gymnastics. (Similarly, ballet on TV has grown and prospered along with a corresponding increase in participation in dance by the viewership.)

Further, TV's technology has changed sports. ". . . tennis earned a permanent place on TV in the 70s, and like other sports the game adapted itself to the medium. A notable concession to TV was the switch to 'optic yellow' balls and colorful shirts from the traditional tennis whites. In turn, TV's interest in tennis served to boost the prize money in tournaments to levels unprecedented for the sport."[3]

> **"These lessons are so vivid and so instructive that thousands of players will improve their game as they themselves practice these methods."**

Millions of viewers may today be seeing what later generations will consider "classics." Among these classics will be sports events which marry television with what communities have always valued and cherished. That is, communities have always liked to entertain themselves by displaying members who are capable of feats of speed, endurance, courage, concentration, coordination, and strength. Not surprisingly, these sports classics have powerful societal impact. So television means that vicarious sports will be intensified in the lives of families. Millions of relatives and friends come together for discussion and drinking and revelry during important TV sports events. This tendency to gather for TV sports makes common such questions as: With whom did you watch the Super Bowl? May I come over to watch the World Series with you? Can we stay up to watch the Olympics?

## Impact On Society

Thanks to TV, the sports fan knows more, thinks faster and embraces more concepts of sports than did the sports fan just after World War II.

Our presidents at press conferences use the language of sports in discussing the strategy of domestic or foreign issues with the sure knowledge that people will clearly understand and accept such semantics. The third grade teacher calls for an instant replay of long division problems.

To keep current in our society—something Americans insist upon as a cultural value—one must know what is happening in television. Since sports are looked at so many hours by such a diversity of people, this is a primary area for staying "au courant." The shaping of concepts and behavior occurs with each program.

The effects can be good, bad or neutral. For instance sports-medicine authorities worry about the influence television has had on spear-tackling in football.[4] On instant replay, a 300 pound, 6½ foot man is shown in slow motion driving his head into the midsection of an opponent. The next week, a 140-pound 16 year old, without special coaching or special body-building, breaks his neck attempting to spear-tackle a high school classmate during a practice scrimmage.

On the other hand viewers learn positive things from television. Both coaches and players, at all levels, are given priceless instruction for practice procedures in basketball by a celebrated former coach. These lessons are so vivid and so instructive that thousands of players will improve their game as they themselves practice these methods. Further, expert commentary often predicts correctly what is about to happen. Athletes might grasp a fuller picture of how special and environmental features are used by world class athletes in a deliberate, long-range plan.

Sportscasts educate viewers about how and why certain things were done, not done or should have been done. This augmented game intelligence makes the viewer, whether he is only a fan or both a fan and performer, much more sensitive to mundane details about a contest. "Sports sense" or game intelligence is taught in each contest. Previously such use of trivial facts would have been only the province of the professional athlete.

The outcome of TV sports exposure may be that viewers will find it easier to accept the emphasis on standards, standings, details and scores. In a society born in revolution and nurtured from a frontier mentality, it has always been culturally important to know exactly where one stands compared with others. Sports on television again found a fertile soil to emphasize what the society has valued.

> **"Perhaps a basic cause of the timeless love of sports is that they provide a direct (or vicarious) satisfaction for humans' particular taste for violence and entertainment."**

When people compete for favorable standing there must be criticism. Sports invites competitive criticism and comparison. Yet as consumers view performances—say in gymnastics—and are taught rudiments and fundamentals, they become more critical. Therefore being intolerant, agitated or uninterested in inferior performances at a local school or college may develop. Some viewers may feel so outclassed that they never give themselves the chance, content to watch rather than participate. Many people in the society may develop more intolerance for inferior performances of any sport and at the same time have less desire to risk doing something in an ordinary way. If millions of people develop this kind of psychology it would have staggering significance to the welfare of the nation.

Watching sports has clearly given us a wider repertoire of possible behaviors and ideas. Sports contests viewed on various continents are powerful geography and civic lessons for all viewers. In a sports program one may learn about the educational facilities in an adjoining state. Next one is being told of the training methods of a team half a world away. Acceptance of different countries and customs can move some people toward thinking as world citizens.

Other indirect lessons are beamed to viewers. Sports on TV suggest the sacrifice and rigor involved in training, planning and practicing in order to be "number one". The lesson is so forceful that any TV sports viewer must be increasingly aware of the difficulties and hardships one elects to undergo in order to be a superlative competitor.

**"Today it may be true that the best recognized person on earth is an athlete, Mohammad Ali."**

At best it demonstrates that athletes all over have a friendly bond and alliance with fellow athletes that transcends race. At worst, however, TV sport may intensify racism. In the U.S. blacks have had their longest, best and greatest opportunity in sports. For the racist this can fit the stereotype that the black is only a physical being who can be exploited for the community entertainment. The racist ignores the fact that quality performance in any sport must be by people who think, concentrate and are masters in out-psyching their opponents.

The role of black athletes illustrates that social customs can be modified. Twenty-five years ago no two white upper middle-class men meeting at their suburban train in Westchester would greet each other by a "skin-slap." Nowadays no one would be shocked to see such a greeting. The general public was first exposed to such greetings by black athletes on TV. The meaning of the more informal, more openly joyous salutation or valediction both molds and reflects the times. Such sociological observations could also be made about numerous aspects of sports on TV including clothing styles (headbands), or spiking the ball after a touchdown.

Among the motives people have for playing games—and obversely for watching them—are aesthetic pleasures. Some find it fascinating to view a human using the body in novel or maximum ways. The person showing athletic perfection is an exciting aesthetic experience. On TV, *body movement* becomes even more *artistic* as it is viewed via camera angles, close-ups, slow-motion and other effects.

Similar changes in customs come from seeing more female athletes on TV. This could lead to markedly different sociological and psychological understandings in America.

Already certain female events outdraw male sports. In the future, ratings of female athletic contests will increase because more people will watch female gymnasts, swimmers, runners, basketball players as well as tennis players or golfers.

Will more female athletics modify the ruthlessness that often accompanies much of present-day athletic competitions? Will we broadcast more co-educational games? If so, will such games cause any alterations in the way citizens view themselves or act toward each other?

In dealing with such questions the viewer must give special weight to the role of televised sports in our lives. Especially to quality and quantity of coverage.

Finally despite the importance of making citizens aware of television in their lives, it is perhaps even more important to make citizens aware of when not to watch television and what options to exert with their time. It follows that despite the enormous opportunities presented for training by sports programs, it is of more importance to get citizens themselves to exercise. As viewers learn how to place television in perspective in their lives, they should realize that life is partly passive contemplation but also abundant active participation.

***REFERENCES***

1. Brown, Les, *The New York Times Encyclopedia of Television,* Times Books, 1977.
2. Ibid.
3. Ibid.
4. Ryan, Dr. Allen, *The Physician and Sports Medicine,* in personal editor communication to the author.

Sports/Worksheet No. 1

## SPORTS VIEWING ANALYSIS

1. Estimate the number of hours you watch sports events on TV during an average week in:

   Fall________________________________

   Winter______________________________

   Spring______________________________

   Summer_____________________________

2. Estimate the number of hours you participate in active sports during an average week in:

   Fall________________________________

   Winter______________________________

   Spring______________________________

   Summer_____________________________

3. Estimate the number of hours you attend live sports events in an average week in:

   Fall________________________________

   Winter______________________________

   Spring______________________________

   Summer_____________________________

4. List the sports you watch on TV in order of preference:

   a.____________________

   b.____________________

   c.____________________

   d.____________________

   e.____________________

5. On a scale of 1 to 7, indicate how violent each of the above sport is in terms of body contact and risk of personal injury:

   a.____________________ 1 2 3 4 5 6 7

   b.____________________ 1 2 3 4 5 6 7

   c.____________________ 1 2 3 4 5 6 7

   d.____________________ 1 2 3 4 5 6 7

   e.____________________ 1 2 3 4 5 6 7

6. Compare your favorite TV sports (4) with your violence scale (5).

*(Worksheet No. 1 Continued)*

7. How would you rate your viewing preference on the violence scale? Do you prefer low or high violence sports?

8. What does your sports viewing say to you about yourself as a person?

9. What does our national sports viewing say about us as a culture?

10. Total your average week's viewing to secure an estimate of how many hours/days each year you spend watching TV sports. Enlarge the total to include all members of your household.

## Sports/Worksheet No. 2

# ANALYSIS OF BEHAVIOR

Observe yourself and household members to gain insight into the effects of viewing television sports. The following questions are guidelines.

1. What are the eating and drinking patterns while viewing? How healthful are the items consumed? List items and quantities.

2. Is there a sense of family togetherness or comaraderie? List the persons who are left out of TV sports viewing in your household.

3. Does the viewing of TV sports in your household create tension, anger or conflict? List specific, typical incidents.

4. In what ways does the viewing of TV sports increase the opportunity for conversations?

5. Do you think the messages of competitiveness and the importance of winning and of being "Number One" influence the thinking and behavior of you or others? In what ways?

6. What are your feelings, attitude and body activity when a player on your team is injured? On the other team?

7. Do any of the effects of viewing go beyond the end of the game, such as anger, aggressiveness, physical inactivity, difficulty in moving into other activities or conversations, driving patterns?

# WORKSHEET NO. 3

## ANALYSIS OF VALUES

List the positive and negative *values, results* or *effects* of TV sports. Evaluate their implications.

| Positive | Negative |
|---|---|
| | |

Chapter 10

# TELEVISION AND MINORITIES

## Minorities on TV

*Brenda Grayson*

*BRENDA GRAYSON is a freelance writer and lecturer on children's television. She has collaborated on the production of children's programs at WBX-TV in Boston and NBC in New York. She is a recipient of a writing award by the Council on Interracial Books for Children (1979).*

In the mid-sixties, long before the creation of "Family Viewing" hours, a television phenomenon existed that brought my entire family happily together in front of our small screen. The big event was the appearance of a black person on a television program. Whether as a walk-on character in a commercial, a bit-part in a drama, or a two-minute act in a variety-entertainment show, "one of us" on the tube was a rare enough occasion to warrant the cessation of all household operations. Mother, father, auntie, and kids would snuggle together, side-by-side in front of the TV, each intently absorbing this major affirmation that blacks were indeed real people who used American products, and existed in other parts of the world. Unlike the network family viewing hours, which are regularly scheduled each evening, our family viewing sessions were impromptu and infrequent. In 1962, for example, only three black faces appeared once every five hours in one television sample.[1]

The same sentiment must have prevailed in other minority households with television sets because blacks were not the only "invisible" people of American T.V. society. Native Americans, Asian Americans and people of Hispanic origin appeared less often than Blacks on television programs. Together, minority groups constituted less than 3% of all characters in televised dramas and comedies in 1974.[2] When ethnic minorities were portrayed, they tended to be cast in exaggerated and stereotyped roles, more as white producers perceived them, than as they perceived themselves.

**" . . . these negative images accumulating day after day [are] teaching self-hate and resulting in destructive self images.[9] For white viewers, such depictions reinforce common misconceptions about how blacks behave."**

Blacks were typically cast as amusement objects or buffoons (e.g. Amos n'Andy, Farina, Stymie and Buckwheat in *Our Gang*), household servants (i.e. Beulah or Rochester on *The Jack Benny Show*), or singers and dancers on variety shows. Native Americans were depicted as blood-thirsty marauders in old hollywood westerns that were often historically inaccurate and lacked understanding of Native American culture. The televised image of the Asian American was that of a passive human being with slanted eyes, bowed head and hands held together in the "ah-so" position. Most likely to work as waiters, laundrymen, gardeners or household servants, the Asian male appeared unattractive and sexless. Asian women were portrayed as docile, submissive, housekeeper types (e.g., *Courtship of Eddie's Father*), or as exotic, sexy background figures patterned after the Geisha Girl image.[3] The popular image of Mexican Americans as crime oriented "banditos" was crystalized in the form of one infamous cartoon advertising campaign character, "The Frito Bandito". Potbellied sporting a wide sombrero, and a pair of pistols slung around his hips, the Frito Bandito robbed innocent victims of their cornchips at gunpoint. His theme song, sung in a spanish accent, was backed by such hispanic flourishes as a strumming guitar and clicking castenets. Along with non-white ethnic minorities, old people and people with disabilities, two large minority groups in America also suffered symptoms of invisibility and gross misrepresentation in American TV society. Old people were restricted to such roles as Grandpappy Amos on *The Real McCoys* or Granny on *The Beverly Hillbillies*. White-haired, taken to forgetfulness and stubborn as a mule, television's old folks contributed little to society apart from tending the kettles and nurturing the young ones at home. Likewise, the major concerns of people with disabilities as portrayed on television tended to be their own illnesses. Handicapped characters were most often portrayed in clinical settings like *Ben Casey* or *Marcus Welby,* medical series in which doctors or nurses played major roles in helping them adjust.

It is my belief that, more than any other mass medium, television has contributed to the misunderstanding and stereotyping of minorities in America. The average American will spend 3,000 days, almost nine years, simply watching television.[4] During that time, many viewers, particularly children, will receive their initial exposure to the lifestyles and cultures of ethnic and other minority groups. Television's greatest influence lies in its ability to shape attitudes and beliefs about people through representation of characters in seemingly real-life situations.

As children, long before we could read or write, we watched television. Today, by the time a child graduates from high school, he will have spent more hours in front of a television set than in a classroom. The cumulative effect of viewing exaggerated and stereotyped portrayals of minorities, with little or no redeeming counterimages on TV may create negative impressions and social behavior expectations that persist through adulthood. Test yourself: close your eyes and visualize any one minority group. Do images of childishness, dependency, pity, or helplessness come to mind when you think of the handicapped? Does the word "Indian" bring to mind feathered war bonnets, moccasins, beads, tee-pees and painted faces? I have found that the image of "The Plains Indian" has remained with me through adulthood, most likely because I have never had any personal interaction with Native Americans. The Plains Indian was the image of Native Americans popularized in television in the late fifties and early sixties. Scantily clad, wearing a feathered war bonnet, he was usually shown riding bareback across the plains on a wild pony. Interestingly enough, upon questioning a group of four year olds as to what an "Indian" was, the most common reply was ". . . . a man who wears a feather in his cap."

Current seasons indicate a new trend in the portrayal of minorities on television. During the 1978-79 season, you might have tuned in one week in February and gotten the impression that at least one minority group, blacks, had taken over prime time.

*The Jeffersons, Good Times,* and *What's Happening* invited viewers into black homes to share black family life. *Welcome Back Kotter* offered insight into multi-ethnic minority experiences in a ghetto high school. *Diff'rent Strokes* delighted audiences with the antics of two black boys being raised by a white real estate agent and his housekeeper, as a last favor to their deceased mother. Seven episodes of *roots-Generation II* captivated mass audiences with the story of generations of one black family growing up under slavery in the old South, while *Backstairs At The White House* faithfully retold the stories of former American presidents through the eyes of Lillian Rogers Parks and her mother, black maids at the White House.

In other recent seasons, Hispanics were featured in *Chico and The Man* and *Chips.* Asian and Pacific Americans were seen as continuing characters in *Hawaii Five-O, Quincy,* and *Barney Miller.* American Indians appeared in *How The West Was Won.*[5] Blacks appeared as supporting characters in *McCloud, Baretta,* and *Policewoman* as well as many others. A CBS special, *The Silent Victory, The Kitty O'Neil Story* featured a deaf woman who became a Hollywood stunt woman. Even the plight of female homosexuals was explored in a network special entitled *A Question of Love* in which a lesbian is faced with the prospect of losing her children because of her sexual preference. Clearly, minorities of any kind are becoming increasingly popular on commercial television. One writer suggests that current television program executives and producers go out of their way to include some ethnic or minority characters in as many programs as possible. Rather than an indication of broadcasting's newfound social consciousness, however, he suggests that the current "ethnicity" craze may simply be a way to "dress up otherwise similar and unimaginative situation dramas . . . with people of exotic skin colors and backgrounds."[6] The major issue today is no longer the number of minority faces on the television screen, but the manner in which minorities are portrayed.

**"The major issue today is no longer the number of minority faces on the television screen, but the manner in which minorities are portrayed."**

Currently, blacks are the most visible non-white minority in television society today. A closer inspection of the portrayal of blacks reveals many trends that are currently being tried with other minority portrayals.

## Trends

Social consciousness might call for thoughtful treatment of black life and issues. Black representation on television might demonstrate to all viewers the extent to which blacks now participate in all levels of the social, political, economic and cultural aspects of American society. However, black shows have been drawing protests from black parents, teachers, psychologists and leaders. Why? Because broadcasters continue to present blacks in a restricted range of roles and formats on television, with most blacks appearing in situation comedies. One of the hottest trends today is the distortion and satirization of black family life in weekly situation comedies. On *The Jeffersons, Good Times, Sanford & Son,* and *What's Happening,* series that deal explicity with black family life, major characters appear as reincarnations of black buffoons of the fifties, caricatures of blacks as perceived by white producers and writers. Comic personalities like George Jefferson, "a splenetic little whip of a man, who bullies like a demented overseer, seldom speaks below a shriek, and worships at the church of ostentation."[7] "Fred Sanford," a bigoted old junk dealer, known to feign heart trouble when asked to work, and willing to do anything to make a fast buck; J.J., the sharp tongued high stepping jester of *Good Times,* and Rerun, the perpetual teen-aged eating machine on *What's Happening,* are the how-to-behave models for black children, and the how-blacks-behave models for white children.

While exaggerated portrayals of whites on situation comedies are offset by the appearance of whites in a variety of shows and formats, blacks retain the situation comedy as the only regularly scheduled vehicle for the depiction of contemporary black lifestyles. Rather than providing a humanistic perspective of this minority's lifestyle and culture, these series often mock it. Comic dimensions are added to images of black families in turmoil. Hostile interactions are sweetened by timely audience laughter and hackneyed displays of affection at show end. In *What's Happening,* Dee, the domineering little sister makes deprecating remarks about her brother Roger and his friends. J. J. constantly chides his sister Thelma about her stupidity and ugliness on *Good Times*. George Jefferson spews forth venomous insults at his maid, Florence, while Fred Sanford curses Aunt Ester, the bible-toting, signifying sister-in-law who insists that he is a *heathen.* In *Baby I'm Back,* one short-lived black series, the children called each other by such endearing names as "toilet-head" and hardly any exchange was free of an insult.[8]

An older sophisticated viewer might question this world in which black females and males are constantly shown expressing hostility toward one another, but to the young, unsophisticated child viewer, this is life as it is in "the real world". Studies have suggested that young children see a series of separate and fragmentary incidents rather than the story of a television film.[9] Thus, while the overall theme and story line of a program might be of social significance and redeeming moral value, young viewers may be tuning in only to the seemingly comic aspects of these abusive verbal attacks. Moreover, as many parents can attest, young children are excellent mimics of what they see on the television screen. Such examples are hardly inspiring. For black children, the total effect of these negative images accumulating day after day is teaching self-hate and resulting in destructive self images.[9] For white viewers, such depictions reinforce common misconceptions about how blacks behave.

As a single parent with a son, one trend that I have found particularly disturbing is the absence of strong, intelligent black father figures on television. Viewers are constantly exposed to a picture of the black family in which the father is either dead or has deserted. Children appear most often in the company of mothers or unsupervised. In *Good Times,* after a season of programs in which the father was depicted as unable to find or hold a job, he was killed in a car accident while away from his family. In *That's My Mama* the father was deceased. In *What's Happening,* the father is not dead, but does not live in the household with his children, and has rarely appeared in any supportive situation vis a vis his children. In *Diff'rent Strokes,* a dying mother entrusts the care of her two black children to a white employer. While many black children live in female-headed households, there is such a figure as a black father, yet these programs give the impression that fatherless black families are the norm. Why are there no strong, intelligent father figures on black programs? Are networks being consciously racist in their presentation of blacks, and other non-white minorities? Or are they being consciously business-minded, assuming that serious and realistic portrayals of minorities in dramatic roles will not sell?

**"In 'Sanford & Son,' and 'What's Happening,' series that deal explicitly with black family life, major characters appear as reincarnations of black buffoons of the fifties, caricatures of blacks as perceived by white producers and writers."**

## Roots

The *Roots* phenomenon (1977) proved that mass television audiences would respond to quality programming about blacks. More than 130 million Americans tuned into at least one episode of *Roots.* The story of seven generations of one black family, *Roots* told the story of black slavery in the U.S. in a manner sympathetic to the black perspective. Both blacks and whites came away from *Roots* perceiving themselves to have more understanding of blacks.[10] Whites indicated that they learned about black historical roots and gained insight into black thinking. Both blacks and whites indicated increased awareness of the strength of the black family under slavery, and of the way tribal identities were stamped out.[11] The impact of *Roots* appears to have been genuinely humanistic.

The success of *Roots I* spurred a number of Roots-type productions; *Roll of Thunder, Hear My Cry* and most recently, *A Woman Called Moses.* Conforming with proven acceptable patterns, these programs had an historical basis, were set in the rural South, and portrayed blacks as struggling victims of the American system. Most were well received by television audiences. An attempt at portraying more contemporary issues, the *King* docudrama, was not well received.

Set in the sixties, during the civil rights struggle, *King* failed miserably in the ratings. *King* was set in a time period that many viewers had lived through and still remembered. One white producer commented, "With *Roots* we were looking back into history; we were not a part of that . . . . with Martin Luther King, the wound is still open, we feel guilty about it."[12] Criticism of the program by King's former associates may have also turned off viewers. The danger of *King's* failure lies in the possibility that it may be used as a rationale for not presenting other dramas about contemporary black issues or public figures. Nevertheless, programs like *King* and *Roots* have been invaluable in communicating the heritage and culture of blacks to both white and non-white audiences. Both presented strong, positive images of blacks that counteracted the weekly parodizing of blacks on situation comedies. Unfortunately, such programs have been limited to the special status, and therefore appear sporatically.

Two other large minority groups are gathering forces to join in the revolt of television's invisible people, the aged and the handicapped. According to the 1970 census report, there are 29 million people over 60 years old in this country. One million are over 85, and 13,000 are centenarians. Americans over 60 spend more time watching television than in any other leisure activity.[13] Yet, the media have ignored old people as a major segment of the population. Old people are considered poor consumers and as such, they are not a popular market with advertisers. In one 1969-71 sample of network prime time programming, the elderly comprised less than 5% of both sexes portrayed, about half their share of the population.[14]

Various programs have dealt with some of the problems of the aged, such as forced retirement and loneliness. *Barney Miller, Fish,* and *The Waltons*. While these programs and others reveal the pain and loneliness of growing old, few have provided images of aging as an exciting and eventual stage of life. In American culture,we seem to believe that from the time we are born we begin to die. By old age, life is but a series of memories. Television provides few models who could demonstrate successful models of aging. Rather than the end of life, viewers could be made aware that old age is another stage of life, with its own advantages and joys. Children might also benefit from more varied and positive portrayals of the elderly. Formerly, personal contact with the aged was provided by grandparents who often shared a household with young families. In today's mobile society, the extended family is a rarity. A child's only real contact with the elderly may be long distance telephone calls to grandma and holiday visits. Television portrayals of the elderly preclude active involvement in community affairs, social activity and even sexual companionship. Television has colored our image of old age a dull uneventful gray. Researchers suggest that stereotyped myths of aging may be limiting the adaptive capacities of the aged. "In the absence of high levels of interpersonal contact, the burden of information (about the aging experience) falls upon the media."[15]

On commercial television, the typical script about a disabled person often begins with a tragic individual, unable to cope, with adjustment problems related to his disability. With the help of a mentor (usually a doctor, nurse or mate) he works hard to overcome mental obstacles and limitations imposed by his handicap, and goes on to lead a meaningful life despite all. In the televised movie *Larry* the institutional psychologist worked tirelessly to teach Larry skills so he might eventually return home. There are approximately fifty million people who have some form of mental and physical disability. Disability labels like blind, deaf, mentally retarded, etc. commonly evoke images of helplessness, childishness, pity, dependency etc.[16] Television might be very influential in changing public attitudes toward people with disabilities by portraying them in a variety of life situations. The image of disabled people might be changed by portraying them in normal roles and ordinary settings—out of doctors' offices and hospitals, away from clinical interactions, and as equals to their peers, rather than as objects of pity.

**"While exaggerated portrayals of whites on situation comedies are offset by the appearance of whites in a variety of shows and formats, blacks retain the situation comedy as the only regularly scheduled vehicle for the depiction of contemporary black lifestyles."**

The success of televised docudramas, miniseries and specials providing thoughtful and sensitive portrayals of minorities is promising. Success breeds imitation in the broadcast industry, so viewers can look for more to come. For instance, NBC will introduce a serious dramatic series, *Harris & Co.,* featuring Bernie Casey as a black single parent. ABC-TV is currently developing a Roots-type miniseries based upon the past of the America Indian. The program will cover two families of Indians living between the late 1700's and 1830. The 1979 presentation of *The Silent Victory, The Story of Kitty O'Neill* on CBS in which the central character is a deaf woman, is another indication that networks are moving in the right direction in the portrayal of the disabled. CBS also recently aired *No Other Love,* the story of a love relationship between retarded young adults.

While broadcasters continue their search for program formulas that will attract mass audiences without reinforcing negative stereotypes of minorities, viewers and public groups remain disturbed by current television fare portraying minorities. The recurrent question arising in classrooms, living rooms and meeting halls remains, "What can we do in the meantime?"

## Parenting

As parent of a five year old, I have been grappling with this question since the day my son discovered Barbarino and Boom-Boom Washington on *Welcome Back Kotter* almost two years ago. At that time, he developed a mysterious limping walk, with hunched shoulders and head thrust forward, which he described as "walking cool, like Barbarino, ma!" Two weeks later, his nursery school teacher informed me that my son was no longer answering to his given name. Apparently he had changed his name to Boom-Boom Washington overnight, and forgotten to tell me.

Naturally, this condition provoked a bit of concern on my part, so I instituted a new game in our household called "Mommy's TV Control Game". For him it was a space fantasy game in which mommy pretended to be a robot who consulted the space center about what he could see on television.

On my level, "Mommy's TV Control" involved conversations with grandparents, babysitters, and others entrusted with the care and supervision of my son, regarding what television programs were suitable for him. Grandparents didn't understand the fuss, but were willing to respect my wishes. The babysitter ignored me, and was eventually replaced. On visits to homes of less enlightened friends, I brought along books and games to occupy the kids, and would politely reject suggestions to "turn on the TV for the kids, so we can talk in peace."

"Mommy's TV Control" also involved supervision of what my son viewed at home. At three years old, he was not familiar enough with television programs to make requests, but I found that clicking the channels in front of him was bound to incite screams of "I want to see that mommy" whenever cartoons, fighting or explosions appeared. Quickly, I learned to peruse the TV listings, pick out a children's special, or other show that I thought might be interesting, and announce with great excitement that we were going to watch TV together tonight. After two years, I find that now, my son has been trained to accept this type of supervision and is quite cooperative. In fact, my own enthusiasm about the shows seems to generate the same feeling in him. Likewise, when my son was three years old, I began the practice of turning off the set after the selected program was over. At first, there were tantrums and tears, but again at five years old, he has become conditioned to the fact that when "Mommy's TV Control" dictates that kids' TV time is over, then its time to do something else. I find it helpful to have something else planned for the child to do immediately after the viewing time in order to distract him. At bedtime we follow TV with a bedtime story. On Saturday mornings, it's light chores or outside playtime.

Selective and controlled viewing, however, are not enough to subvert some of the more deadly messages about minorities being conveyed on television. Parent-child discussion has proven most effective for me. Despite limitations in his ability to understand certain television content due to his age, talking to my son about minority characters, and situations on TV helps him to understand what is happening, and helps me to understand what he is seeing. In one incident, for example, where Fred Sanford rants and raves about how ugly Aunt Ester is on *Sanford and Son,* I noticed how intently he was watching. "Do you think he looks like a grandpa?" I asked. "Yep," he replied. "Do you think grandpa ever acts like that?" I then asked. "No mommy, grandpa's not on T.V.!" he explained. Immediately following the show, we talked about the difference between how people on television act and how people that we know act in real life. I doubt that his five year old mind understood everything we discussed, but I view this discussion as a continuous and cumulative process. More than anything else, "Mommy's TV Control" involves communication with my child, talk and more talk, knowing what he's feeling and thinking while he's watching, indicating related situations with minorities in real life, and pointing out the non-stereotyped examples of minorities on TV where they occur. For example, we searched several superhero programs on Saturday morning one week, looking for a black superhero. When we finally found two, "Superstretch and Microwoman" we were both excited. While I wasn't exactly ecstatic about the characters themselves, I felt rewarded when my son explained to me "See mommy, superheros come in all colors!"

Teaching critical viewing skills is an ongoing process, that must start at an early age and increase in sophistication as the child's developmental capacities increase.

Parents interested in reducing the impact of television portrayals of minorities on their youngsters might begin with these steps:

1. Institute a program of selective viewing excluding those programs that portray minorities in distorted and stereotyped roles.
2. Look for programs that portray ethnic and other minorities in positive, realistic and sensitive manners (e.g. specials, miniseries, afterschool specials.)
3. View these programs with youngsters, discussing their relevance to real life people and situations familiar to the youngster.
4. Provide reading materials, projects, trips to films, museums, etc. that might be related to the subject matter of the program, providing exposure beyond fictionalized TV presentations.
5. Get children into reality situations with minority persons, elderly people, and others.

### *REFERENCES*

1. Regina Lowenstein, Lawrence Plotkin, and Douglas Pugh, "The Frequency of Appearance of Negroes on Television," (The Committee on Integration, New York Society For Ethical Culture, 1964 p. 4)
2. "SAG Documents Use of Women and Minorities in Prime Time Television," Screen Actors Guild news release, October 31, 1974
3. "Window Dressing on The Set: Women and Minorities in Television," U. S. Civil Rights Commission, August 1977, p. 8
4. Grant Noble, *Children in Front of The Small Screen,* Sage Publications, 1976 p. 7
5. "Window Dressing on The Set: An Update" U. S. Civil Rights Commission, Jan 1979, p. 2
6. Howard Suber, "Ethnic Window Dressing on The Tube" Newsday,
7. Frank Rich, "Blacks on T.V., A Disturbing Image." Time, March 27, 1978
8. Pamela Douglas, "The Bleached World of Black T.V." Human Behavior, Dec. 1978 p. 63
9. Grant Noble, "Children In Front of The Small Screen," p. 94
10. Howard, J., Rothbart, G., Sloan L., "The Response to ROOTS: A National Survey," J of Bdcstng, Summer 1978, Vol 22:3, p. 285
11. Ibid.
12. Alexander Keneas, " 'KING', a flawed but important 'docudrama' " Newsday, Feb 10, 1978
13. R.H. Davis, "Television and The Image of The Aged," Television Quarterly, 12:37-41, Winter 1975
14. C. Arnoff, "Old Age in Prime Time," J. of Communications 24:86-87 Autumn 1974

15. Beth B.Hess, "Stereotypes of The Aged," J. of Communications 24:79, Autumn 1974
16. Douglas Biklen, Robert Bogden and Burton Blatt, "Label Jars, Not People" in *Promise and Performance, Children With Special Needs.* ACT Guide to T. V. Programming for Children, Vol I, Bollinger Publishing Co. 1977.

# Minorities/Worksheet No. 1

Children often use information they get from TV in deciding what kinds of persons are important in our society. *(See Chapter 3.)* They especially notice whether or not:

a. That type of person is seen at all, or how frequently.

b. That type of person is shown in positions of authority and is shown respect.

In coming to conclusions about individuals, children begin to make assumptions about the groups those individuals represent.

**Exercise:**

1. Watch the television programs your children most often view. For each program, and for the commercials, keep a record of the following:
   a. What minority persons are shown and how often?
   b. Which minority persons are not shown at all?
   c. What types of persons are most often shown in authority or leadership roles that gain the respect of other people?
   d. What types of persons are most often shown in secondary roles and bit parts?

2. Review your findings and decide which programs you feel are most appropriate for your children.

3. If your children now view programs you feel they should not view, watch the programs once with the children, pointing out and discussing what you find objectionable.

4. With your children, watch the programs you think are appropriate, pointing out and discussing the positive values you find in the programs and why you think those values are important.

## Minorities/Worksheet No. 2

Select a TV series in which minority characters are in major roles. List the main characters in the program.

Character
_______________

Race
_______________

Role
_______________

Evaluate each character from a low of 1 to a high of 7 in terms of:

Powerful
_______________

Likeable
_______________

Successful
_______________

Bad/good
_______________

Status
_______________

Object of amusement
_______________

Solid citizen
_______________

## Minorities/Worksheet No. 3

In a television commercial, the first scene is of a bare-chested white male putting on a shirt. Next, there is a similar scene of a black male. In the third scene, the two men are together in a disco, "boogieing on down," with an Asian-looking woman between them.

1. Think about the commercial and discuss the following with another person:
   a. Why do you think the commercial was cast the way it was? For example, why was an Asian-looking woman chosen? How would the situation be changed if the woman is white? If she is black?

   b. If you didn't know this was a shirt commercial, what would you say was being advertised?

   c. What are some of the other messages you find in the commercial?

2. Repeat this same technique with other commercials you find on your TV screen.

# Minorities/Worksheet No. 4

With your past TV viewing experiences in mind, select questions from the following list and discuss your responses with another person or with your family.

1. How favorably or positively are persons in general, of all races and groups, portrayed on television?
2. Which types of persons are treated most favorably? Which are treated least favorably?
3. The following are often listed as U.S. minority groups: Native Americans, Hispanics, Blacks, Asian Americans, Jews, old people, handicapped persons. What others would you add? What changes or deletions would you make in the list?
4. Which minority groups are never or almost never represented on TV? Why do you think that happens?
5. Is the phrase "minority group" negative, positive or neutral for you?
6. Do you believe it is harmful to:
   —never show persons of a minority group on TV?
   —show persons of a minority group most often in a negative way?
7. If so, do you think it is more harmful to show members of a minority group in a negative way than it is to show members of the majority group in a negative way? Why?
8. Did you watch *Roots I, Roots II, and King?* Why or why not? Why do you think *Roots* got very high audience ratings and King got very low ratings?
9. Which program you have seen on TV seemed most representative of the "black way of life"? Why?
10. In what way would *All in the Family* be a different show if scripts and direction remained the same but the characters were all Asian Americans? Would it be as widely viewed?
11. What TV programs can you remember seeing that portrayed Native Americans living in the 1970's?
12. Did you think of Chico in *Chico and the Man* as accurately representing Hispanic Americans?
13. If a TV series was developed featuring persons with serious disabilities such as mental retardation, amputees, or stroke-damaged persons, would you watch regularly and how would you feel—happy, pitying, embarrassed, sickened, interested? Would you allow children to watch?
14. Judging from TV, what impression do you have of senior citizens?
15. If you had the opportunity to create a TV series, describe what it would be like. What subjects would it deal with? Would it be a funny or serious program? What kind of persons would play leading roles?
16. What, if any, persons do you see on TV that you identify with? (Seeing them is a bit like seeing yourself.)
17. What persons and behaviors do you see on TV that you would want your children to imitate?
18. Can majority group persons, who make almost all television industry decisions and almost all the programs, create programs that accurately portray people of a minority group? Why or why not?
19. Given television's reliance on large audiences and its purpose of making money, would ethnic minority producers tend to create quite different TV programs than are on the air now, or quite similar ones?

## Minorities/Worksheet No. 5

1. Select a minority group you have no personal contact with. For that group write a description of what you think a typical individual of that minority group would probably be like. Consider such things as what kind of job, how well educated, how well they take care of their children, how trustworthy, what kinds of foods they eat, what activities they like, how satisfied they are with their lives, etc..

2. Since you have had no direct experience with the person you are describing, try to figure out where your opinions and impressions of that person came from.

3. If you can find TV programs where one or more members of the minority group you selected are on the screen, watch several such programs.

4. Re-read your description of the person and answer the following questions:
   a. How accurate do you think your description is?
   b. How closely does your description of the person compare with how such persons are shown on television?
   c. How much of the information you used in describing the person may have come from your past TV viewing?

5. If you feel your present information about persons of other groups is not accurate or is incomplete, list some ways you can begin to correct that situation.

## Minorities/Worksheet No. 6

Television very often presents stereotyped portrayals of ethnic minority individuals, having them behave in a way that fits the impression many people have about what that minority group is like.

The following are some common stereotypic minority roles found on TV:

____________________ Sidekick role, always second in command.

____________________ Servant role.

____________________ Dumb, slow witted, can't remember instructions.

____________________ Amusement object or buffoon.

____________________ Poorly educated or speaks English poorly.

____________________ Linked with criminal activity.

____________________ Comes from a bad home.

____________________ Is poor or a transient.

____________________ Can't run own life, needs help.

____________________ Is a helpless victim of people with power.

1. Using the above list, watch TV programs in which you find individuals you can identify as minority persons. Make a check by items on the list each time you find that stereotype on the screen. Consider why the producer may have used that stereotype. (Humor, quick identification, etc.) Try to notice when someone is presented in a ridiculous way. Add other stereotypes to the list as you notice them.

2. Do the same with programs featuring all non-minority persons. What difference do you find? What similarities?

## Minorities/Worksheet No. 7

If you are a member of an ethnic minority group, consider the following:

1. What are some of the most important values or special heritage of your group that you think should be passed on to children?

2. How do you make sure children learn about them?

3. What ways do you have for celebrating the positive feelings you have about your group?

4. List some ways that changing how television is used in your home might help protect your heritage and help in passing it on to children.

## Minorities/Worksheet No. 8

Changing how you use TV. Helping to change the television system.

1. Create a set of guidelines for how you will use TV in a responsible way, including details on:
    a. How much you will watch.
    b. How you will decide what to watch.
    c. What positive and negative things you will look for.
    d. A plan for discussing your viewing experiences with other persons.
    e. How you find sources (other than TV) for getting accurate information about groups you have no direct knowledge of.

2. Read Chapter 8 in this book to learn how viewers can help change the television system. List some of the things you can begin to do immediately.

# Chapter 11

# TELEVISION AND THEOLOGY

*George C. Conklin*
*Linda W. McFadden*

*GEORGE C. CONKLIN is Director, Center for Media Studies at the Pacific School of Religion, Berkeley, California. He is a T-A-T trainer.*

*LINDA W. MCFADDEN is a candidate for Masters of Divinity Degree at Pacific School of Religion, Berkeley, California. She has been Professor of English at Virginia Commonwealth University and at North Carolina Central University.*

Our transactions with television are often puzzling to us. Television's array of images holds endless fascination for us quite apart from the medium's ability to entertain, to influence, or to inform us. We may sometimes ask ourselves as we reach wearily for the "off" button, "Why did I watch *that?*" or think, "I guess I just wasted the whole evening." These familiar experiences reflect the fact that we sometimes watch television without consciously choosing one program over another. We are often bewildered or chagrined to discover that we have watched from beginning to end something we didn't like or enjoy.

We suggest that there is in fact a reason for our ambivalent love/hate affair with televison and that there is more involved in our viewing habits than entertainment, information, boredom-chasing, or avoidance of interaction with others. Our central thesis is that our interaction with ourselves and with the world—including our attaction or repulsion toward advertising, sex, violence and other television content—can be understood in terms of our human condition. The approach of Television Awareness Training is based primarily upon psychological and sociological understnding of human behavior. Here we will reflect on use and abuse of television within the perspective of Christian theology.

## Theological Assumptions

Central to Christian faith is the belief that the ultimate reality of the universe is God, who created the world and all it contains. As creature, humankind is intended to live in relationship with all Creation and with the Creator. Because we are created in the image of the Creator, we also have the capacity to create. And so humans create—cities, works of art, codes of law, languages, systems of belief, technologies. Our very ability to have developed television—both the technology and the human social systems which shape it—reflects our divinely-given creativity.

Created in an act of divine love, we are stewards of our lives and of the Creation, intended to live in relationship and dialogue with the Creator. Although we, as creatures, are finite, bounded by the limitations of time and space and of our physical bodies, we can transcend our finitude by being in dialogue with God. But we are not compelled to be in relatedness to God. We have a choice. Created in freedom, we are able to choose relationship to God or to turn away.

The human inclination to turn away from the Source of our being rather than to live in dialogue and partnership with God is what Christian tradition has called *sin.* In turning away from the Creator, we cut ourselves off from the Source of our being, the Source of our wholeness and creativity, the Source of our ability to be in relationship. In this cut-off-ness, we experience anxiety and rootlessness, the sense that we have no "home" in the Creation and that we have no essential worth.

When we choose to turn away from the Source or Ground of our being, we rupture the primary relationship in our lives. Turning away from God is a distortion of our existence. In choosing to leave the unconditional relationship for which we were created, we carry with us a desire to be loved unconditionally, to have ultimate meaning, to know that we are loved. The Original Sin of broken relationship with the Creator is worked out in human history in complex, manifold ways. The unfulfilled, passionate longing for the ultimate which belongs to our relationship for God creates within us fear, anxiety, rage—a nameless terror in the night. Because of this, we reach out to possess, to dominate, to control, to exploit, to consume things or persons we believe can satisfy our emptiness.

In our desperate quest for worth and belonging and control of our destinies, we use our divinely given creativity to fashion gods. We elevate some aspect of our finite reality—such as race or sex or affluence or power or another person—and bow down before it. Our idols cannot, of course, satisfy our being; the finite can never satisfy infinite longing, but we persevere. We love our idols for the totality of their claims on us and curse them when we discover that they have left unsatisfied our craving to be unconditionally valued.

Our use of television, like every other facet of human life, reflects both our essential creativity and our fallen nature. The existence of the complex technologies which permit television to exist as well as the content of television, reflects human creativity. But the fact that this powerful medium is used to play upon our deepest longings for love, security, and belonging for the sake of promoting consumption of toothpaste, automobiles and life insurance constitutes a betrayal both of those who create television and of those who view it. Our paradoxical transactions with television are not really new in human history. St. Paul's plaint, "For I do not the good I want, but the evil I do not want is what I do" (Rom. 7:19, RSV) echoes in our own perplexity about our love/hate relationship with the medium. We attempt to find ultimate meaning for our existence in television, but such meaning is not there to be found. Yet we continue to seek—and to know anxiety—cursing both television and ourselves because we do not find the meaning we seek.

Placing television and its uses in a Christian theological perspective helps us to see that our stewardship of television is an issue of human community. All of us—sponsors, producers, and viewers—share responsibility for what television is and how it is used. It has been easy—and commonplace—to point the finger of blame at "them," the makers of television systems and their images, claiming that "they" do things to "us," the viewers. It is conversely commonplace for sponsors and producers to claim that they are "giving people what they want." But both "they" and "we" share our common human experience. We all bring our rootlessness and sinful turning away from God to our interaction with each other. The stories, myths, promises and dreams of Television Land are not other than ourselves.

The producers, writers, talent and sponsors of televison are no less victims of the situation than are the viewers. The assurances of being valued and loved to which the audience aspires are also powerful motivators for the makers of the dream. The dream-makers seek to wear the right things, know the right things, have the right symbols of power, be the right persons—exactly as we, the viewers, do. We share the same dream. It is a measure of the effectiveness of our unrelenting, aggressive seeking of transcendent value that we honor a relative hierarchy of finite values. Ironically, we do homage to those dream-makers who are most adept at beguiling us through their skillful casting of the finite in the guise of the transcendent.

## Salvation

Christianity affirms that sin is part of our human nature, but holds the promise of salvation from our brokenness. We believe that God, in an act of self-revelation, entered the world in the person of Jesus Christ. His life and death hold for humankind the possibility of the restoration of dialogue / communion / relationship between Creator and creature, a healing which we cannot bring about through any effort of our own. In Christ, the powers and false worship of the world are met and ended not with a challenge of greater power but with the transcendent power of humility and vulnerability. The central message in Christ's incarnation is: Trust God's love and call to relationship as the Source of all your being. Do not place your trust in the values, beliefs, actions and things of the world in their claim of salvation for you.

The question "What must I do to be saved?" finds its way into television in several formulations:

How can I obtain love?
How can I be valued?
How can I matter to others?
What must I do in order to be acceptable?

Television's answers also are manifold. News programs suggest *who* is important and *what* makes a person important or interesting. The emphasis on economic or political power, physical or intellectual prowess, or behavior which departs from cultural norms tends to reinforce in us the idea that power, prowess or eccentricity make us interesting. Dramatic and adventure programs provide role models of noteworthy behaviors which are portrayed as obtaining the attention, affirmation and affection of others. Advertisements demonstrate countless things which we may add to ourselves in order to be valued, loved and secure. Television suggests that salvation from our longing for love and security lies in things performed, worn, used, applied, driven, learned or experienced. Ironically, the way is always clear for more products, new or old, with the same promises stated anew. There is an inherent futility in seeking ultimate meaning or love from things, ideas, experiences, or even people within the order of Creation, but this fact does not preclude our renewing our search.

The question "What must I do to be saved?" presupposes that a person has recognized her or his own need and wishes to act to change that situation. Such a question cannot even be asked until the futility of the quest for love, value and security through that which can be purchased, consumed or acquired has been recognized. Christian faith teaches that there is nothing we can *do* to deserve or receive God's love. Our worth, value and ability to love and be loved at the every center of our being is a given, part of God's Creation. We can respond to God's love, but we must surrender our quest for domination and control of the Creation before we can return to relatedness with our Creator.

Television is demonic to the extent that it suggests that happiness and belonging *are* ultimately obtainable through our own efforts. We cannot accept God's gift of salvation until we choose to be free of our striving and attachment to material possessions. Television helps us not to see this, and we are guilty of complicity. We willingly accept the premise that whiter teeth or a snazzier car will advance our quest. The message of success, power and affluence proclaimed by television is a secular gospel in direct competition with the Christian gospel. Television promises an easeful "good life," a paradise on earth which is always just beyond our grasp. Its false promises are a seductive lure away from the way of eternal life.

## Transcendence

The dream of going beyond the limitations of time, space and mortality is woven throughout human history. The wish to fly through the air, to travel in time, to live forever, to be all-knowing or all-powerful has been recorded in song, fable, art and literature for as long as human culture has existed. In recent decades, human technology has brought many of these dreams to realization through developments in aviation and space exploration, in communication and in medical science. Indeed, it is the quest for transcendence which has provided the impetus for the research and exploration which has pushed the frontiers of human limitation in some astounding ways.

Television technology makes the possibility of transcending human finitude appear within grasp. Transcendence is depicted primarily in two ways. It may result from human activity, particularly from science and technology, or from contact with powers, forces or beings from outside our time, space or solar system. Television's myth of transcendence is based upon science, which is portrayed as giving humans the ability to travel backward or forward in time, venture to other galaxies, or be rebuilt into super beings. The fantasies once enacted only in art and literature are now acted out with familiar actors and actresses appearing as reconstructed humans or visitors to other worlds. The vivid images of such a telecast have every appearance of reality, and it is left to the viewers' experience of reality to help them distinguish fact from fantasy.

The difficulty with these tantalizing fantasies is that there is no transcendence in television's transcendence. Transcendence is trivialized, stripped of its place in the divine-human relationship. From a Christian point of view, transcendence of our finitude is a gift of God. In Genesis, God is depicted as surveying Creation in search of some creature who will be in dialogue and relationship with the Creator. Not finding any such creature, God creates humankind specifically for this special relationship. Out of the entire order of Creation, only humans are called to be in communication and relationship with the Creator. It is out of this relationship, which is the gift of God's grace, that we transcend our creaturely limitations. Neither the entrance of superior extraterrestial beings into our universe nor our own efforts to be above or outside the Creation can surpass the power that is ours by virtue of our relatedness to God.

## Storytelling

Storytelling rests at the center of our awareness of who we are, what our history has been and what our purpose in life is. Christian theology has recently rediscovered the importance of storytelling in the process of discovering who God is and who we are. The Jewish and Christian traditions experience God's self-revelation to humankind as story/event. Our scriptures and much of our religious tradition are stories told from a human perspective of the drama of God, the Creator, acting in history, the Creation. Storytelling is a divine gift which is more than simple diversion or recollection. It lies at the center of communication with other persons and with God. Through the telling of stories which embody our individual and cultural experience of the transcendent, we seek the Holy, that place where we know and believe that our stories and God's story intersect. The telling and re-telling of stories helps us to discover the reality of God's love for us, the central purpose in our lives, and reconciliation with the Power which created us.

For centuries, stories have been passed on through myriad forms of creative activity. They have been spoken, sung, danced, written, painted or enacted by every human culture. Now television has become one of the most powerful and pervasive storytellers in human history. As a people we now commit years and years of our lives to the stories of a medium which has existed for only thirty years, a mere moment in the span of human history. There are stories which touch, in varying degrees, the realities of the world—the news and documentaries. There are stories which draw on the themes of human life as we know it—drama and soap operas. Our hopes and fantasies for the future are caught up in science fiction, advertisements and even quiz programs. The average American adult now devotes more than *four hours of* every day to the stories told or suggested in the dramas, advertisements, news programs, documentaries, sports broadcasts and other formats of television.

Our personal stories are of ultimate importance for each of us. We find meaning in them and seek opportunity for greater meaning. Stories hold for us the fascination of helping us discover who we are and who we might become. We bring to our television viewing not only a desire for diversion or entertainment, but this quest to discover who we are, what promise life holds for us, how to be ultimately valued. Both the viewers and the creators of television cooperate in the shaping of contemporary stories.

Television has become mythogenic; it creates new myths and stories, changes old ones and ignores others. Sometimes it presents anew the great and enduring stories of our tradition. More often, however, television exploits, impoverishes or ignores the central Story of our history. Christmas and Easter in particular have been transformed by television from times of Christian observance to occasions for the spinning of new myths and stories, particularly by those programs intended for children. These stories may draw upon the language and imagery of the Christian story, but without the central point of reference of God's self-revelatory dialogue with humankind. These polished presentations, featuring winsome bunnies or reindeer and fresh-scrubbed children, may emphasize the values of caring and sharing, but they do not point beyond the clever stories to the Source of the gift of love, to a God who loves us infinitely.

## Religion on Television

With its emphasis on storytelling, salvation and transcendence, television has many of the attributes of a religion. It tells us what it is like to be out of relationship, how to restore relationship, and what to do in order to be saved. The creators of the dream are not different from the audience. They, too, struggle with the purpose and meaning of life, and they communicate their understandings throughout the content and structures of television. Television reflects the spiritual experience of those who shape it. Much of television as we now experience it reflects the anxiety and fear we feel in our separation from God, and our search for transcendence in the finitude of Creation.

The core of the Christian faith stands over and against television today. In Jesus' life and teaching we see how he confronted the structures, powers, and actions of his society. He stood with those who were powerless, exploited, and dehumanized; he stood with women; he stood with children; he stood with the minority outcasts of first-century Palestine. His preaching of the acceptance of God's justice and love was not simply one among many ideologies to be considered. It called for a yes or no, not consent but decision. A decision in freedom must be unconditioned by techniques. Jesus said that all of Creation was accountable to the Creator. The nature of such a decision rules out any manipulation of our human fears, anxieties or feelings of cut-offness to obtain endorsement of a toothpaste or a belief. Were television to communicate this facet of the Gospel, television itself would be called into question. The medium does not allow its power to be used to question power.

Television shapes our perception of reality—our values, mores, and attitudes. The medium is often called a "mirror of society." Critical reflection on the nature of the Church and the manner in which religion, religious practice, and religious leaders are portrayed on television, however, suggests that the mirror possesses the curved planes of a carnival mirror, reflecting a caricature of religious faith and practice rather than reality.

A most curious phenomenon is the portrayal of religion, church, faith and religious practice on television. Those programs presented by religious groups are clustered on Sunday mornings when the audience is the smallest. Censorship of Gospel proclamation has existed throughout the history of broadcasting. Standards and practices of proper and improper opinion are exercised by industry executives.

Religion on television is trivialized. In the past, television's codes of self-regulation required that religion and the portrayal of religious leaders be treated with respect. Today, a minister, a sister, or a church building is likely to be used as part of a dramatic story line or even for a commercial. The church and its leaders do participate in society, are part of the fabric of life, do consume products and services. But the use of religious symbols or vocation or architecture to authenticate a product or add realism to a plot is a betrayal of the significance of religious garb or the role of a church building in the life of a congregation and community. The depiction of clergy and religious in commercials which suggest that salvation is found in consumption of goods is a distortion of the vocation of those who seek to minister in God's name.

An element of the proclamation of the Good News is the holding up of a mirror, in which all of our human existence is seen with clarity, for what it truly is. At the core of our human questions about love and alienation lie theological questions stemming from our human condition. The emotions and questions of our finitude include alienation, estrangement, guilt, penitence, renewal, redemption. These themes are frequently treated on television, in dramas, news and advertisements. Rarely, however, is the transcendant nature of these questions conveyed.

## Proclamation

The gift of God's love not only transforms our lives, but it also impels us to share the "Good News" of God's love revealed in Christ with others. Proclamation of this Gospel is central to our faith. Since television is such a powerful and pervasive means of disseminating information, it would appear to be an important vehicle for Christian proclamation. The first known broadcast of music and the human voice was on Christmas Eve, 1906. It consisted of Christmas music and a reading from St. Luke's Gospel. The use of radio, and in more recent years television, to proclaim the Gospel has been shaped in large measure by the technical and social structures of broadcasting.

There are unresolved theological tensions between the message of the Christian Gospel and the medium of television. Can a Gospel which assumes dialogue be communicated through a medium whose current technology broadcasts a monologue, inviting no response from the audience except conformity to suggested patterns of behavior? Can the news of God's initiative to heal our pain and separateness be heard amid the din of television's message that salvation and transcendence come from the things of this world? Where, in the midst of television's beautifully told stories and charming myths, is there room for the message that the ultimate story is the story of God's love affair with humanity? The use of television for proclamation is not simply a matter of obtaining air time for the broadcasting of church services or of churches owning radio and television stations. The nature of the Gospel is such that it calls into question the system of values proclaimed by television, and, indeed, ultimately calls into question the structures—technical, social, economic, and political—which shape television as it presently exists.

## Conclusion

Television participates in our turned-awayness from God. It often dehumanizes people; it presents the idols of society in dazzling imagery; its monologue says that ultimate worth is to be found in the things of the world. Its visage is one of friendly, open, earnest communication, a "communication" with the appearance of I-Thou relation which in reality almost always makes objects of us all. Part of the proclamation of the Gospel is to name and call to account the uses we make of our God-given freedom and of Creation. A theological reflection on television must begin with ourselves and the uses we make of television, must continue with the uses it makes of us and of all Creation, and must result ultimately in our choice of a new relationship with this aspect of Creation.

# Theology/Worksheet No. 1

In the article, Conklin and McFadden make the statement, "With its emphasis on storytelling, salvation, and transcendence, television has many of the attributes of a religion. It tells us what it is like to be out of relationship, how to restore relationship, and what to do in order to be saved."

For the purpose of this exercise assume that the TV screen *does* speak for and illustrate a religion or theology. Using knowledge stored from previous television viewing, plus observations gained from watching two hours of current programming, write answers to the following questions:

1. What god or gods are you being asked to worship?

2. Who are the prophets and what are they saying?

3. Who are the Christ figures?

4. What would be an appropriate list of ten commandments for television's religion or theology?

# Theology/Worksheet No. 2

Television tells stories about the world, about people and how they think, feel and behave. Many of the stories deal with clashes between good and evil as seen by writers, directors and persons in the cast. Some stories involve rather trivial conflicts, some involve life and death situations, many force persons to decide what is important. Television tells such stories both in programs and commericals.

For this exercise think of television's stories as being *parables* that communicate messages about who we are as humans. Watch two hours or more of varied programming, deciding for each program:

1. What are the basic messages of the parables?

2. Which television parables are in some way complementary to the parables of your own theology?

3. Which are contradictory?

# Theology/Worksheet No. 3

We all have needs and hungers. Each of us can best list them for ourselves, but some commonly identified ones are those such as—a need to feel at home in the world, have a sense of purpose and belonging, to be loved and be able to love, be accepted, act out of a sense of right and wrong, be part of community, be secure, have a sense of well being, be saved, be entertained. Actually, it is our needs and hungers that take us to the TV screen. We seek something there, perhaps even some form of salvation and transcendence.

Television deals constantly in needs and hungers. Programs show us people in search and finding solutions. Commercials identify problems and demonstrate how things or activities will set us free and satisfy our needs.

For this exercise, make two lists:

1. List the needs and hungers you feel and can identify in yourself.

2. List the needs and hungers that television seems to most often express.

# Theology/Worksheet No. 4

Most theologies include an acknowledgement of good and evil in ourselves and our world and provide value standards that help individuals define what is good and what is evil. Watch at least one situation comedy episode and at least one action adventure episode and using your own understanding of good and evil, write down instances where they appear in the programs.

| **Good** | **Evil** |
| --- | --- |
| | |

# Theology/Worksheet No. 5

**List some of the ways television can help create community. Consider the implications of whether the act of viewing makes you feel good or bad, whether specific programs make you feel good or bad, and whether you ever pick up ideas, behaviors or expressions which are useful to you in relationships.**

# Theology/Worksheet No. 6

The ending sentence of the article "Television and Reality" says, "A theological reflection on television must begin with ourselves and the uses we make of television, must continue with the uses it makes of us and of all Creation, and must result ultimately in our choice of a new relationship with this aspect of Creation."

With that statement in mind draw up a set of guidelines for yourself on using TV in a way that is in keeping with your own theology.

Include criteria on:

—How you decide whether or not to watch TV (stewardship of time)

—How you decide what programs to watch

—How you respond to and evaluate implications of programs

—How you use what you are learning about yourself and your relationship with TV in continuing to change how you use television.

A person has a number of ethical questions to decide. How shall I use my time? What kind of entertainment will nourish my mind and spirit? What programs will I support by viewing? What products will I purchase as a result of television commercials? How will television programs change my values or beliefs or relationships with others? What else might I be doing—who else might I be with—if I were not viewing television?

Each of us is of great value. We don't need to allow ourselves to be sold to the highest bidder at the network rating auction. We can make choices for ourselves and our families in selecting what television programs we watch which will be enjoyable, healthful, informational and will enable and encourage deeper personal relationships.

*Nelson Price,*
THE SEED, *(Birmingham, Ala.)*

The church should be educating the people to the languages and techniques by which their lives are being shaped.

*Wm. Kuhns*

"These magnificent electronic extensions of ourselves can teach, and heal and inspire, if we use them not for the ruthless pursuit of the least common denominator but for their highest human potential."

*The Public Trust*
*The Carnegie Corp.*

Chapter 12

# TELEVISION'S SOAP OPERAS AND GAME SHOWS

*Clare Lynch O'Brien*

*CLARE LYNCH O'BRIEN is the educational Consultant to ITT's* Big Blue Marble, *an award-winning children's television series. After working as a classroom teacher, Mrs. O'Brien spent nine years in publishing, developing print and audio-visual materials for young children. In addition to her work in television production, she writes and lectures on the subject of children and television.*

Against a backdrop of bathtub rings and diaper rash, daytime TV viewers are treated to a steady diet of passion, romance, crime and fantasy unparalleled in any dramatic form—or for that matter in life. With dramatic pacing that swings from borderline hysteria on game shows to the languid energy in some of the daytime dramas, an estimated 20 to 40 million viewers trip from rags to riches or get lost in small-town USA where villains and victims act out improbable dramas in settings as common and diverse as the audiences' homes. One finds, over time, that commercials on the soaps and game shows offer essentially the same promise—real life setting with enough fantasy to make it seem exciting. Each show holds out a pretense of real life but only as a backdrop for the game or drama.

One commercial for Calgon Bath Oil Beads shows a four-way split screen depicting aggravating household difficulties—an ugly spill on a newly cleaned floor, a crying baby, noise and commotion—and so forth. Now cut to a bathroom (modeled upon a fantasy) and to our housewife, transformed by her bath from anguished to serene. The fact that the real life drama, split screen and all, continues outside the bathroom door matters not. It is the promise of escape that holds daytime TV viewers and this promise is consistent in the commercials, serials, and game shows.

Soap operas, the half-hour, hour, and now 90-minute dramas, dominate daytime TV which begins at 10 am and continues until 4:30 each Monday to Friday. Actually, the serials began long before television. During the Depression, and after, 15-minute dramas occupied radio time from ten in the morning until six in the evening. At that time, twice as many radio listeners tuned in radio serials than watch TV serials today. Yet the appeal of TV serials, called "soap operas", after the products which sponsor them, borders on the fanatic today. Along with an audience that continues to expand both in size and demography, there are 19 magazines devoted to daytime drama, including several *weekly* newsletters that give plot summaries for fans who have missed a particular episode. Many newspapers have weekly or daily columns summarizing the plots. Soap stars have fan clubs, write advice columns, or dispense favorite recipes. In some viewers' minds, the actors and characters are so interchangable that villains report receiving hate mail, characters who are sick receive "Get Well" cards, and newlyweds receive gifts. People name babies after stars, and some viewers have gone into periods of grief and depression when a TV character "dies."

**"It is the promise of escape that holds daytime TV viewers, and this promise is consistent in the commercials, serials, and game shows."**

The appeal of soaps has increased over the years with several significant demographic changes. While the audience was once largely comprised of housewives 18-40, soaps are now attracting men, male and female viewers under 18, and working women. Producers are capitalizing on the audience shift and creating younger characters and youth-oriented story lines, which in turn is attracting more young viewers. People who were formerly embarrassed to admit their interest in soaps have come "out of the closet" with the likes of Van Cliburn, Thurgood Marshall, Dan Wakefield, and Carol Burnett admitting that they work their schedules around a favorite soap opera. Some students avoid midday classes and fill the student union with "Ryan's Hope" fans. Stewardesses leave notes to one another in the attendants' lounge—"Mary lost the Baby" or "Jack is sterile"—so that one may not have a tortured layover in Rome without the latest story development. Thirty percent of the *All My Children* audience are male, 20% of the *Ryan's Hope* audience are male.

What are these serials and what is their appeal?

## Settings

The early soaps which include *Search for Tomorrow* and *Love of Life* (1951), *The Guiding Light* (1952), and *As The World Turns* (1956) are all set in small towns where traditional values are adhered to, no matter how preposterous the story line becomes. The more recent soaps, like *The Young and the Restless* (1973) and *Days of Our Lives* (1965) are set in small towns but have more contemporary themes.

Only one soap has tried, with any

success, to use a large city setting. ABC introduced *Ryan's Hope* (1975), with a religious Irish American Catholic family in New York City.

The various settings give each soap a different feeling, but the conflicts and storylines are the same—perhaps identical—from time to time.

## Relationships

The drama of soaps seems to revolve around relationships that are rarely permanent and never uncomplicated. It is not unusual for a woman in her 30's to have been married four or five times, as when Rachael Davis Matthews Clark

**"The only happy, rewarding, trouble-free relationships on the soaps are the ones that exist in the memories of the seven or eight widows and widowers."**

Frame added Corey to her string of surnames *(Another World)*. Lisa Coleman *(As the World Turns)* eloped with Bob Hughes, divorced him to marry another man, then decided she wanted Bob back, but she married a wealthy Chicago businessman, had an affair with and became pregnant by the married Dr. Shea, bore his illegitimate son, married him and became his widow shortly thereafter. She then decided she wanted Bob back. Instead she met and married a lawyer, Grant Coleman.

There are marriages for money, convenience, paternity, on the rebound, and for almost everything but love and stability. Marriages are threatened by deception, infertility, impotence, jealousy, conflicting careers, former lovers, and lately, organized crime. Those marriages not beset by weaknesses of the flesh are usually attacked by the cruel hand of fate, debilitating illness or terminal disease.

Incest and oedipal conflicts dominate some programs and almost everyone who is pregnant is not living with the baby's father.

The only happy, rewarding, trouble-free relationships on the soaps are the ones that exist in the memories of the seven or eight widows and widowers.

The newer relationships have hints of incest. Lovers find out they are half sister and brother, *(Somerset)* daughters attempt affairs with or actually marry their mother's husbands. *(Love of Life, Young and the Restless)*. Narcissism is the hallmark of most characters and no subject seems taboo. One story currently on the air in afterschool hours centers around a pornographic filmmaker who is blackmailing one of his former leading ladies who is now engaged to the local District Attorney. The filmmaker has a print of a film, which the actress made when she was down on her luck and in need of money. He promises to destroy the film if she sleeps with him. She does and he has a secret videotape camera recording that. It seems likely that the filmmaker will be killed, once there are enough potential suspects. This soap is famous for its long court trials.

## Children

The children of these complicated relationships seem relatively unphased by the mess of their parents' lives. But they manage to become embroiled in other, seemingly unrelated, predicaments, usually caused by fate. Probably the most accident-prone child on any soap is Phillip (Charlie) Tyler who is the son of Tara Martin Tyler Brent and Phillip Brent. Little Phillip was conceived by Tara and Phil after they "married themselves" in a small deserted chapel in a snowstorm a day before Phillip was to leave for Vietnam. Presumed dead, Phillip couldn't claim his child so Tara married Chuck Tyler to give the boy a name. When Phil, not dead, but an MIA, returned to Pine Valley, confusion reigned. Most knew who little Phillip's real father was, due to Tara's delerious revelations during her brush with death in childbirth. (In soap operas, truth tends to be revealed in delerium.) Little Phillip's true paternity was revealed, not surprisingly, when in a life-threatening illness he required blood transfusions. Soap opera children whose paternity is ambiguous always require blood transfusions. Since then Little Phillip has fallen out of a

**"For many, soaps are a replacement for real life contact."**

treehouse, nearly killing himself, been sent out West with severe asthma, and come home to Pine Valley to face two other life-threatening illnesses.

The overall impression of children on soaps is as props to the drama. Most often they appear as complications to an adult romance. They are sweet and adorable one day, and then shoplift the next. They can go away to boarding school at seven (most do) and then return three months later all ready for their first affair. Not only does little of it make sense, none of it projects much about the real needs of children. When one combines these soap opera images with childrens' portrayals on daytime commercials—sources of tension headaches, greasy fingerprints, and diaper rash—we see that children don't fare too well anywhere on daytime TV.

## Family Relationships

While mothers and fathers are separating and divorcing, children are having accidents and getting rare diseases, siblings are acting out some of the most complicated dramas. Two sisters on *The Young and the Restless* continually seduce each other's boyfriends and husbands. Currently one is carrying her sister's husband's baby. Since the paternity is not clear to both women, we need only wait for the birth and the blood transfusions so that the real father can discover his paternity, seek out a relationship with his child, and complicate the marriages forever after. This same theme is on *Search for Tomorrow, The Guiding Light, As the World Turns,* and *General Hospital.*

Mothers and daughters seem to love, sleep with and marry the same man. Somehow the men are supposed to be plausible choices for both women. Despite their differences, they carry on normal mother-daughter lives.

## Men and Women

Men on soap operas often fill a fantasy ideal. They are father figures extraordinaire, having limitless daytime hours to devote to listening to women's problems. They console and support, but women usually solve most of the difficulties, unlike prime time. Men are both extremes—very good and very evil. Male villains are unrelentingly nasty. Soap writers don't seem to think it's necessary to give any psychological justification for male villainy. Most are just simply evil. Female villains usually are understood within the context of whatever misfortune has led to their meanness.

Some villainesses are simply schemers, but because most have problem pasts, they enjoy some sympathy. Most male villains enjoy no sympathy from other characters or the audience.

The good female characters are usually mother or grandmother types. Their goodness is not limited to their own family. They provide shelter to everyone in need. While there are any number of "good" male characters, there are no "earth father" types to match the all-good female counterparts.

Along with villains and saints, a third category of character contributes to soap opera drama: the victim. More often female than male, the victim is manipulated by a stronger character or fate's fickle finger.

## Careers

White collar professions dominate the soaps; most are doctors and lawyers. If there is a disproportionate number of doctors and lawyers, another career distortion exists. Women not only have terrific jobs, they get them without any reasonable preparation. When Nicole Drake was widowed, she became co-anchor for the evening news. Tara Martin Tyler Brent became a teacher, even though she never finished one year of college. Pat Randolph divorced her husband and when looking for her first job, took over an executive position in a publishing company. Meteoric career rises are not unusual for women at all.

Careers are often the subject of marital conflicts. Mary Ryan's career gets in the way of her marriage and child. Kathy Phillips often feels that her career is in contest with her marriage because she is more successful than her husband, Scott.

**" . . . most say they find soaps engrossing in the way that gossip is. They like the contact with problems without the responsibility to do something about them."**

## Illness and Disease

The ailments people get on soap operas are scarcely believable. According to daytime TV, pregnancy is one of the most hazardous health conditions known to women. Most pregnancies end in miscarriage unless the child is unwanted or illegitimate. They then come to term. Complications of delivery suggest that there have been no medical advances in the last one hundred years or so. Miscarriages are often punishments for deception (Heather Webber in *General Hospital*) or indiscretion (Chris Foster in *The Young and the Restless*). The result of miscarriage is often a nervous breakdown (Erica Brent, *All My Children*) or the inability to ever have children again (Heather Webber, *General Hospital*).

Nervous breakdowns are common on soaps. Leslie Brooks had one because of her rival sister's schemes. Ann Tyler Martin had one when her retarded daughter died. Erica Kane Brent had one after a miscarriage. Amnesia in soapland is like the common cold. Alcoholism is a problem on almost every soap. Unlike real life, where these problems are painful and difficult to cure, on soaps they appear to resolve themselves with ease. Except with Kevin Thompson who comes to town a mystery man, was thought to be a playboy, marries a lovely woman, turns out to be a secret, occasional alcoholic and dies of cirrhosis of the liver in a few months.

Cures for other diseases are nothing short of miraculous. Cancer is almost always cured. Blindness is reversed. Terminal blood diseases go into remission. As if ordinary diseases are not enough to cope with, producers use little known diseases to make daytime life more treacherous. Two of the most exciting were Extra-Vascular Supplementary Propensity and Abona Fever. Soap opera studios are scarry places to be. Diseases are not usually used on soaps to kill people off; characters die from accidents or are murdered.

If soap operas don't take the details of life too seriously they certainly take a sober look at moral issues. Abortions are never decided on lightly. More often than not they are contemplated seriously and then decided against. Many characters are concerned with "doing the right thing." While it happens all the time, lying is told to be morally wrong.

If one believes that "to err is human, to forgive is divine," soap opera characters are as close to heaven as anyone. No deed is too dastardly to be forgiven eventually. And many characters make seeing something from another's point of view seem quite easy.

The popularity of soap operas among people who stay home is not hard to imagine. For many, soaps are a replacement for real life contact. Among those who must make a real effort to watch, like working people and students, most say they find soaps engrossing in the way that gossip is. They like the contact with problems without the responsibility to do something about them.

Almost no subject is off-limits on daytime drama. Long before night-time would go near the subjects, daytime serials treated abortion, child abuse, teenage alcoholism, and drug abuse. Recently a storyline featuring a homosexual has been developed. Realism is not one of the hallmarks of soap operas, but producers have the opportunity to treat certain aspects of life more realistically. Grief from a death can take several months or years on a soap, where it can almost never be honestly represented in a one-hour drama on night-time TV.

**"When Jimmy Carter was Governor of Georgia, he appeared as a mystery guest and stumped the panel."**

Soap operas are enormously profitable for both networks and sponsors. Proctor and Gamble has always been the biggest supporter of daytime TV. In 1977, they spent $141 million in daytime advertising alone. Along with P & G, the top five advertisers include General Foods, American Home Products, Bristol Myers and Lever Brothers. The list has not changed significantly in the last ten years.

The audience that watches soaps has not changed much over the last ten years, either. One trend, first noticed in 1977, is the decline in daytime TV viewing among women in general, the greatest decline is among women over 50. While this disturbs both the sponsors and networks, no one is clear on the cause of the decline. Those who are concerned that there is some disenchantment with specific programs or program types are experimenting with various program changes to determine influences. Advertisers are looking at changes in households to see if an increase in women who work might be the cause.

## Game Shows

While soap operas have developed in a steady fashion, increasing in both popularity and production quality, the other half of daytime TV, the game show, has had a very different history. One cannot characterize the development of game shows as a progression, but rather as a series of different directions and formats. The earliest format was the panel show, two examples of which are *What's My Line* and *I've Got a Secret.* The primary appeal of these programs was neither the payoff nor the game. Small sums were awarded as prizes, and the games were simple and repetitive. The big attraction was generally the panel of celebrities. *What's My Line* began in 1950 and remained on the air for nearly twenty years. The panel included the urbane and witty publisher of Random House, Bennet Cerf, the acerbic columnist, Dorothy Kilgallen, Arlene Francis, and a fourth panelist who changed from time to time. As host, John Daly gave the panelists the task of guessing the occupation of each of three contestants and one "Mystery Guest." When Jimmy Carter was governor of Georgia, he appeared as a mystery guest and stumped the panel. The success of this show launched the team of Goodson and Todman, the company which has dominated the production of game shows ever since.

*I've Got A Secret* which aired from 1952 to 1962, was essentially the same guessing game show with a different celebrity panel. Instead of guessing occupations, Bill Cullen, Henry Morgan and Jayne Meadows guessed the contestant's secret which had already been divulged to the audience. Hosted by Gary Moore and produced by Goodson and Todman, *I've Got A Secret* proved that virtual duplication of programs, airing simultaneously, does not preclude audience appeal. Introduced at more or less the same time, 1951, *You Bet Your Life* was another game show with small prize money and huge popularity. Developed primarily as a comedy vehicle for the popular humorist, Groucho Marx, contestants on the show were insulted or otherwise worked into one of Groucho's wisecracks or jokes. Female contestants were frequenty shapely and beautiful and usually not very bright. The appeal of these shows, and several others like them, was the personality of the celebrity host or panelists. One wonders whether or not some of them would have a television career were it not for these shows. Also aired in the fifties were two of the most unusual daytime programs ever to appear on television, *Strike It Rich* and *Queen For A Day.* Combining the elements usually found in soap operas, these shows had contestants who qualified for prizes because of some personal tragedy. *Strike It Rich* featured a "heartline," a telephone that lit up to indicate that an appropriate "prize" was found for a particular contestant. Usually these prizes were donated by someone who heard the sad story on the program. Other general sorts of prizes were available—refrigerators, medicines, clothes and so forth and these were given away with the appropriate plug for the manufacturer and product. In order to become the big winner the participant told his or her story, usually about a serious problem—a disabled spouse, hungry children, invalid parents, broken-down appliances, or whatever. The heartline would light up and a wheelchair, food, a hospital bed, or a new refrigerator would be awarded. On *Queen For A Day,* five women would compete for the most pathetic story. Premiering in 1955, *Queen For A Day* was the number one daytime show for nearly all of the twenty years that it was on the air. Jack Bailey, as host, with appropriately sympathetic clucking,

**"Sure *Queen* was vulgar and sleazy and filled with bathos and bad taste. That was why it was successful. It was exactly what the general public wanted."**

drew the tragic story out of each contestant. Poverty, illness, abandonment, and more children than would fit in a shoe were the most popular themes of these stories. Few women got through their woeful tales without wrenching sobs, but that seemed to be part of the appeal of the show. The audience members would applaud the contestant of their choice and the applause was registered on an applause meter. The woman with the most applause was crowned with a rhinestone tiara, wrapped in an ermine-trimmed, velvet robe and waltzed around the stage by Jack Bailey. Then came her prizes. A night club tour of New York City courtesy of Night Tours International (plug); a mink coat from Fabulous Furs of Los Angeles. (Just the thing to change the life of every indigent woman.) To say that the prizes had nothing to do with the problem is understatement. But the comment it makes on how the producers felt about human feelings is significant. Les Brown, in his *New York Times Encyclopedia of Television,* cited an article published in 1966 in which the program's producer, Howard Blake, said: "Sure *Queen* was vulgar and sleazy and filled with bathos and bad taste. That was why it was so successful. It was exactly what the general public wanted." For whatever its appeal, it is safe to say that there has never been another show quite like *Queen For A Day.* Public humiliation for money was not to end with *Queen For A Day* however; it reemerged in the sixties after the most significant event in the history of game shows, the quiz show scandal in the fall of 1959. *The $64,000 Question* premiered in 1955. Together with *Twenty One* and a few others, it was part of a small group of programs which required keen intelligence and thorough knowledge of a subject in order to win. Aired in prime time, these programs were very popular. Loyal followers were shocked to learn that several of their favorite contestants were "briefed" by the producers before appearing on *Twenty One.* The most significant evidence brought to light in the investigation that followed came from Charles Van Doren, an English instructor from Columbia University, when he confessed that he received help with the questions before each broadcast. Van Doren testified that he had been told by the producer that his opponent, Herbert Stempel, was an unpopular champion because he was not lively enough. After being assured by the producer that briefing a contestant was a common practice on game shows, Van Doren went on to win $129,000. The audience increased each week which pleased the sponsors. Once the investigation began, no program escaped scrutiny. There were two significant consequences of the quiz show scandal. Quiz shows virtually disappeared from television and network policies were established which resulted in more careful controls of program practices. Revlon, which sponsored *The $64,000 Question,* was implicated in the scandal when contestants testified that Revlon had provided answers to attractive contestants to keep them on the air. Since the late fifties networks have been extremely cautious about a sponsor's involvement in programming, particularly where money and prizes are involved. It was several years before game shows returned to television with the same degree of enthusiasm. Programs that emphasized knowledge never reappeared with the exception of *Jeopardy* and the *G. E. College Bowl.*

**" . . . Alice, are you going to risk that beautiful 14 karet gold Benrus watch, encrusted with genuine diamond chips, designed to make you feel and look elegant—or—will you risk it to take the contents of one of those boxes on the stage?"**

Next came the era of screaming participants, bright sets with flashing lights and more excitement than can possibly be good for a person so early in the morning. Most of these programs revolved around the buying and selling of goods: *The Price is Right, Let's Make a Deal, Jackpot, Treasure Hunt,* and *Supermarket Sweepstakes. Let's Make a Deal* was both the most successful and the most absurd of these programs. The audience members dressed up in ludicrous costumes or wore strange body signs to attract the attention of Monty Hall who would invite their participation in a game which required the contestant to gamble his holdings for the chance to win more valuable prizes. The program appeared to be a 30-minute commercial because of the way the questions were phrased. Monty Hall would yell, "Alice, are you going to risk that beautiful 14 karet gold Benrus watch, encrusted with genuine diamond chips, designed to make you feel and look elegant—or—will you risk it to take the contents of one of those boxes on the stage?" Often, she would risk the watch. The box might have an entirely new dining room set in it or it might have 200 live roaches and a can of Raid. Viewers evidently liked the suspense. *Let's Make a Deal* stayed on the air for twelve years.

Humiliation for money and prizes returned in the late sixties, this time with an extra added attraction of sexual inuendo. On *The Newlywed Game* and *The Dating Game* contestants were asked questions designed to embarrass them or their mate. "On your honeymoon, what will your husband say you should have worn to bed?" Taste and a sense of limits were never considered a desirable trait for a contestant on these programs. The prizes weren't particularly good, raising a question about what motivated these people to embarrass themselves on network television.

Game shows today are really a combination of all the earlier formats. In the public humiliation area, we have *The Gong Show* on which contestants go to great lengths to embarrass themselves in an amateur talent contest. So-called celebrity panelists, such as Rex Reed, skilled in the art of the stinging rebuke, slash the contestant into ribbons. All this for audience amusement. *The $20,000 Pyramid, Hollywood Squares,* and *The Match Game* also featured celebrity panelists, but in a slightly more humane role. The remaining programs, *Card Sharks, High Rollers, Wheel of Fortune,* combine chance and very little knowledge in the pursuit of cash and prizes. If soap operas are profitable, costing $170,000 to $220,000 each week to produce in return for $130,000 in advertising revenues each day, game shows are a veritable gold mine. Five half-hour game shows, which require no significant rehearsal time, are taped in one day. Studio and production costs are quite low. Talent which is paid for shows aired rather than days worked, make five days pay for one day's work. The sponsors do equally well because of all the promotional credit. Airlines, hotels, limosine services and a host of manufacturers are involved in trade outs, and exhange of goods and services for commercial time. For many game show viewers, the programs are 30-minute commercials, interrupted from time to time by a semblance of a game. Contestants are usually chosen for their vitality. Game show producers do not want dull participants. Viewers obviously enjoy playing along. While no single game show has ever matched the success of some of the soap operas, one or two have aired for nearly twenty years. It is neither fair nor accurate to suggest that all daytime viewing consists of soaps and game shows. Many local stations have talk shows like *Midday Live* and *Panorama* or syndicated shows like *Donahue, Not for Women Only* and *Dinah Shore.* The success of these programs, particularly *Donahue,* suggests that daytime viewers are interested in more intelligent and thoughtful programming. Phil Donahue hosts a 60-minute audience participation talk show that deals with some of the more serious concerns of TV viewers, even some that are covered on the soap operas. The audience response suggests that the viewers' own lives are filled with conflicts that are interesting and challenging. Their relationships may not contain the dramatic peaks of soap opera characters, but they are thoughtfully considered. For the main part of daytime TV however, it seems that the use of fantasy eases the boredom of rings around the bathtub and collar and just may make these chores bearable.

## SOAP OPERAS

1. List positive values and effects of soap operas.

2. List reasons why you and others watch soap operas.

3. Discuss what psychological needs in viewers soap operas fulfill.

4. Evaluate how soap operas enable you and other viewers to deal with life situations and personal problems more creatively and constructively.

5. Evaluate your (and others') conversations which are related to soap operas. Do the soaps allow you to become more involved or less involved in the lives of real persons?

6. Do you see any dangers in applying soap opera solutions to real life problems?

7. To what degree do you think the lives of your friends and acquaintances are like the lives of soap opera persons?

8. If you are a regular viewer of soap operas, what would your sense of loss be if all soaps went off the air tomorrow? Describe in a few words what you have lost. How would you replace the loss?

Soap Operas and Game Shows/Worksheet No. 2

## SOAP OPERA MESSAGES

Soaps carry messages about who people are. One of the ways to get an idea of who a person is to find out what they are concerned about. Someone's idea of what are important personal or life problems can give you a very good picture of that person's values and beliefs. Select one soap opera that you are going to watch and analyze.

1. Before the show starts, write down five answers to each of the following questions:
   a. What are women really concerned about?

   b. What are men really concerned about?

   c. What are children really concerned about?

2. Now watch the program and the commercials. What do the programs and commercials propose as answers to the questions?
   a. What are women really concerned about?

   b. What are men really concerned about?

   c. What are children really concerned about?

3. Compare your two sets of answers. What are the similarities? What are the differences?

4. Soaps have been called contemporary folk legends that tell us truths about what is important to our culture. What do you think?

## SOAP OPERA ISSUES

Soap operas on television are mostly problem-centered programs. That is, they tend to concentrate on the problems their characters are encountering for most of their plot material. Analyze several soap operas in terms of the types of problems they deal with: Personal (interpersonal relationships, unhappy relationships, disappointment, etc.) and social (women's rights, minority rights, energy, world hunger, etc.)

| Personal issues | Social issues |
|---|---|
| | |

## GAME SHOW VIOLENCE AND SEX

Violence and sexuality are things that concern many people about all of television. What kind of violations (put-downs, slams, roasts, etc.) do you see on game shows? What kind of sexuality (innuendos, flirtations, suggestiveness, etc.) do you see? Select a program and count the number of personal violations and sexual references you see in a half-hour game show, using the tally sheet below. Note where possible the nature of the incident in a word or brief phrase. Compare your perception with those of others. Indicate with a plus +, or minus −, whether you consider these positive or negative.

**Violence and Sexuality in Game Shows**

| Personal violations | Sexual references |
|---|---|
| | |

## GAME SHOWS AND SELLING

One of the purposes of game shows is to sell products, not just in the commercials, but through the promotional announcements made for products that are the prizes. Watch a game show and count the number of such promotions in it and the number of specific mentions of products or brand names.

Name of show ________________________ Network ______ Time of day______

List products mentioned:

## ANALYZING DAYTIME PROGRAMMING ON YOUR LOCAL STATIONS

Begin analyzing daytime programming in your area by filling out the chart below with the following categories of programs. S = Soaps, G = Game Shows, N = News, P = other public affairs, M = Movies, C = Children's, D = Drama, S = Sitcom. For each half-hour on each station, put a letter in the blank on the chart based on this week's *TV Guide*.

Daytime TV worksheet for ____________________ (date)

| | NBC | ABC | CBS | IND | PBS | OTHERS |
|---|---|---|---|---|---|---|
| 6:00 | | | | | | |
| 6:30 | | | | | | |
| 7:00 | | | | | | |
| 7:30 | | | | | | |
| 8:00 | | | | | | |
| 8:30 | | | | | | |
| 9:00 | | | | | | |
| 9:30 | | | | | | |
| 10:00 | | | | | | |
| 10:30 | | | | | | |
| 11:00 | | | | | | |
| 11:30 | | | | | | |
| 12:00 | | | | | | |
| 12:30 | | | | | | |
| 1:00 | | | | | | |
| 1:30 | | | | | | |
| 2:00 | | | | | | |
| 2:30 | | | | | | |
| 3:00 | | | | | | |
| 3:30 | | | | | | |
| 4:00 | | | | | | |

After you have filled out the chart for one day for all stations, or for one week for one of the stations, do the following.

A. Mark the shows that come from the network by circling them.

B. Look over the uncircled local shows. Do they reflect the diverse needs of people in your community who watch television during the day? Do the network shows do that?

*(Worsheet No. 6 Continued)*

C. Write a letter to the program director of one of the stations, indicating you have done this analysis, and let him or her know what your conclusions are, whether they are positive or negative.

D. Design your ideal daytime program line-up for one station. Which network shows would you carry? Which local shows? What kind of additional shows would you put on? Is the order of the schedule good? Should it be changed? List the programs by exact title (by type, if no such show exists) that you think should be added to the line-up.

The average set is on 6.82 hours a day.
*George Comstock*
Television and Human Behavior

---

"There is no suggestion that networks or individual stations should operate as philanthropies. But I can find nothing in the Bill of Rights or the Communications Act which says that they must increase their net profits each year, lest the republic collapse."
*Fred Friendly—1966*
*letter of resignation to CBS quoting*
*Ed R. Murrow*

---

"We remember the Egyptians for the pyramids, and the Greeks for their graceful stone temples. How shall Americans be remembered? As exporters of sensationalism and salaciousness? Or as builders of magical, electronic tabernacles that can in an instant erase the limitations of time and geography, and make us into one people. The choice is in our hands, and the time is now."
The Public Trust
*The Carnegie Commission*

---

Children see twenty thousand commercials each year.
*George Comstock*
Television and Human Behavior

---

Sixty percent of the mothers of first graders place absolutely no restrictions on the amount of time their children watch television.
*George Comstock*
Television and Human Behavior

---

# Television Awareness Training

The Viewer's Guide for Family & Community

## PART 2

# Education in Century Twenty-One

*Alberta E. Siegel*

*ALBERTA E. SIEGEL is Professor of Psychology in the Department of Psychiatry and Behavioral Sciences at Stanford University. She was a member of the Surgeon General's Scientific Advisory Committee on Television and Social Behavior.*

American psychology has been dominated by the study of learning. The processes by which rats, mice, monkeys, and even human beings acquire new behaviors have been carefully studied in literally thousands of experiments in hundreds of psychological laboratories around the land. Our most respected theorists have been our learning theorists. Experiments on learning dominate our textbooks and our lecture courses.

American social thought has been dominated by faith in education. Americans see education as the path to the good life, as a social elevator, as the essential qualification for a job, as the solution to various social ills. We invest not only our money but also, and more importantly, our time, in a massive educational effort supported by local government, state government, now the federal government, and also by private foundations and trusts.

Given the preoccupation of American psychologists with learning, and the preoccupation of all Americans with education, it is easy for a psychologist to equate learning and education. If something can be learned, then it must be the task of the schools to teach it.

## Growing Expectations for Our Schools

In the last decade much of psychological research has been devoted to demonstrating the importance of early learning. There is now a substantial research literature showing how central the learning is that occurs in the first few years of life. The response of many Americans to these findings has been to extend the educational establishment downward into early childhood. If learning at age four is important, then we'd better have schools for four-year-olds. We are identifying important learning that occurs in the second and third year of life; already we hear suggestions that this means that toddlers should be schooled. And as Americans ponder the new flood of researches on infancy, showing how much is acquired during the very first year of life, there is talk about the mother as a teacher and the crib as a school.

In the nineteenth century, America's social thinkers tended to believe that most behavior was innate. People were competitive because of a competitive instinct. Acquisitiveness was an instinct. So were generosity and nobility. The behavioral differences among individuals were not ignored by these social thinkers; they regarded these differences as innate also. Blacks were thought to be inherently and intrinsically different from whites, and females were thought to be instinctively different from males. Given these stamped-in differences among people, differences which were not based on learning and therefore had nothing to do with education, the task of schools was relatively simple. A few skills were acknowledged to be based on learning and the task of the schools was to concentrate on developing these few skills. Reading and writing must be learned and thus it was the task of the schools to teach these skills. No one thought that a child would spontaneously come to understand geography and American history. But he could learn to understand them if given the proper instruction.

Throughout the twentieth century, American social thought has come to assert that most behavior is learned. We no longer speak of a competitive instinct: we believe people learn to be competitive. Acquisitiveness is seen to be a learned habit. Even generosity and nobility are thought to be fostered and enhanced by benign learning settings. More, we now see behavioral differences as based on learning. No longer do we see black-white differences as innate and inherent; now we regard them as the product of different learning environments. There are even suggestions that females differ from males not becuase of instinct but because of learning.

I must pause to note that no serious social thinker believes totally in either genetic endowment or social learning as an explanation for all behavior. All of us recognize, and all of us teach our Psychology I students and our Educational Psychology students, that behavior results from an interaction of endowment with experience, that the human being is jointly a product of his biological constitution and his life events.

But the increasing emphasis on learning in the twentieth century has created an awesome set of tasks for the schools. For as soon as Americans find that something is learned, they want the schools to teach it. And they want it taught in the

Alberta E. Siegel, "Educating Females and Males to be Alive and Well in Century Twenty-One" in CHANGING EDUCATION: Alternatives From Educational Research, edited by M. C. Wittrock, © 1973, pp. 155-168. Reprinted by permission of American Educational Research Association, Washington, D.C.

morally correct version. If both competitiveness and cooperativeness are acquired through experience, then we'd better arrange a child's school experiences so he learns the correct balance of competition and cooperation. If children do indeed learn to be generous, then let's start devising ways to teach them this trait in our schools. And if differences between blacks and whites are the result of learning, then it is the task of our schools to teach blacks and whites to be equal.

## Learning Outside of School

By equating schooling with learning, we have come to think of our schools as the site of the solutions for all our social problems. It is this thinking I wish to examine here.

Schools are a relatively recent cultural invention. Some schools for a few privileged young people have existed for centuries. The idea of universal education is a very new idea. Until recently, schools were primarily for males, for freemen, and for the wealthy. This remains true in many parts of the world today. In America, schooling for females began more recently than schooling for males, and it is only even more recently that serious efforts were made to offer schooling for blacks, for Native American Indians, and for other minorities. Now we are seeing the opening up of collegiate education to ethnic minorities. And there are fresh efforts to open up professional education to females, in medical schools, law schools, and business schools. So we are still living through and working through the creation of equal educational opportunities for all Americans.

Schools are new, and universal schooling is even newer. But learning is ancient. Take language, for example. It is evident that children *learn* to speak. French children learn to speak French while their cousins across the channel are learning to speak English. The language a child learns depends on the learning environment in which he lives. But language learning was occurring long before there were schools.

Sherwood Washburn estimates that man has been using a phonetic code for communication for the past 40,000 years. This estimate is based on secular changes in brain size and configuration, as inferred from human fossils. At the time man began to use language, all human beings lived in small nomadic groups which made their living by hunting and gathering. Such societies are almost entirely extinct today, so it is easy for us to forget that 99 percent of the span of human life on this planet has been the era of the hunters and gatherers. Only in the last one percent of his time on earth has man had agriculture. And industrialization has occurred in the last tiny fraction of that brief period. The world population was smaller in the past than it is at present, of course, but even so 95 percent of all the humans who have ever lived have been hunters and gatherers. It is worth our thinking about how children learned language in the hunting and gathering group.

Our best guess would be that they learned speech the same way they do now: through observation and imitation of the speech of those they loved. What impelled this learning is the same motivation that underlies speech acquisition in today's toddler: the wish to maintain a close emotional exchange with the adult to whom he is attached. Probably youngsters today learn language more readily than was the case 30,000 or 40,000 years ago, for today's youngsters benefit from thousands of years of natural selection for linguistic capacity. There seems little doubt that selective pressures favored the communicator—he had a better chance to survive to adulthood and to reproduce—and that over the ages the contributors to the gene pool have increasingly been those whose nervous systems were best wired for speech acquisition. By now most human infants are born with the potential to develop a central nervous system that will permit social speech. Only moderately supportive social systems are needed to assure the acquisition of language. That such systems are still needed, however, is evidenced by the tragic histories of children when they are reared outside of families, for whom no close bonds form to adults. These children are almost invariably retarded in language skills. Even in the best orphanages—the models devised by dedicated pediatricians and social workers, the orphanages in which infants are well fed, well bathed, healthy, and motorically advanced—speech lags behind the orphans' other skills.

We are talking about a skill—the use of language—which most people link with schooling. An individual who speaks ineffectively is thought to be poorly educated. Cultivated speech is said to mark the literate person. But the most reasonable guess is that this skill is acquired in learning environments outside the school, learning environments which have existed for thousands of years, and learning environments which affect the child during what we call the "*pre*-school" years. Schools simply teach the refinements and sophistications of language: reading and writing.

If the point needed to be made any more vividly, I could mention that some specialists in nonverbal communication estimate that 70 percent of all human communication in face-to-face interaction occurs over extra-lexical channels. Like other primates, human beings convey meaning to one another through posture, gesture, facial expression, and vocal intonation. Infants begin to acquire skills in nonverbal communication very early in life. Their gestures and postures display emotions in the first months of life, and the social smile occurs by two or three months. By the time a child is a year old he has a large repertoire of facial expressions and postures by which he communicates with others around him, and also he is adept at reading the meaning of their postures, gestures, and vocal intonations. There is no doubt that these behaviors are learned. There is no doubt that the learning builds on a biological substrate which goes back into man's history into the millennia before humans had lexical speech. There is no doubt that humans were communicating gesturally before any society had schools. And there is little doubt that skill in nonverbal signalling is essential for social success.

For our present purpose it is important only to note that a child can readily learn to read and write only if he has already mastered speaking, listening, and nonlexical signalling. And he will add the sophisticated language skills to his repertoire most readily if his parents value reading and writing, just as he learned gestural communication, speech, and listening in the context of the emotional exchange with his parents and siblings.

The remarks I've been making about language, both lexical and extra-lexical, could perhaps be made as well about other skills we now know to be learned. The learning begins in the very early years. It occurs in the learning environment which is significant for the child because of his strong emotional bonds with the other persons in that environment. It occurs largely through observation and imitation.

It is the family which is the child's most critical learning environment. We have tended to think that schools are where learning occurs. We have tended to say that if something is learned it is the task of the schools to teach it. In fact, some of the most important learning that humans achieve occurs in homes rather than schools. And if that learning fails to occur the child who enters school at age six is impaired, perhaps irremediably, in his potential for learning the skills that schools must teach. Our options are either to extend schooling institutions downward into infancy, or to strengthen families to continue to serve as primary learning environments.

If children are to grow up to be both alive and well in century twenty-one, it will be because we have managed to keep the family alive and well in century twenty. If we achieve that goal in the latter third of our century, then the children we have reared will have the skills and the motivations to achieve that goal in century twenty-one.

When our ancestors were living as hunters and gatherers, there were no professional educators and educational researchers to discuss how to meet the needs of the coming generation. Education occurred in the context of living, as a byproduct of the work and social life of the community. All institutions shared in the responsibility for socialization of the young. Children learned through observation, imitation, and practice, as well as occasionally through formal instruction. The legends and myths told around the fire conveyed the community's moral insights. Work skills were learned through apprenticeship. Children learned what it means to be male and what it means to be female by observing various males and females in their daily lives. Children learned what it means to be young, to be middle-aged, and to be old by their acquaintance with the youthful and the elderly. Learning occurred because the social institutions functioned to permit children to learn. Our children today are born with the learning capabilities of the children in hunting and gathering societies. They can learn readily through observation and imitation, a fact which has been documented by the researches of Professor Albert Bandura and his students.

What can we learn from the hunting and gathering way of life—which was the life style of 95 percent of our ancestors, the life style in the millennia of human evolution from which we have inherited our genetic capacities and potentialities—which will be helpful with today's young savages?

What we can learn is to question the distinction between educational institutions and other social institutions. And we can learn that most socialization is unintentional, a byproduct of the life of the community. When we acknowledge that the child is socialized by all the institutions in which he has a part, then we may begin to examine all of them for the positive contributions they make to the next generation.

## TV Learning Environment

Recently I have been involved in this way of thinking about television. The American commercial television enterprise is profoundly committed to the notion that TV exists to entertain and to communicate the news.

As a child psychologist, I am struck by the fact that American children spend more hours watching television than they spend in school. I can't fail to believe that such a pervasive activity has lasting effects on children. The evidence for these effects is coming in, and is becoming increasingly convincing. In the meantime, we have common sense as a guide. It seems only good sense to believe that TV watching must affect children, since they spend much time in front of the tube and do so voluntarily.

What has impressed me is the profound unwillingness of most people to think of TV as a learning experience. This unwillingness was evident in the work of the Surgeon General's Scientific Advisory Committee on Television and Social Behavior. Many Americans are deeply committed to the view that learning is serious, learning occurs in schools, learning is what we elect school boards to worry about. In contrast, television is fun, television is entertainment, television isn't linked to the schools in any meaningful way, and we don't need to think of the TV set as an instructor with the child as its pupil.

When an anthropologist contemplates the hunting and gathering societies and how they socialized their young, he sees no schools and no school boards. What he sees are learning environments, and youngsters being socialized in these. When he contemplates American life today, can he doubt that TV produces a learning environment for our youngsters? Can he doubt its socializing role?

I believe that our society could benefit if we began thinking of many institutions in terms of their functioning in the socialization of the next generation.

## Community Learning Environment

The community is a learning environment. The child learns from the people around him, as he meets them in supermarkets and banks, as he visits their apartments and houses, as he encounters them in churches and other organizations. He learns what it means to grow old by observing the elderly in his community. He learns about sickness by watching what happens when people in his community fall ill and need medical attention and nursing care. He learns about infancy not only when he is himself an infant but through observing the infants around him and how they are regarded by adults and other children. He learns about marriage through visiting different families, watching what happens when a couple is divorced, observing the effects of a death on a family. He learns about racism by observing who lives where, in what kinds of homes, and by observing who works in what kinds of jobs.

I haven't served on a Surgeon General's Committee for the real estate activity in our country, as I have served on a Surgeon General's Committee to study television. But I think it is a fair guess that the real estate operators are even less aware of their role as educators than the television people are. Most television writers and producers have a dim awareness that there are children watching, though they don't think about children very much and they don't know much about children's thought processes. But are real estate operators thinking about children at all as they develop our new neighborhoods and rearrange older ones? How are they altering the community as a learning environment when they develop separate living facilities for the aged? When retirement communities are built away from cities and towns, we not only change the tax base for our schools, we also change the experience base for our children. No longer will the child confront the realities of aging in his daily life. No longer will he have casual everyday opportunities to know a grandfatherly or grandmotherly older person who loves him in the especially clearsighted and indulgent way that older people love the very young.

I selected this example, rather than the more obvious one of ethnic segregation in housing, just because I think it is slightly less obvious. We all know that it is housing patterns that have segregated our communities. The fact that schools are asked to adjust their arrangements in order to compensate for the inequities perpetrated by another social institution—the building and real estate industries—is just an example of my general point that Americans ask schools to bear most of the burdens of educating our young while ignoring the educative functions of other social institutions.

## Effects of Environments on Children

My general point is that a modern understanding of the psychology of learning compels us to examine all our social institutions as learning environments. And a modern understanding of the central importance of the family in the child's learning compels us to examine all our social institutions to see ways in which they support and strengthen the family, or ways they may compete with it and inadvertently be destructive to the family.

It is naive and arrogant to say that if something is learned then the schools must teach it. Language is but one example of many a skill which is learned outside of school. Realistically, we must examine every social institution to see the contributions it can make to the child's learning. These contributions will be of two kinds: the direct contributions, made through serving as a learning environment for children, and the indirect contributions, made through support to the family.

For example, television makes direct contributions to children's learning by providing them with imagery about social roles, by providing models of language use, by displaying examples of human conflict and how it is resolved, by vaunting certain products that children are urged to use in order to grow big, strong, and powerful, and so forth. Television is indirectly involved in the welfare of children by all that it does bearing on families. In some nations, this fact is recognized in the practice of shutting down the TV transmitters during the dinner period: TV does not compete with parents for the child's attention at the traditional time of family togetherness. Television could be helpful to children if the magnificent teaching potential of the medium were used to teach mothers about mothering. This electronic teacher that brings an integrated message to eye and ear, and that reaches into 96 percent of the homes of this nation, could be a powerful aid to the pediatrician and the nurse in showing mothers how to nurse their babies, how to guard their health, how to encourage social communication early in life, and so forth.

Similarly, the housing industry makes direct contributions to children's learning by creating environments in which children can or cannot meet people at all stages in the life cycle, can or cannot observe people at work, can or cannot easily meet one another to play together, can or cannot explore their neighborhoods without being exposed to mutilation by automobiles. At the same time, it makes indirect contributions to children's lives by the ways it affects families. The housing industry helps to determine how long the daily commute run is for men, and thus how many waking hours a child will spend in the company of his father. The housing industry helps to determine whether a family has privacy, whether family recreation is accessible, and so forth.

## School as a Socializing Institution

Now so far I doubt that I've written much with which most educators would disagree. Most educators are likely to appreciate the significance of early learning, to recognize the importance of the family to the child, to agree that all social institutions have effects on children, and that we could serve children's welfare by examining these effects.

But now I want to turn this argument back on the schools. For schools are themselves learning environments, and they are also social institutions that can aid the family or can contribute to its further deterioration. We have asked what the television industry is doing to create favorable learning environments for children and adolescents, and what its practices are doing to strengthen families. We may ask the same question of the education establishment.

## Support of the Family

Since schools are set up to educate children, it may seem foolish to ask what schools are doing to enhance learning opportunities for children. But I want to ask that question not about classroom practices but about schools as a social institution. Like every other social institution, the school constitutes a learning environment for children, and thus may affect them directly, and it has effects on the family, which may be either supportive or competing.

An example of the ways schools affect families is the relation of schools to geographical mobility of families. In my opinion, geographical moves have the potential of being destructive to family life. It is because of the dispersal of our population all over the landscape that children now grow up in homes with only two adult relatives (at most) who care about them, far distant from their grandparents and aunts and uncles who might otherwise enrich and diversify their lives and might relieve and aid their parents in time of sickness and family stress. It is the middle class in this country which is geographically mobile, as the businessman father climbs the executive ladder in a major corporation by locating and relocating at various branch offices, or as the academician father moves from one university to another to climb the academic ladder from instructor to professor.

What can schools do about geographical mobility? Here it is important to think of the school as an employer rather than as a classroom. Can schools arrange their employment practices so that it is not necessary for their employees to relocate in order to advance? Can they give the most careful consideration of the qualifications of their own employees before going outside to fill leadership positions? Can they arrange instructional opportunities in their own communities so school employees need not travel in order to advance their own educations? As employers, schools are in a position to be either helpful or destructive to the family lives of their own employees, and in this way to contribute importantly to those we want to be alive and well in century twenty-one.

A second example has to do with the health of the family. Our nation is in a crisis of medical care. We tend to think of schools as contributing to the solution of this problem through schooling: by guiding more youngsters into the health professions, by encouraging students to consider careers in nursing, medical social work, dietetics, physical therapy, rehabilitation, and the like. And we tend to think of the need to establish curricula in these fields in our high schools and junior colleges, as well as in our colleges and universities.

But the schools are also health insurers. The educational establishment constitutes one of the largest networks of employers in the United States. What kind of health care arrangements do schools offer for employees? As an enlightened institution, a school system can be a model health insurer, insisting that school employees are covered for preventive medicine as well as for treatment of illness, insisting that the employee's family receives the same thorough medical protection that is given to the employee personally. I mention this example not to get off onto the subject of medical care but to provide a common-sense example of a supportive role the schools can play to aid family life.

## Socialization of the Sexes

My remaining examples will have to do with the changing roles of females and males. There has been a lot of talk about ways that schools may participate in the current revolution in women's roles. I'd like to focus my concern on what the schools can do as socializing institutions and as institutions which support family life.

The industrial revolution and subsequent technological revolutions have profoundly altered the lives of males. Because men's work is increasingly mechanized, brain power and personality attributes are increasingly lengthy and specialized as the contribution the individual man makes in his work. This means increasingly length and specialized education for men. And whereas in the hunting and gathering eras and in the agricultural era, men worked out of their homes, now men's work occurs in factories and offices sometimes far distant from their homes. For today's urban and suburban male, the home is the place where he sleeps at night, where he performs brief morning ablutions and takes a hasty breakfast, and where he spends his evenings for a few brief hours of family life.

The effects of the industrial revolution have been felt by women much more recently. They include a change in the ways food and clothing are manufactured and processed, and thus a change in the work that must be done within the home. They also include an increasing emphasis on a money economy, with the result that many women need to earn money to meet the families' needs. Also significant are twentieth-century advances in contraception and in medical care of mothers and children, with the results that women can plan their families, few of them die in childbirth, and few of their children die. All of this is new, and so is the modern emphasis on a small family. A woman's life expectation today is 20 years longer than it was in 1900. These changes in women's lives come together in the increased employment of women outside the home. Only in this generation are the majority of adult women likely to be working outside the home, but the proportion of women in the working force has been increasing steadily throughout the century and is likely to continue to increase as we inch towards century twenty-one.

How should schools respond to these radical changes in the lives of women? The tendency has been to make direct responses: to alter the curriculum and rewrite the textbooks. This is all to the good.

Certainly it is appropriate to launch guidance programs for women to encourage them to enter careers in science, law, medicine, and engineering, as well as in librarianship, nursing, teaching, and social work. Certainly we do need a revolution in admissions policies at the level of higher education, to enable young women to synchronize their education with their personal and family lives. We must find ways to enable a college woman to transfer from one institution to another when she marries and her husband relocates. We must find ways to offer a college education to women with families, enabling them to come to college on a part-time basis and at hours suited to family life. All of this is so obvious it is easy to overlook the fact that many institutions of higher education have hardly started to make these changes. Among other direct responses, probably it is worthwhile to scrutinize our children's readers, to see whether Jane is destined to be a homebound drudge and Dick is on his way to growing up as a male chauvinist. And certainly it is important to expand the horizons of our history books, to see that children learn more about our past than the political history of white males.

But my guess is that the most important significant responses are going to be the indirect ones. And this is my point of distinction between education and learning. It is not as educational institutions but as environments for learning and as social institutions that our schools are likely to have their greatest usefulness in enabling individuals to adapt to the changing roles of men and women. Indeed, most of the direct responses have already been made: more than any other social institution in our society, our schools have created opportunities for females to develop their abilities. It is precisely because we have so many well-educated women that we are impressed by the barriers to them in science, medicine, and law. It is precisely because our schools have already shown how capable girls are of achievement that we wonder why they have no more opportunities for using their talents in business, finance, higher education, politics, and so on.

What can we do to create schools which are environments for learning and are social institutions that promote family strength and that aid males and females in adapting to the requirements of century twenty-one? I have already mentioned policies which reduce the pressures to geographical mobility and thus aid the extended family to stay together. And I have mentioned policies which enhance the medical and psychiatric services to families with a member who is employed by the schools. A third contribution schools can make is to avoid sex-role stereotyping in employment.

If children learn through observation, as I believe our research shows they do, then our children are going to learn about sex roles by observing what kinds of jobs are held by males and females in the institutions which are open to them. Children will learn about sex-role stereotypes in our churches, by watching TV, in our banks and supermarkets. They will also learn in our schools. Who are the secretaries? Who are the custodians? Who are the shop teachers? Who are the nurses? Who are the principals? Who are the kindergarten teachers? Who are the librarians? As an employing institution, our schools can serve to reduce sex-role stereotyping by giving children the opportunity to meet physicians who are females and librarians who are males, the opportunity to take classes from female science teachers and male chefs.

Here I am suggesting that we put our money where our mouth is. All the high school guidance lectures and pamphlets about women in science are not going to have the effect of knowing one woman scientist who is happy and effective in her work. All the talk in the world about career opportunities for women in engineering is as nothing compared to the opportunity to meet and work with one woman who is a contributing engineer.

A fourth suggestion is based on the observation that people marry people similar to themselves. A psychologist often marries another psychologist, a lawyer is likely to marry another lawyer, and teacher is very probably going to marry another teacher. How can our schools as social institutions contribute to family stability and family strength when both parents are in the same profession? As employers they can minimize the barriers to the employment of spouses.

Schools should reexamine their nepotism policies for their cost and benefits. We have all been preoccupied with the dangers in employing both a woman and her husband. We fear bloc voting. We fear favoritism in promotions. We fear jealousy of the family earning two salaries. In this preoccupation, we have failed to consider the dangers in not employing both a woman and her husband. We have ignored the strains on a marriage when the two partners have to commute some distance in order to serve in independent institutions. We have ignored the even more profound strains when one partner is unable to use the education he or she has received and when that partner sees the marriage as blocking his or her professional growth. And we have also ignored the benefits of employing both a wife and her husband. When a couple are both teachers in the same school, the students in that school have the opportunity to observe a marriage relation which includes intellectual respect. They have the opportunity to learn to know a woman who is both a wife and a professional person and to observe that she fulfills both roles effectively. They also have the opportunity to observe the strains that some women experience in combining marriage with professional work, and thus to learn that this challenge is not for every woman. When both a wife and a husband are employed in the same institution, that institution becomes a richer learning environment for the youngsters being socialized in it. And that institution is contributing to family stability by providing an opportunity for a marriage between professionals to work comfortably. These are very real benefits, and we ought to consider these benefits in our cost-benefits reanalysis of nepotism. In my view there's nothing wrong with nepotism as long as you keep it in the family.

My fifth suggestion comes from everyone who has made a serious study of women in our current era. Part-time jobs are the way society can use women's educational achievements today while not interfering with women's work as wives and mothers. Our schools should be especially sensitive to this need. Schools are concerned with children, and it is folly for them to be engaged in employment practices which interfere with a woman's ability to be a fine mother to her children. The part-time job permits a woman to achieve independently in the world of work while continuing to give her children and her husband the attention and care they require from her. As employers, schools can be pioneers in creating more and more part-time positions for women. Especially valuable would be jobs whose hours coincide with the hours children are in school, so that employed mothers may be at home when their children are. Needless to say, I am urging part-time jobs not only for teachers, administrators, and other professional staff but also for service employees, clerical workers.

With part-time jobs available, the schools will become interesting learning environments in which the male and female students may observe both women and men in a variety of work activities. Their acquaintance with women workers will not be limited to older women whose children have grown nor to the very young women who have not yet had children, but rather they will become acquainted with women from all stages of life and thus will be enriched in their understanding of what it means to be a woman in an industrial era.

## Summary

How can we educate our children to be alive and well in century twenty-one? To begin with, we must recognize that their learning equipment is the result of evolutionary selection within the context of hunting and gathering society. They learn from all their experiences, not just from those labeled "education," and they learn through observation, imitation, and practice. This means we must examine all our social institutions as learning environments for youngsters. Second, we must recognize that the most important learning occurs within the family, from the people with whom the child has strong emotional bonds. Other social institutions can aid children to be alive and well in century twenty-one by aiding the family. This means examining our social institutions for the unintended consequences that technological advances may have had for the family. Is there a sort of pollution of the family caused by industrialization, electronic communication, geographical dispersal, and the other marvels of the twentieth century? Schools can set an example for all social institutions by examining not only the direct educative influences they have on children but also the indirect ones. Females and males will be more likely to be alive and well in century twenty-one if our schools reform all aspects of the learning environment they provide for children, and if they modify all their practices to bring them into line with the continuing viability and strength of the family.

# The Networks Cry Havoc

*Les Brown*

*LES BROWN is the television correspondent for The New York Times.*

Never in 30 years of commercial television has there been a season so chaotic, so paradoxical and so marked with convulsive change.

It has been the best of times: Advertisers have scrambled to buy air time as never before, creating a sellers' market and driving up prices for commercial spots to the highest levels in history. Each of the networks last year achieved more than $1 billion in sales.

It has been the worst of times: Few hits have emerged from the shows introduced in September, a record number of weekly series have been canceled (although virtually none actually lost money) and the revolving doors in the network executive suites have not stopped spinning since August. Two network administrations and two news administrations have been toppled, with high-ranking executives jettisoned faster than the Nielsen-rated turkeys in the topsy-turvy network ranking race.

Adding to the turmoil, viewing appears to be declining during both the daytime and the nighttime hours, and prospects for the year ahead are unsettling. Wall Street predicts that the networks' extraordinary profit growth of the last two years will "normalize" in 1978, and the signs have already appeared in price-cutting for the first quarter.

"This season has been a balloon," a network salesman remarked last fall, "nice and fat but not made of very much." And now the balloon is starting to lose air.

For CBS and NBC it has been a season of profits without honor. The boom in revenues belies their essential place: they are running second and third in the ratings, well behind the leader, ABC, with few programming triumphs since September and a critical shortage of successful shows to provide a foundation for next season's schedule.

For this situation a parade of executives has paid the extreme penalty—among them Herbert S. Schlosser, president and chief executive officer of the National Broadcasting Company; John A. Schneider, president of the CBS Broadcast Group; Robert J. Wussler, president of CBS-TV, and Robert T. Howard, president of NBC-TV. Some veteran observers of the industry have always found more drama behind the television screen than on it, and this season's second act closed with a dazzling surprise twist.

Mired in third place, NBC engineered a coup last month by spiriting away ABC's programming ace, Fred Silverman. The price was Mr. Schlosser's job and a record salary of close to $1 million a year.

Mr. Silverman may be short on experience as administrator of a large company—one that encompasses, in addition to the television network, NBC News, NBC Radio and all the television and radio stations owned by NBC. But most Wall Street analysts believe that, at worst, Mr. Silverman will do less damage to NBC as its chief than he would have done next fall as its opponent in program strategy.

Moreover, the analysts believe, Mr. Silverman's presence at NBC will tend to block ABC's raids on stations affiliated with NBC. They won't want to switch allegiances, at least for a time.

Whether he was responsible for it or not, the developments in network television during Mr. Silverman's two years as ABC's head of programs—he had spent 12 years with CBS—were the chief cause of all the turbulence today: a change in the balance of power.

ABC's rise to the top in the national audience ratings, after more than 20 years as the also-ran network, has left third place open for either NBC or CBS—with a good deal more than corporate pride at stake.

The third-place network usually lacks bargaining leverage with advertisers and program producers, and is least equipped to retrench successfully in a possible downturn in the television economy. But more significant, the pretax profit differential between the first-place and the third-place networks can exceed $100 million a year. That spread would include the considerable earnings of the network-owned stations. Each has five in major cities, and their audience size is frequently determined by the parent network's popularity.

"Everyone knows you can be a blooming idiot and still make money in television today, but the incremental profits come from those who know what they're doing," said Robert M. Howitt, analyst for the First Manhattan Company.

A year ago CBS made more than the others. Edgar H. Griffiths, president of RCA Inc., the parent company of NBC, detected a $50 million profit difference between CBS and NBC and orderd the gap closed. It was indeed closed in 1977, but by ABC.

Figures compiled for 1977 by Broadcast Advertisers Reports, which checks on whether the commercials purchased are aired, put ABC's estimated 1977 revenues some $65 million ahead of NBC's. Of that difference, about 15 percent would be paid out in advertising agency commissions and a smaller percentage in compensation to affiliated stations for carrying programs. But the remainder represents pretax profits, the margin of difference between the first-place network and the third.

In addition, ABC Inc. would get tens of millions more in incremental profits from greater success of the ABC-owned stations, all benefiting from the popularity of such ABC shows as *Laverne and Shirley, Happy Days, Love Boat, Charlie's Angels, Welcome Back, Kotter* and *Barney Miller*.

The focus of the three-network race tends to be prime time, the peak viewing hours between 8 and 11 p.m., which account for half the profits.

Although all three networks have the same amount of air time to sell in that period—six minutes an hour, except for movies, where seven minutes are allowed—the rates are governed by the size and demographic composition of the audience; in short, by the ratings for the programs.

The networks' costs for weekly series do not vary greatly, but the prices they receive for commercial spots range widely. In this season's economy, a 30-second spot in a low-rated show might go for $45,000. In a high-rated show, the 30-second rate could be $100,000.

All but a few of the prime-time series are produced in Hollywood by the major film companies or independent producers. When a network "buys" a series, it actually licenses it for two runs—an original run and a rerun—at a price that usually falls somewhat below the production cost.

Producers are expected to recover the remainder of their investments and make their profits from overseas sales and the subsequent syndication of the series to local stations in the United States. If a series fails in its first or second season on a network, there is rarely an opportunity for subsequent syndication.

Typically, half-hour series are licensed to the networks for $140,000 to $180,000 an episode; the range for one-hour shows is usually $350,000 to $400,000. The difference often depends on how much location shooting is involved and whether foreign locales are used.

The profitability of successful shows is told by this example:

ABC's high-rated situation comedy, *Three's Company,* commands about $90,000 for each 30-second spot, while spots on NBC's relatively low-rated *C.P.O. Sharkey* sell for around $50,000 each.

Both programs cost the networks approximately $160,000 an episode. But *Three's Company* grosses $540,000 from its six 30-second commercial units each time it is aired, while *C.P.O. Sharkey* takes in $300,000. Subtracting the cost of the program and the amounts paid in agency commissions and station compensation, the ABC half-hour comedy has a weekly net of $220,000 and the NBC show, only $50,000.

In television economics the source of greatest expense to the networks is program failure, but paradoxically in these bullish times even the flops make money. The networks are heading for a record in program cancellations. ABC, CBS and NBC together have already canceled more than a score of series, although none of the departed shows were, in practical terms, in the red.

One of ABC's biggest flops was a one-hour Monday night entry called *The San Pedro Beach Bums,* but with the advertisers' surging demand for air time, the program was able to charge $50,000 a half minute, so the show cleared close to $100,000 for ABC each evening it played.

But as an official of the network remarked: "Think what we could have made if we had a hit in that hour."

If a network cancels a series before it has completed its run, it must usually pay for episodes it will never broadcast. Often, as in the case of *Young Dan'l Boone,* which CBS dropped after three episodes last fall, this can be more than $2 million worth of wasted programs. Nevertheless, networks will cancel modestly profitable programs.

"If you don't make the change, you go down like a stone," explained David C. Adams, vice-chairman of NBC. "You make your competitors stronger because the audience simply goes over to them. You hurt the programs that follow the failures in the schedule. And you stand in danger of destroying the whole evening."

Replacements for failed programs must be sold to advertisers, affiliated stations and the viewers, and this fresh round of advertising and promotion also drives up the costs.

Keeping a program that clearly isn't making the grade may also tempt affiliated stations to substitute a more potent movie or syndicated show, thereby reducing the network audience.

It is the audience and not programs or air time, that the networks really sell. The key for advertisers is cost-per-thousand-households delivered. Some advertisers, having no age or sex requirements, buy an undifferentiated audience but a majority—perhaps three-fourths of those who buy network television—will pay a premium to reach a target audience of customers most likely to buy their products.

Companies that make cosmetics or other products marketed to women have paid as much as $8 a thousand this season to reach 18-to-49-year-old women.

These rates contrast with the basic rates for households (undifferentiated audience), which have ranged from $3 to $4.20 a thousand, depending on the network and the time of year.

In earlier days of commercial television, advertisers sponsored whole programs and even assumed production responsibilities. This carried high risks—an unsuccessful program was inevitably a poor investment and often resulted in market-share losses for the product being advertised.

Since the mid-1960's, however, the advertiser has been on much safer ground. Few sponsor programs; instead they have been dealing in one-minute or 30-second "participations"—spots spread over a wide number of programs, often on more than one network. Not only does this minimize the gamble, it also assures reach and frequency.

The networks have provided major advertisers with a further safeguard by guaranteeing circulation. Thus, an advertiser who has spent $10 million with a network to reach women aged 18 to 49 at the price of $8 a thousand will be assured, through "make-good" spots, of receiving full value.

"Make-good" spots are half-minutes over and above those in an advertiser's original contract that are provided by a network when its programs have failed to deliver the intended audience.

Network advertising is purchased in two modes: "up front," or six to nine months in advance, often for a full year's worth of spots, and "scatter," or short-term arrangements that might be made on a day's notice.

The advertisers that buy up front have the advantage of staking out specific programs and time periods at an assured price, even if subsequent advertiser demand should drive rates higher. The scatter buyers tend to play the market, and they frequently come away with bargains, although they risk being shut out of network television when the demand intensifies, as it has the last two years.

Network computers figure prominently in the buying and selling of television time. They are programmed with projected audience ratings and demographics for all the forthcoming shows. When an advertiser presents his requirements to a network, the computer provides several possible schedules of programs on which his spots might appear.

If an advertiser accepts a plan, the contract usually calls for make-good spots if necessary. But if the ratings for the programs should exceed the computer's projections, the advertiser enjoys a bonus at no extra payment.

While this practice has served to make network buying comfortable for advertisers, it has enlarged the problems of a network that has a disappointing season. The lower a network's ratings, the more make-good spots it will have to provide, and of course each make-good reduces the network's inventory of salable advertising time.

One of the joys of being the leading network is having to provide fewer make-good spots. And because the leading network is more likely than the others to provide the advertiser with bonus audience, its rates are often 10 percent higher than the other networks'.

**"Series have been canceled, audience has declined, executives have fallen like chaff— never has the industry made so much profit and been in such a turmoil."**

Ultimately, as the network running first in the ratings, ABC winds up with more spots to sell than its rivals increasing its profitability.

Network profits in 1977 would have been even higher had program costs not risen as much as 40 percent, largely as a result of a quick program turnover and the competitive programming tactics of the networks known as "stunting."

This involves the displacement of regular programs with specials, movies or mini-series intended to blunt a rival network's success. It also includes the expansion of ordinary series into special two-hour presentations and such surprise maneuvers as ABC's scheduling of *Washington: Behind Closed Doors,* the political mini-series based on John D. Ehrlichman's novel, "The Company," two weeks before the official opening of the season.

NBC's troubles in running third were exacerbated by its program costs—the highest among the three networks because of its concentration on mini-series and specials.

The sobering conclusion reached by all the networks from this season's experience is that regular weekly series make more economic sense than do bombardments of so-called "event" programming. In addition to costing less and providing a network schedule with stability and an intensely loyal following, the weekly series has an extremely valuable byproduct: the rerun.

The cost of a rerun that is aired in the same television year as the first-run episode is approximately 20 percent of the original licensing fee. The cost represents essentially the residuals that must be paid the actors, writers, directors and other creative talent for the second use of their work. Advertising rates do not decline by anywhere near a proportionate amount, however.

During the rerun seasons—late spring and summer—the rates drop to two-thirds of those of the peak season. There is unquestionably more profit in reruns than in first-runs, and that, finally, is the greatest advantage in being the No. 1 network.

Frederick S. Pierce, president of ABC Television, recently outlined to a group of Wall Street analysts what ABC-TV could do to keep profits buoyant in the event of an advertising recession. The key ploy was to cut costs by reducing the number of first-run episodes and reverting to reruns of its most popular shows. With a large flock of hit series, ABC could sustain itself nicely.

The other two networks, however, each with only a handful of successful weekly shows, would risk falling further behind ABC if they had to resort to reruns of their flops.

Wall Street's media analysts have been predicting, not a recession, but a return this year to the "normal" pre-1976 growth patterns for the networks—in other words, a decline in the percentages of profit increase.

"This will be a good year for network television," said Ernest S. Levenstein of E. F. Hutton, "But 1977 was an excellent year, and 1976 was super." He defined "excellent" as a year when revenues rose 20 percent and "super" as one in which they exceeded 20 percent. "I think we're closer to a growth of 10 or 15 percent in 1978," Mr. Levenstein said.

Beth Dater, the analyst for Fiduciary Trust, regards the last two years as "extraordinary" and predicts revenue gains at the networks of only 8 to 10 percent this year.

William Suter of Merrill Lynch, Pierce, Fenner and Smith, who predicts an 11 to 14 percent growth, expects the uncertainties of the changes at the networks and the apparent decline in daytime viewing to hold down prices and defer large advertising decisions to 1979.

Mr. Howitt of First Manhattan said, "The stock market has not evaluated these companies [ABC, CBS and NBC] as if the last two years were maintainable." He, too, predicted a lessening of advertiser demand for air time this year and a general flattening of prices.

These forecasts, which gain some support from a softening of the network television market in the current quarter with a consequent lowering of prices, contribute to a climate of nervousness and even panic in the network skyscrapers along the Avenue of the Americas. Price cutting has already begun, and in some instances of short-term "scatter" buys, the household cost-per-thousand has dropped to $2.50.

The three networks, which earned an estimated $400 million last year on sales of $3.3 billion, have entered a retrenchment period because business is slipping back to merely good. All three may be expected next season to settle into the old-fashioned, economical groove of 22 weeks of first-run programs, a scattering of specials, a minimum of stunting and, to the extent possible for NBC and CBS, a long and lucrative period of reruns.

# PBS: For Better or Worse

*Clare Lynch O'Brien*

*CLARE LYNCH O'BRIEN is the educational consultant to* ITT's Big Blue Marble. *In addition to her work in television production, she writes and lectures on the subject of children and television.*

Even among the staunchest advocates of American public television, confidence in the system is a rarity. For as many persons who are critical of the system, there are ten who can't even begin to understand how it functions. If the viewing public is confused, it is not surprising. In its short, twelve-year history public television has gone through dozens of changes in an attempt to establish a workable alternative to commercial television, what the Carnegie Commission on Educational Television described in its 1967 plan for public television as ". . . all that is of human interest and importance which is not at the moment appropriate or available for support by advertising." Unravelling the problems that ensued since the '67 mandate became the goal of Carnegie II, the name given to the second major study of public television which was presented in January of 1979. Among the many recommendations of Carnegie II was the dissolution of the Corporation for Public Broadcasting, the same bureaucratic system that was the invention of Carnegie's earlier study. Such organizational ironies are at the heart of many facets of public television. Making sense out of them is no easy task.

What is now known as PBS (Public Broadcasting Service), the 264 public stations, actually began in 1952 when the FCC determined that certain channels be set aside for educational broadcasting. Each of the stations was totally independent and in spite of the fact that many produced programs suitable for wider telecast, local programs were broadcast on one station. Adequate funding for programming remains a problem today, but in the early 50's it was the undoing of many educational stations. Money had to be raised locally and even stations in large cities found it difficult to acquire subsistence level financing. Many credit the Ford Foundation for keeping public stations functioning in the early years. In fact, its support was significant to both the young stations as well as to what became the public television system. Since 1951 the foundation contributed nearly $300 million to public television, more money than it received from any other source.

Some of the larger cities, principally Washington, New York and Los Angeles, did not have any VHF stations available, which in effect kept these cities out of public television for nearly ten years. In addition to the fact that each of the existing stations was firmly rooted in local concerns, public stations were and are licensed to different groups whose goals don't conveniently mesh. Some stations are licensed to state authorities, some to colleges, some to non-profit civic corporations, and a few to Boards of Education. The combination of regional differences and what is viewed by some as incompatible programming goals has contributed to the development of strong and often disagreeable factions among the stations. The general difficulty in raising money made the factions even more self-protective. Public stations remained in this state of uncomfortable disarray until the Public Television Act of 1967 renamed the system "public television" and a new era of slightly different uncomfortable disarray began.

It was the first Carnegie Commission report, *Public Television: A Program for Action,* which led to the Public Television Act. Funded by the Carnegie Corporation, a commission of fifteen members spent two years investigating educational television. The commission, which represented the interests of education, communication, broadcasting, business and the arts, made several recommendations. It suggested the creation of a corporation, The Corporation for Public Broadcasting (CPB), which would raise and distribute funds for the system, administer federal appropriations for public television, allocate money for national programming and take responsibility for the interconnection of the many stations. Not all stations agreed that becoming part of a national system was desirable. In fact, traces of that resistance remain evident today.

Another important role of CPB is to protect the system from government intervention. The issue of freedom has been hotly debated. Because the fifteen members of CPB are presidential appointees, many believe the membership incapable of being nonpartisan, and therefore unable to protect freedom. This opinion is not without evidence to support it. Presidents Johnson and Nixon both had a difficult time drawing a line between government interests and television. Johnson had particular problems with anything in print or on television that ran counter to his plan for the Vietnam War. Perhaps because of his personal interests in broadcasting, Johnson was a great supporter of the Public Television Act, and in fact, helped its speedy passage in Congress. The first chairman of

CPB, Frank Pace, Jr., was then duly appointed by Johnson. Pace was a former Secretary of the Army and former executive of General Dynamics, a company with several military interests. In 1968, a British Documentary called *Inside North Vietnam* was scheduled for broadcast on NET. The film showed aspects of the Vietnam war never seen on American television, and proffered a clearly negative view of American participation in the war. Prior to broadcast, and even before viewing the film, Congressmen sent letters to NET pressuring executives to cancel the broadcast. Some letters contained veiled threats about the future of NET management and at least one suggested that he would never again vote for public television appropriations. CPB provided little of the insulation from government pressure that had been promised by the Public Television Act. NET ran the program in spite of the resistance from the government, but the experience was a bitter pill that few have forgotten.

Nixon's opinion of television journalism is well known. Instead of exerting pressure through Congress, he chose to be even more direct. He vetoed two separate public television appropriations, making it "perfectly clear" that he did not like the direction public television was taking. Programs considered to be anti-administration, like *The Great American Dream Machine* that satirized government, incensed Nixon, who had a personal taste for action adventure series and situation comedies. With federal funds as leverage, the White House Office of Telecommunications issued a statement about what would be required of public television if it wanted continued government support. Among other things it would have to put an emphasis on local affairs and leave public affairs programming to the networks. It would also have to raise $2.50 for every federal dollar offered. CPB offered little support to public television stations in these situations and station managers grew increasingly disgusted with a system they had little confidence in to begin with. In 1973 the situation came to a head when an organization of station managements engaged in an active fight against CPB control of public television. What followed was an agreement between PBS and CPB to form a partnership by which a gradually increasing percentage of federal funds would go directly to the stations. CPB lost a good deal of its power but it continued to support some public television programs and provide seed money for new projects.

The new seventeen-member Carnegie Commission, chaired by Dr. William McGill, president of Columbia University, issued a blueprint for the future of public television and radio in January of 1979. It urged a federal government commitment to a *larger, better financed,* and *more independent* public television system. Specifically, it called for the Government to triple its financial support by 1985. It suggested a $553 million budget for 1979 with approximately one third coming from federal funds. As an additional revenue source, the Commission suggested that a fee be imposed on commercial broadcasters for use of the public airwaves. With banner profits every year, such a plan seems easily affordable by commercial broadcasters, but, needless to say, they opposed the idea.

**"Instead of asking what will sell, we have to look out there and say, 'What need will it fill?' "**

The study, which took 18 months and cost $1 million to complete, also calls for the elimination of the CPB. The new organizational plan consists of two groups, a *Public Telecommunications Trust* and a *Program Services Endowment.* The trust would replace CPB. Its responsibility would be to both set goals for the system and to evaluate its performance. Members of the trust would have no voice whatsoever in programming decisions. The second organization, the endowment, is envisioned as "a safe place for nurturing creative activity," where new and experimental forms of programs would be developed and financed. Distribution of funds would be made directly to stations under a more generous formula of two federal dollars for every three raised locally. Monies would be designated for programming rather than facilities. The changes are aimed at significantly increasing money for programs, further protecting the system from political interference and eliminating some of the many bureaucratic blocks to the development of new programs. A new selection process for board members would be designed so that appointments did not become political plums. How successful the plan to " . . . create a free press sponsored by the government" might be remains to be seen. The CPB President called the study "thoughtful and informed" but gave little indication of his support. At a televised press conference shortly after the release of Carnegie II, President Carter said that he was impressed by the report. When pressed further to describe what he liked in particular about it, he explained that he had not actually read it. When he does get around to it, he might decide that tripling federal support to public television is inconsistent with his anti-inflationary themes.

A continual flaw with the public television system is the cost of running the management system and the pathetically low portion of the annual budget that is directed to production.

In a series of articles published in the *New York Times* in 1977, Les Brown reported the failure of increased government support to improve the quality of programming. "More than 8 of every 10 federal dollars go to the support of public television's own bureaucracies and to activities unrelated to national programming." While little more than 10% of the budget is used for programs, the remainder is consumed by the staffs of PBS and CPB who often duplicate each other's work. The rest goes to operational costs and the expense of equipment. It is nothing less than appalling that such a ratio of management expenses to program output be permitted. To illustrate the sort of management waste that exists, Mr. Brown reported a meeting in Mexico with the Mexican network, Televista, to discuss program exchanges and coproductions. Sixteen executives made the trip at public television's expense.

## Programming

For whatever criticism that can be leveled against public television, it does provide an alternative to commercial television programs. A nagging question is whether or not it provides enough in the area of children's programs. Public television has been the only respite for parents who worry about what children view. *Mister Rogers' Neighborhood,* a program for pre-school children, fosters self-esteem in youngsters, something commercial television thrives on debasing with its particularly inappropriate violence and self-depreciating humor. PBS's current use of the *Neighborhood* programs is one of the more intelligent and efficient things they do. At the moment there is an inventory of 500 *Mister Rogers' Neighborhood* programs. Understanding that a new generation of pre-school children appears every three years, the programs are recycled. In 1979, Fred Rogers will produce 10 new half hours to develop new ideas or refresh old ones. This is a sensible approach to any kind of children's program whose content would become stated or stale.

In the education-entertainment area, *Sesame Street,* and to a lesser degree, *The Electric Company* proved that children could be entertained and learn something at the same time. Whether or not the programs have achieved their stated goals, or if their goals were appropriate to begin with, is a source of constant debate. The important point, in terms of this discussion, is that the idea was worthy of experimentation and only in public television could this experiment be undertaken.

Minority groups have been seriously neglected in commercial children's television. In public children's TV, special programs have been developed for minority youngsters, including handicapped children. *Villa Allegre, Que Pasa USA?* and *Carrascolendas* addressed a significant Spanish-dominant American population, and provided positive Latin role models. *Feeling Free,* the first program featuring a group of handicapped youngsters, aired on PBS in the spring of 1978. *Zoom* provided an opportunity for children to participate in what has always been a passive activity.

The greatest flaw in public broadcasting for children has been its failure to provide enough money for producers to develop programs adequately. With the exception of Children's Television Workshop, which seems to have no trouble raising money for its projects, many children's programs disappear before they have the time to either build an audience or realize their purpose. *Feeling Free,* which was a series designed to introduce young children to their disabled peers, only received enough money to produce six half-hour programs. It is ludicrous to imagine that anything more than a superficial treatment of the subject might be accomplished in six half-hours. The responsibility for raising money for new programs rests with program producers who cannot reasonably be expected to produce programs and raise money at the same time. After production, when the creative momentum is interrupted for fund raising, financial support for the production staff is rarely available to help them raise money for new programs. More often than not, producers go on to other projects and important programs that have been given expensive seed grants die in infancy.

**" . . . What remains is the bureaucratic nightmare which is at the heart of public television."**

Lack of adequate financing is the most frequent excuse given for the proportionately high number of foreign imports on public television, a frequent criticism leveled by both viewers and American producers. While many of the imports are excellent, critics argue that we have the resources to develop our own programs and that American public television should support American, not foreign, productions. To be sure there have been a number of excellent American productions. The thirteen-hour WNET production of *The Adams Chronicles* pleased both the audience and critics, proving that a fine American production could equal the quality and scope of the British programs. American programs seem to take an agonizingly long time to produce and are usually over budget. Public television management, discouraged by high production costs, find it easier and cheaper to import finished programs than create new ones.

The government funding process is so hopelessly bureaucratic that it discourages all but the most stouthearted producers. Productions are funded in "steps," with hugh delays between each step. First one must obtain a planning grant which pays for a survey to determine whether or not there is a need for a particular program. A second step, a research and development grant, supports the development of scripts and any research that might be required for the project. If these two stages are successfully completed, one must apply for a pilot grant. Finally, a production grant is necessary. At every step, a proposal must be written. Often these proposals take longer to write than the scripts.

The reviewing process for proposals is lengthy and tedious. For *The Adams Chronicles* scholars debated endlessly whether the scripts should be true to the *Adams Papers* or historically accurate. Such debates take a painfully long time and discourage many producers. Through all these steps and proposals, production budgets become obsolete. Actors and crews lose interest and go on to other projects. *The Adams Chronicles* took nearly five years to complete. Few producers can sustain interest for a project that drags on so endlessly.

WGBH recently presented *The Scarlet Letter,* a production that took four years to complete, was significantly over budget, and received generally negative reviews. As with the *Chronicles, The Scarlet Letter* had some consultant scholars, each with a different perspective. The project went through a series of proposals and script reviews, several writers, and in the end didn't work. While the excuse for this "step" arrangement is that it protects public funds, so much money is wasted in revisions and delays that it becomes increasingly difficult to defend the system.

One of the mandates of public television is to provide alternative programming. For the most part, it does that quite well. Few disagree that programs on PBS are unlikely to be broadcast on commercial television. A question remains on how adequately the existing PBS schedule reflects the interest of the total viewing public. In a given week WNET, the PBS station in New York City broadcasts nearly 100 hours of programs. Of those 100 hours, 36½ are devoted to children's programs which is more than one third of the total broadcast time. Thirty, or slightly more minutes each day, is devoted to "How to" shows like *Yoga, Julia Child* or *Crocket's Victory Garden.* Local programs are broadcast between 36 and 60 minutes each day. After children's programs, the next largest number of broadcast hours are devoted to news commentary and analysis, debate or discussion shows. Approximately 7½ hours are devoted to drama and 3 to a movie. In the week reviewed, both dramas and movies were British. Music programs, all classical, accounted for another 3 hours. The remaining 9 hours were divided more or less equally between tennis, pro soccer and a science program.

What does the selection tell about the audience? It assumes that many are children between the ages of 3 and 13 years of age. There were no programs for teenagers. There was one parenting program broadcast mid-day mid-week, leading one to assume that parents are at home and don't work. The schedule also assumes a well-educated, literate audience. It also suggests that the audience can't get enough of Dick Cavett. His program was broadcast three times in one day. There was no minority program for adults. The minority programs for children were Spanish. The only program for the handicapped was captioned *Zoom.* Why don't more drama or music programs reflect a broader range of tastes? Why are the concerns of teenagers ignored by public television? Is it really necessary to run *Sesame Street* and *Electric Company* twice each day? It is too easy to blame inadequate funding for a program schedule that is so unbalanced. Any new plan for public television should include a serious look at programming objective.

## Funding for Public Television

Government funding for public television provides a financial base for the system, however limited. Foundations, including Ford, Markle, and Mellon support public television. The National Endowments for the Arts and the Humanities both fund certain projects. Stations run subscription drives and hold annual auctions to raise money for operating costs. A host of businesses, large and small, provide program grants and support for individual stations. At a particularly lean time for public television, the early 70's, a combination of events released a flood of money for public television productions. During the oil crisis, increased oil prices, reports of illegal gifts and bribery, and apparently excessive oil company profits, soured public opinion toward the big oil companies. To turn public opinion, Mobil, Exxon, Gulf and Atlantic Richfield spent some of their rich profits on public television. Several of PBS's most lavish and ambitious projects were supported by oil companies. Exxon presented *Theater in America,* Atlantic Richfield, *The Adams Chronicles,* and Mobil, *Masterpiece Theater.* So generous were the underwriting grants that producers and station managers made regular appointments to solicit funds for new projects. The stations could boast a rather elite audience, better educated and presumably more affluent. This became the public television "pitch" in fundraising.

PBS exercises caution, sometimes bordering on paranoia, about corporate contributions. Concern about an organization influencing program content allows a station to accept a corporation's cash, but not a program produced by them or any subsidiary. In the fall of 1978, a James Michener documentary, "Poland: The Will to Be," was offered by WCET in Los Angeles. The program was scheduled to run on a Sunday evening. On the Wednesday prior to broadcast, PBS notified its stations that the program was produced by Emlen House, a wholly-owned subsidiary of Mrs. Paul's Kitchens, the company that underwrote the program. Several stations, including WNET in New York, refused to run the program and substituted another at the last minute. PBS left the decision to run the program or not with the stations:

> "PBS has decided to distribute the program notwithstanding this relationship (underwriter-producer) in light of the problems which would arise for local station schedules should the program not be fed."

Program director of WNET in New York, Robert Kotlowitz, criticized PBS for leaving the decision to run the program up to the individual stations. His attitude reflects the puzzling sort of ambivalence many stations have about autonomy and the role of PBS management.

This particular series, *James Michener's World,* was produced with Reader's Digest Films until the production company went out of business. The series' producers continued the project with Emlen House Productions. Rather than scrap "Poland: The Will to Be," several less wasteful alternatives could have been considered. The station could have run an announcement with the opening and closing billboard, alerting the audience to the relationship between the underwriter and producer. Any audience interested in the documentary could evaluate the relationship. Program managers, presumably capable of making judgments about program content, could have previewed the film. If the viewing didn't leave them with an irresistable craving for fish sticks, they might have assumed that the relationship between the underwriter and producer was a clean one and the program suitable for public television. The basis for such funding policies is obviously the best interest of the public and one cannot take conflict of interest concerns lightly. However, it is difficult to understand how public television cannot draw from the considerable management pool of PBS and CPB, some people who could make individual judgments in cases like these. This sort of hassle is hardly worth it to a company which gets a much better return on money spent on commercial television.

## Producing for Public Television

Many producers claim that despite frustrations of working in the public television system, they prefer it to commercial television. The rewards of producing a program without concern for ratings or advertisers outweigh the frustrating delays. The opportunity to develop a series like *Nova* or *The Adams Chronicles* does not arise in commercial television and the prospect of turning out a spin-off of *Laverne and Shirley* would make many shudder. Some talk about the lack of job security. When a staff is hired for a program, the members are usually employees of the station. If funding dries up for that project, some are let go and others move on to other programs. Jobs are no more secure in commercial television. The only difference is the reason for the cut.

Generally, public televison producers have no equity in either their production or any secondary uses of their ideas. They are considered employees of the particular station and ideas conceived while in the employ of a station belong to it, not them. *Zoom* was created by a producer at WGBH. Royalties from books and records were not shared by the producer. Financial incentives to work in commercial television are greater. For the most part, salaries are higher and royalties or residuals give a producer the luxury to develop new ideas between programs. Despite the inequity in compensation, public television stations attract excellent producers. WGBH in Boston produces one-third of the total number of programs shown on PBS. The atmosphere is relaxed and casual and management is said to be supportive. In an interview with Lynde McCormick of *The Christian Science Monitor,* Sylvia Davis, Executive Producer of Special Projects for WGBH said, "Instead of asking what will sell, we have to look out there and say, 'What need can we fill?' " If an executive from one of the three networks delivered a line like that to a Normal Lear, he'd better have oxygen handy. Networks seem to revel in duplication of the obvious. They copy current movies, foreign television, old movies and each other. It wasn't enough to have one rip-off of the National Lampoon's *Animal House,* each of three networks came up with one.

While huge amounts of time and money are wasted in funding delays, producers can rarely complain of inadequate time restricting the programming end. While their commercial counterparts are churning out a program a week, the pace is more leisurely in public television.

## The Future of Public Television

The potential impact of Carnegie II on public television's problems is not clear. Some of the recommendations from the report are not new. Critics have complained about the preponderance of British dramatic series for years, yet only four American series have been broadcast to date. If the key to progress is funds, it remains to be seen whether Congress will come up with enough federal money to free the system from some of its financial traps. There is no indication that the dreadful step arrangement will be shortened or dropped. With all the reforms and all of the promises, what remains is the bureaucratic nightmare which is at the heart of public television's problems. It is hard to imagine how two new structures, no matter how elegant their charters, will ever deliver to the public more than PBS and CPB has. Simply saying that they will do better is not enough. One would hope that public television could continue to provide a serious alternative to commercial television, but attempt to reach a wider range of viewing tastes, no matter what the future of funding.

# Cable and Pay TV on Eve of Technological Revolution

*Les Brown*

*LES BROWN is Television Correspondent for The New York Times.*

The commercial television industry has weathered in its three decades countless storms of controversy, public pressure and regulatory and legislative sanctions, and it has come through all of them essentially unharmed.

But television's extraordinary resiliency will be put to a more rigorous test in the 1980's, because what is building up now is a technological wave that will be sweeping in formidable new industries that some believe will drastically change the American television system.

Several new forms of electronic communications have already become flourishing businesses expecting significant growth in the next five or six years—cable television, pay television, satellite transmission, portable video cameras, home video recorders and video games.

There is even more to come: the video disk, fiber optics, electronic data transmission and several forms of computer-linked television.

In Washington, where there is a strong sense of a television revolution in the offing, the House Subcommittee on Communications is attempting to recreate a free and open market for the emerging technologies on the theory that the public interest is served by diverse, competing media.

## Troublesome Issues Foreseen

Other policymakers, however, foresee numerous troublesome issues rising from the new technologies, such as invasions of privacy, monopolistic control of electronic communications in a community and the entry into homes of pornography and extremist propaganda.

How soon the wave hits will depend, it is generally agreed, on consumer acceptance of the new devices. "It's when technological forces turn into economic forces that you have the big bang," a Wall Street analyst said.

The networks are expressing skepticism about the effects the new industries will have on their business, but they concede that changes are inevitable and that their own giddy period of steady audience growth will probably be over in the 80's.

Vincent Wasilewski, president of the National Association of Broadcasters, also minimizes the effects the new industries will have on commercial broadcasting before 1985: "I don't detect a great public demand for change."

## Unavoidable And Soon

However many experts in the field, citing the fact that the new developments are affecting every aspect of the television process—production, distribution and display on the home screen—contend that change is unavoidable and that it must come fairly soon.

What the developments are leading to primarily, these experts indicate, are vast increases in the number of viewing channels available to viewers and cheaper and more efficient national distribution of programming. More channels and cheaper distribution would inevitably mean more networks, and these may be expected to cut into the audiences for ABC, CBS and NBC, as well as the existing local television stations.

Melvin A. Goldberg, a research vice president of ABC-TV, said in a recent speech to the American Association of Public Opinion Research: "Time is television's basic commodity. It can be divided but not expanded. To the extent that these new technologies take people away from watching programs, broadcasters must be concerned."

The marriage of two-way cable televison to the computer, making it possible to change viewers for programs they order with the press of a button, is likely to result in a wide variety of specialized programs because they would not require mass audiences. One million viewers paying $2.50 for an opera, for example, could be more than enough to justify the telecast.

## Cultural Revolution Predicted

"We will be seeing not just a technological revolution," Gustave Hauser, president of Warner Cable, one of the largest cable systems in the country said, "but also a cultural revolution. People will be learning to use television differently and to expect different things from it."

Warner is the parent of Qube, the remarkable two-way system having its tryout in Columbus, Ohio. Qube permits viewers to be polled, order products through television and purchase movies, college courses and cultural and sporting events not offered on conventional television.

Whether Qube, which involved a $20 million investment by Warner, can develop into a profitable business will probably not be known for another year. If it should succeed, expectations are that other large companies will enter the field spreading the technology to major cities.

The various Qube installations could be interconnected to form a number of networks by means of domestic satellites.

### Effects Of Satellites

The satellites are regarded by experts as the surest instruments of change for the business of television. They not only are altering the methods of distributing television and radio programming, but they are also opening national distribution—the almost exclusive province of the networks—to all comers.

Moreover, they have already met the test of market acceptance. Traffic has steadily been increasing on the two domestic satellite systems—Western Union's Westar and RCA's Satcom. Westar's two satellites can handle between them 24 television transmissions at a time, and Satcom's pair have the capacity for 48 simultaneous transmissions.

Among their regular users are the Public Broadcasting Service, the Mutual Broadcasting System, the Christian Broadcasting Network, the Independent Television News Association, Home Box Office, and the Robert Wold Company, an organization that sets up temporary networks and arranges regional transmissions of sporting events.

Foreseeably, in the 80's the variety of part-time or ad hoc networks fostered by the satellites will loosen the full-time dependency of affiliated stations on those networks.

### Options To Sell Own Ads

Film companies may elect to eliminate the network as middleman and send motion pictures directly to the stations by satellite, after having sold the commercial spots in the films themselves. Advertisers would have the ability to lease the satellite to send out programs of their own choosing instead of relying on the networks' choices.

According to a number of experts, satellites will make their full impact on commercial television, liberating individual stations from network dominance, when there is a broad proliferation of earth stations—the special receiving antennae for satellites—around the country.

A large earth station, which is a parabolic dish aimed at a specific satellite, cost $100,000 to build and install a few years ago. But compact six to eight meter dishes are being engineered now at lower cost, and they are described by one expert as "no more than the price of two Cadillacs."

The cost of satellite transmission itself is expected to come down substantially. Charles Jackson, technology specialist on the staff of the House Communications Subcommittee, predicts that in time the expense of distributing a program nationally will be no more than $100 an hour. The drop in price will come, he suggests, when each of the transponders on the satellite can be subdivided to carry four signals at a time instead of one.

### Only Half Way There

"From what we know of the developments that are coming," he said, "we're seeing only half of the technological revolution now."

Although the three major networks use the satellites frequently to relay news reports and sports coverage to their transmission centers, they remain holdouts in the use of satellites for the distribution of their programs to stations.

Network officials maintain that their current tariffs for telephone land lines are economically sound and that there would be no substantial savings for them to switch to the satellite mode. But in the view of outside experts, the networks' reluctance to switch from the old

### TECHNOLOGY GLOSSARY

Cable television: The technology of distributing television signals to homes by wire instead of over the air.

Two-way cable: Cable installations in which two wires are used, one carrying signals from the transmission center, the other taking signals from the television set. Sophisticated forms of bidirectional cable permit pictures to be sent in both directions.

Pay-cable: Pay-television by means of cable, which essentially provides subscribers with new movies, sporting events and special variety programs.

Qube: Trade name for a form of twoway cable whose special feature is the union of a cable with a system of polling computers.

Fiber Optics: Light-wave technology in which a hair-thin flexible glass fiber can be substituted for the copper wire in cable systems to substantially increase the number of channels.

Satellites: Orbiting space vehicles, 22,300 miles above the equator, used in place of terrestrial lines to relay television signals and other communications services—telephone and teletype—over long distances.

Home video recorders: Videotape recording and playback devices that can record off the air up to four hours of programming for later viewing while the viewer is away, asleep or watching another channel.

Video disk: The video counterpart of the phonograph record, expected to come onto the market later this year. It stores visual and audio matter that can be displayed on the television set through a special turntable.

Electronic data transmission: Systems developed in Britain and France that, by means of a decoder attachment to the television set, permit the viewer to call up a variety of printed matter, such as news bulletins, stock quotations and sports results.

technology to the new is based chiefly on their wariness about changing the dependency of their affiliates on the network transmission lines.

There is growing sentiment in Washington now for a wide-open marketplace, relatively unhindered by regulation, in which the new developments and the existing television system could compete freely.

### 'It Won't Even Take 10 Years'

Broadcast regulation is based on the idea of frequency scarcity. Representative Lionel Van Deerlin, Democrat of California, chairman of the House subcommittee and spearhead of the new bill, believes with certainty that there will be an abundance of television channels in the 80's.

For this reason, he has proposed in the bill to relieve television operators of most of their license responsibilities after 10 years. "It won't even take 10 years," Mr. Van Deerlin said. "The scarcity problem will be over a lot sooner than that."

The expansion of channels is expected to come about in one or all of at least three ways:

- Through cable television which now has systems offering 36 channels and, in its modification with fiber-operatics technology, can increase the number to thousands;
- Through satellites transmitting directly to homes equipped with special, relatively inexpensive antennae;
- Through a new device developed by Texas Instruments and financed by the Federal Communications Commission, known as the TI/Fun. It divides the electromagnetic spectrum more efficiently than present technology to allow for the creation of substantially more stations on the VHF and UHF television bands now in use.

### Cable Is Penetrating Suburbs

It appears that cable-television will be the most immediate source of new outlets and networks. Spurred by consumer demand for such pay-television networks as Home Box Office and Viacoms's Showtime, cable is beginning to make significant penetration in the suburban communities of major cities.

William J. Donnelly, a vice president of Young & Rubicam Advertising, who has been concentrating on cable and other emerging media, has predicted that cable television would be taken seriously as a national advertising medium when it reached into 30 percent of the country's television homes. The figure is currently 17 percent.

Thirty percent is not a figure pulled whimsically from the air, Mr. Donnelly points out: "It was the magic number that made television a mass medium and that later made color television matter to advertisers."

A 1976 study of cable by Arthur D. Little Inc. had forecast a 30 percent penetration by 1985, but Mr. Donnelly, citing the recently accelerated growth of the medium, predicts the critical mass will be reached in 1981.

### Networks Don't See Jeopardy

The networks are, of course, fully aware of the technological developments, but they contend the present system of television is in no jeopardy.

Gene F. Jankowski, president of the CBS Broadcast Group, in a recent interview derided the prognostications of doom for the networks:

"People are making a set of assumptions about the 1980's. It's all still speculation, but they have set up a situation that becomes, in their minds, fact. These new industries are interesting, and we might want to enter some of them ourselves, but they are small businesses and not a serious threat to us. Networking in the 80's will be very similar to what it is today."

To illustrate his belief, he cited that CBS recently committed $35 million to license "Gone With the Wind" for 20 years. NBC similarly paid $21.5 million to license "The Sound of Music" for 22 years.

### Struggle For Advertising Revenues

At NBC, Alfred Ordover, the executive in charge of corporate planning, said that according to his studies the expansion of cable to the 30 percent mark would more seriously affect small stations than it would the networks. Moreover, he suggested, any future network that might be formed would face a struggle for the advertising dollars that now support the three networks comfortably.

Asked how the network was planning to meet the changes promised by the technology, Mr. Ordover replied: "Rational analysis says nothing is going to happen. But if all these things do come to pass in a large way, defying rational analysis, we have no plan."

# It Was Cool to See All Those People Shot

*Jim Klobuchar*

*JIM KLOBUCHAR is a reporter with the* Minneapolis Star.

The kid was a playground warhawk. He punched out other kids who didn't give him the run of the yard, and he talked like a hood. For a while he worked a little protection racket. The ones who let him run the show were safe; the ones who objected got hammered.

His teacher and principal brought him into a social worker, who in trying to uncover clues to the 6-year-old boy's behavior asked him about the things he liked and didn't like. The talk got around to television, which in the fashion of times occupied an impressive part of the boy's life.

"Wasn't it cool when they shot those people and they fell into the hole?" the kid said.

The boy had been watching *Holocaust* the night before.

The social worker, Len Colson of one of the suburban school systems, has conditioned himself not to be surprised by 6-year-old kids, but this one stunned him.

Kids sometimes say outrageous things deliberately to attract attention and to set them apart, he recognized.

"But this boy seemed really to mean what he said," Colson concluded, "and it left me pretty shaken because of the raw ugliness of the implications—not that this was an ugly kid but that he was actually turned on by an act as terrible as that. The problem wasn't the production, which was worthy. But we're seeing more and more in the behavior of aggressive and assaultive children that there is a cumulative power in what the kids are watching on television, at a time when they really aren't old enough to tell reality on television from fantasy."

Alarm over the potentially corruptive power of television violence is hardly original, particularly in its impact on young minds just emerging from fairy tales or still immersed in them.

No overwhelming progress has been made so far in slowing down some of its more wrenching attacks, and if Colson's evaluations are right, the damaging impact of it on the mentality of children is mounting.

He conceded it can be overstated.

"Most kids are going to get through the elementary grade part of their growing up without being turned into problem cases," he said. "But we find an awful lot of television-related stuff in the cases of children referred to us for assaultive behavior, the acting out kids. Many of them watch the more violent programs. There certainly can be and are other reasons for their behavior. Yet it has to disturb a teacher, as it did one I know, when you turn loose a bunch of kindergarten kids for free time with building blocks and see a lot of them preferring to run around as gunslingers and musclemen."

All right, but how is the kid today different from the pre-television kid who played cops-and-robbers and cowboys-and-crooks, and liked the idea of walking up to the dinner table with a gun in his holster?

"The difference is degree and the awesome power of suggestion and persuasion that the television set in the living room has," Colson said. "Unlike the movies kids saw in another generation, it's there every hour, every day. Kids learn to see the world through it and a lot of their attitudes are shaped by it. We might have seen Hopalong Cassidy and Roy Rogers in the theaters or on TV 20 years ago, but now a shootout isn't a guy getting winged by the hero. A shootout is spurting blood, and 20 bullets banging into the victim, in slomo or stop action.

"I don't know if all that is very healthy for a 6-year-old kid, who has problems enough with right-and-wrong without being confused by the new wide open dramatic liberties which sometimes make it impossible for the kid to know how much of that brutality is bad and how much of it is okay. If he's allowed to watch all of it he's going to learn that it's okay sometimes to gun down the guy next door, especially if he's been playing around with your wife. Now this might be adult drama and it might even be good adult drama, but you can't believe it's the right drama for the 6-year-old when he tells the social worker the next day: 'I'm gonna kill him.' "

Len Colson thinks the best solution for the parents is to join in a boycott of the sponsors supporting the more brutal shows.

But in his own house he does it by exercising a less-than-lenient controling hand on the dial.

I don't know of a better way. The problem is that an increasing number of parents find themselves, on the advice of their counsellors, seeking love and understanding in all its works, and that includes the love and understanding of their children. And they temporize on the TV dial. So there, at least, we may be seeking love in the house but losing a lot of direction.

This article first appeared in the *Minneapolis Star*.
Reprinted by permission.

# Different Forms of Violence

*Erich Fromm*

*ERICH FROMM is author of a dozen books, including the best-selling* The Art of Loving and The Heart of Man.

In contrast to malignant forms of destructiveness, I want to discuss here some other forms of violence. Not that I plan to deal with them exhaustively, but I believe that to deal with less pathological manifestations of violence might be helpful for the understanding of the severely pathological and malignant forms of destructiveness. The distinction between various types of violence is based on the distinction between their respective unconscious motivations; for only the understanding of the unconscious dynamics of behavior permits us to understand the behavior itself, its roots, its course, and the energy with which it is charged.

The most normal and nonpathological form of violence is *playful violence*. We find it in those forms in which violence is exercised in the pursuit of displaying skill, not in the pursuit of destruction, not motivated by hate or destructiveness. Examples of this playful violence can be found in many instances: from the war games of primitive tribes to the Zen Buddhist art of sword fighting. In all such games of fighting it is not the aim to kill; even if the outcome is the death of the opponent it is, as it were, the opponent's fault for having "stood in the wrong spot." Naturally, if we speak of the absence of the wish to destroy in playful violence, this refers only to the ideal type of such games. In reality one would often find unconscious aggression and destructiveness hidden behind the explicit logic of the game. But even this being so, the main motivation in this type of violence is the display of skill, not destructiveness.

Of much greater practical significance than playful violence is *reactive violence*. By reactive violence I understand that violence which is employed in the defense of life, freedom, dignity, property—one's own or that of others. It is rooted in fear, and for this very reason it is probably the most frequent form of violence; the fear can be real or imagined, conscious or unconscious. This type of violence is in the service of life, not of death; its aim is preservation, not destruction. It is not entirely the outcome of irrational passions, but to some extent of rational calculation; hence it also implies a certain proportionality between end and means. It has been argued that from a higher spiritual plane killing—even in defense—is never morally right. But most of those who hold this conviction admit that violence in the defense of life is of a different nature than violence which aims at destructiveness for its own sake.

## Reactive Violence Has an Unreal Base

Very often the feeling of being threatened and the resulting reactive violence are not based upon reality, but on the manipulation of man's mind; political and religious leaders persuade their adherents that they are threatened by an enemy, and thus arouse the subjective response of reactive hostility. Hence the distinction between just and unjust wars, which is upheld by capitalist and Communist governments as well as by the Roman Catholic Church, is a most questionable one, since usually each side succeeds in presenting its position as a defense against attack. There is hardly a case of an aggressive war which could not be couched in terms of defense. The question of who claimed defense rightly is usually decided by the victors, and sometimes only much later by more objective historians. The tendency of pretending that any war is a defensive one shows two things. First of all that the majority of people, at least in most civilized countries, cannot be made to kill and to die unless they are first convinced that they are doing so in order to defend their lives and freedom; second, it shows that it is not difficult to persuade millions of people that they are in danger of being attacked, and hence that they are called upon to defend themselves. Such persuasion depends most of all on a lack of independent thinking and feeling, and on the emotional dependence of the vast majority of people on their political leaders. Provided there is this dependence, almost anything presented with force and persuasion will be accepted as real. The psychological results of the acceptance of a belief in an alleged threat are, of course, the same as those of a real threat. People *feel* threatened, and in order to defend themselves are willing to kill and to destroy. In the cause of paranoid delusions of persecution we find the same mechanism, only not on a group basis, but on an individual one. In both instances, subjectively the person feels

"Different Forms of Violence" from THE HEART OF MAN by Erich Fromm
Volume Twelve of the Religious Perspective Series
edited by Ruth Nanda Anshen

in danger and reacts aggressively.

Another aspect of reactive violence is the kind of violence which is produced by *frustration.* We find aggressive behavior in animals, children, and adults, when a wish or a need is frustrated. Such aggressive behavior constitutes an attempt, although often a futile one, to attain the frustrated aim through the use of violence. It is clearly an aggression in the service of life, and not one for the sake of destruction. Since frustration of needs and desires has been an almost universal occurrence in most societies until today, there is no reason to be surprised that violence and aggression are constantly produced and exhibited.

Related to the aggression resulting from frustration is hostility engendered by *envy* and *jealousy.* Both jealousy and envy constitute a special kind of frustration. They are caused by the fact that B has an object which A desires, or is loved by a person whose love A desires. Hate and hostility is aroused in A against B who receives that which A wants, and cannot have. Envy and jealousy are frustrations, accentuated by the fact that not only does A not get what he wants, but that another person is favored instead. The story of Cain, unloved through no fault of his own, who kills the favored brother, and the story of Joseph and his brothers, are classical versions of jealousy and envy. Psychoanalytic literature offers a wealth of clinical data on these same phenomena.

## Violence Seeks Revenge

Another type of violence related to reactive violence but already a step further in the direction of pathology is *revengeful violence.* In reactive violence the aim is to avert the threatened injury, for this reason such violence serves the biological function of survival. In revengeful violence, on the other hand, the injury has already been done, and hence the violence has no function of defense. It has the irrational function of undoing magically what has been done realistically. We find revengeful violence in individuals as well as among primitive and civilized groups. In analyzing the irrational nature of this type of violence we can go a step further. The revenge motive is in inverse proportion to the strength and productiveness of a group or of an individual. The impotent and the cripple have only one recourse to restore their self-esteem if it has been shattered by having been injured: to take revenge according to the *lex talionis:* "an eye for an eye." On the other hand the person who lives productively has no, or little, such need. Even if he has been hurt, insulted, injured, the very process of living productively makes him forget the injury of the past. The ability to produce proves to be stronger than the wish for revenge. The truth of this analysis can be easily established by empirical data on the individual and on the social scale. Psychoanalytic material demonstrates that the mature, productive person is less motivated by the desire for revenge than the neurotic person who has difficulties in living independently and fully, and who is often prone to stake his whole existence on the wish for revenge. In severe psychopathology, revenge becomes the dominant aim of his life, since without revenge not only self-esteem, but the sense of self and of identity, threaten to collapse. Similarly we find that in the most backward groups (in the economic or cultural and emotional aspects) the sense of revenge (for example, for a past national defeat) seems to be strongest. Thus the lower middle classes, which are those most deprived in industralized nations, are in many countries the focus of revenge feelings, just as they are the focus of racialist and nationalist feelings. By means of a "projective questionnaire" it would be easy to establish the correlation between the intensity of revenge feelings and economic and cultural impoverishment. More complicated probably is the understanding of revenge among primitive societies. Many primitive societies have intense and even institutionalized feelings and patterns of revenge, and the whole group feels obliged to avenge the injury inflicted on one of its members. It is likely that two factors play a decisive role here. The first is much the same as the one mentioned above: the atmosphere of psychic scarcity which pervades the primitive group and which makes revenge a necessary means of restitution for a loss. The second is narcissism, a phenomenon which is discussed at length in Chapter 4. Suffice it to say here that in view of the intense narcissism with which the primitive group is endowed, any insult to its self-image is so devastating that it will quite naturally arouse intense hostility.

## Violence: A Shattering of Faith

Closely related to revengeful violence is a source of destructiveness which is due to the *shattering of faith* which often occurs in the life of a child. What is meant here by the "shattering of faith"?

A child starts life with faith in goodness, love, justice. The infant has faith in his mother's breasts, in her readiness to cover him when he is cold, to comfort him when he is sick. This faith can be faith in father, mother, in a grandparent, or in any other person close to him; it can be expressed as faith in God. In many individuals this faith is shattered at an early age. The child hears father lying in an important matter; he sees his cowardly fright of mother, ready to betray him (the child) in order to appease her; he witnesses the parents' sexual intercourse, and may experience father as a brutal beast; he is unhappy or frightened, and neither one of the parents, who are allegedly so concerned for him, notices it, or even if he tells them, pays any attention. There are any number of times when the original faith in love, truthfulness, justice of the parents is shattered. Sometimes, in children who are brought up religiously, the loss of faith refers directly to God. A child experiences the death of a little bird he loves, or of a friend, or of a sister, and his faith in God as being good and just is shattered. But it does not make much difference whether it is faith in a person or in God which is shattered. It is always the faith in life, in the possibility of trusting it, of having confidence in it, which is broken. It is of course true that every child goes through a number of disillusionments; but what matters is the sharpness and severity of a particular disappointment. Often this first and crucial experience of shattering of faith takes place at an early age: at four, five, six, or even much earlier, at a period of life about which there is little memory. Often the final shattering of faith takes place at a much later age. Being betrayed by a friend, by a sweetheart, by a teacher, by a religious or political leader in whom one had trust. Seldom is it one single occurrence, but rather a number of small experiences which accumulatively shatter a person's faith. The reactions to such experiences vary. One person may react by losing the dependency on the particular person who has disappointed him, by becoming more independent himself and being able to find new friends, teachers, or loved ones whom he trusts and in whom he has faith. This is the most desirable reaction to early disappointments. In many other instances the outcome is that the person remains skeptical, hopes for a miracle that will restore his faith, tests people, and when disappointed in turn by them tests still others or throws himself into the arms of a powerful authority (the Church, or a political party, or a leader) to regain his faith. Often he overcomes his despair at having lost faith in life by a frantic pursuit of worldly aims—money, power, or prestige.

The reaction which is important in the context of violence is still another one. The deeply deceived and disappointed person can also begin to hate life. If there is nothing and nobody to believe in, if one's faith in goodness and justice has all been a foolish illusion, if life is ruled by the Devil rather than by God—then, indeed, life becomes hateful; one can no longer bear the pain of disappointment. One wishes to prove that life is evil, that men are evil, that oneself is evil. The disappointed believer and lover of life thus will be turned into a cynic and a destroyer. This destructiveness is one of despair; disappointment in life has led to hate of life.

In my clinical experience these deepseated experiences of loss of faith are frequent, and often constitute the most significant *leitmotiv* in the life of a person. The same holds true in social life, where leaders in whom one trusted prove to be evil or incompetent. If the reaction is not one of greater independence, it is often one of cynicism or destructiveness.

While all these forms of violence are still in the service of life realistically, magically, or at least as the result of damage to or disappointment in life, the next form to be discussed, *compensatory violence,* is a more pathological form, even though less drastically so than necrophilia, which is discussed in Chapter 3 of my book *The Heart of Man.*

## Violence as a Substitute

By compensatory violence I understand violence as a *substitute* for productive activity occurring in an impotent person. In order to understand the term "impotence" as it is used here, we must review some preliminary considerations. While man is the object of natural and social forces which rule him, he is at the same time not *only* the object of circumstances. He has the will, the capacity, and the freedom to transform and to change the world—within certain limits. What matters here is not the scope of will and freedom, but the fact that man cannot tolerate absolute passivity. He is driven to make his imprint on the world, to transform and to change, and not only *to be* transformed and changed. This human need is expressed in the early cave drawings, in all the arts, in work, and in sexuality. All these activities are the result of man's capacity to direct his will toward a goal and to sustain his effort until the goal is reached. The capacity to thus use his powers is *potency.* (Sexual potency is only one of the forms of potency.) If for reasons of weakness, anxiety, incompetence, etc., man is not able to *act,* if he is impotent, he suffers; this suffering due to impotence is rooted in the very fact that the human equilibrium has been disturbed, that man cannot accept the state of complete powerlessness without attempting to restore his capacity to act. But can he, and how? One way is to submit to and identify with a person or group having power. By this symbolic participation in another person's life, man has the illusion of acting, when in reality he only submits to and becomes a part of those who act. The other way, and this is the one which interests us most in this context, is man's power to destroy.

To create life is to transcend one's status as a creature that is thrown into life as dice are thrown out of a cup. But to destroy life also means to transcend it and to escape the unbearable suffering of complete passivity. To create life requires certain qualities which the impotent person lacks. To destroy life requires only one quality—the use of force. The impotent man, if he has a pistol, a knife, or a strong arm, can transcend life by destroying it in others or in himself. He thus *takes revenge on life for negating itself to him.* Compensatory violence is precisely that violence which has its roots in and which compensates for impotence. The man who cannot create wants to destroy. In creating and in destroying he transcends his role as a mere creature. Camus expressed this idea succinctly when he had Caligula say: "I live, I kill, I exercise the rapturous power of a destroyer, compared with which the power of a creator is merest child's play." This is the violence of the cripple, of those to whom life has denied the capacity for any positive expression of their specifically human powers. They need to destroy precisely because they are human, since being human means transcending thing-ness.

## Violence is to Control

Closely related to compensatory violence is the drive for complete and absolute control over a living being, animal or man. This drive is the essence of *sadism.* In sadism, as I have pointed out in *Escape from Freedom,* the wish to inflict pain on others is not the essence. All the different forms of sadism which we can observe go back to one essential impulse, namely, to have complete mastery over another person, to make of him a helpless object of our will, to become his god, to do with him as one pleases. To humiliate him, to enslave him, are means toward this end, and the most radical aim is to make him suffer, since there is no greater power over another person than that of forcing him to undergo suffering without his being able to defend himself. The pleasure in complete domination over another person (or other animate creature) is the very essence of the sadistic drive. Another way of formulating the same thought is to say the aim of sadism is to transform a man into a thing, something animate into something inanimate, since by complete and absolute control the living loses one essential quality of life—freedom.

Only if one has fully experienced the intensity and frequency of destructive and sadistic violence in individuals and in masses can one understand that compensatory violence is not something superficial, the result of evil influences, bad habits, and so on. It is a power in man as intense and strong as his wish to live. It is so strong precisely because it constitutes the revolt of life against its being crippled; man has a potential for destructive and sadistic violence because he is human, because he is not a thing, and because he must try to destroy life if he cannot create it. The Colosseum in Rome, in which thousands of impotent people got their greatest pleasure by seeing men devoured by beasts, or killing each other, is the great monument to sadism.

From these considerations follows something else. Compensatory violence is the result of unlived and crippled life, and its necessary result. It can be suppressed by fear of punishment, it can even be deflected by spectacles and amusements of all kinds. Yet it remains as a potential in its full strength, and whenever the suppressing forces weaken, it becomes manifest. The only cure for compensatory destructiveness is the development of the creative potential in man, his capacity to make productive use of his human powers. Only if man ceases to be crippled will he cease to be a destroyer and a sadist, and only conditions in which man can be interested in life can do away with those impulses which make the past and present history of man so shameful. Compensatory violence is not, like reactive violence, in the service of life; it is the pathological *substitute* for life; it indicates the crippling and emptiness of life. But in its very negation of life it still demonstrates man's need to be alive and not to be a cripple.

## Finally: Violence as Blood Lust

There is one last type of violence which needs to be described: *archaic "blood thirst."* This is not the violence of the cripple; it is the blood thirst of the man who is still completely enveloped in his tie to nature. His is a passion for killing as a way to transcend life, inasmuch as he is afraid of moving forward and of being fully human (a choice I shall discuss later). In the man who seeks an answer to life by regressing to the pre-individual state of existence, by becoming like an animal and thus being freed from the burden of reason, *blood* becomes the essence of life; to shed blood is to feel alive, to be strong, to be unique, to be above all others. Killing becomes the great intoxication, the great self-affirmation on the most archaic level. Conversely, to be killed is the only logical alternative to killing. This is the balance of life in the archaic sense: to kill as many as one can, and when one's life is thus satiated with blood, one is ready to be killed. Killing in this sense is not essentially love of death. It is affirmation and transcendence of life on the level of deepest regression. We can observe this thirst for blood in individuals; sometimes in their fantasies or dreams, sometimes in severe mental sickness or in murder. We can observe it in a minority in times of war—international or civil—when the normal social inhibitions have been removed. We observe it in archaic society, in which killing (or being killed) is the polarity which governs life. We can observe this in phenomena like the human sacrifices of the Aztecs, in the blood revenge practiced in places like Montenegro or Corsica, in the role of blood as a sacrifice to God in the Old Testament. One of the most lucid descriptions of this joy of killing is to be found in G. Flaubert's short story *The Legend of St. Julian the Hospitaler*. Flaubert describes a man about whom it is prophesied at birth that he will become a great conqueror and a great saint; he grew up as a normal child until one day he discovered the excitement of killing. At the church services he had observed several times a little mouse scurrying from a hole in the wall; it angered him; he was determined to rid himself of it. "So, having closed the door and having sprinkled some cake crumbs on the altar steps, he posted himself in front of the hole, with a stick in his hand. After a very long time a small pink nose appeared, then the whole mouse. He struck a light blow, and stood aghast over this tiny body which no longer moved. A drop of blood stained the flagstone. He wiped it away quickly with his sleeve, threw the mouse outside and said nothing to anyone." Later, when strangling a bird, "the bird's writhing made his heart thump, filling him with a savage, tumultuous delight." Having experienced the exultation of shedding blood, he became obsessed with killing animals. No animal was too strong or too swift to escape being killed by him. Shedding blood became the utmost affirmation of himself as the one way to transcend all life. For years his only passion and only excitement was killing animals. He returned at night "covered with blood and mud, and reeking with the odor of wild beasts. He became like them." He almost attained the aim of being transformed into an animal, yet being human he could not attain it. A voice told him that he would eventually kill his father and mother. Frightened, he fled his castle, stopped killing animals, and instead became a feared and famous leader of troops. As a reward for one of his greatest victories he was given the hand of an extraordinarily beautiful and loving woman. He stopped being a warrior, settled down with her to what could be a life of bliss—yet he is bored and depressed. One day he began hunting again, but a strange force made his shots impotent. "Then all the animals that he had hunted reappeared and formed a tight circle around him. Some sat on their haunches, others stood erect. Julian, in their midst, was frozen with terror, incapable of the slightest movement." He decided to return to his wife and to his castle; in the meantime his old parents had arrived there and had been given by his wife her own bed; mistaking them for his wife and a lover, he slew them both. When he had attained the depth of regression, the great turn came. He became, indeed, a saint, devoting his life to the poor and the sick, and eventually embracing a leper to give him warmth; "Julian ascended toward the blue expanses, face to face with our Lord Jesus, who bore him to heaven."

Flaubert describes in this story the essence of blood thirst. It is the intoxication with life in its most archaic form; hence a person, after having reached this most archaic level of relatedness to life, can return to the highest level of development, to that of the affirmation of life by his humanity. It is important to see that this thirst for killing, as I observed earlier, is not the same as the love of death, which is described in a later chapter of *The Heart of Man.* Blood is experienced as the essence of life; to shed the blood of another is to fertilize mother earth with what she needs to be fertile. (Compare the Aztec belief in the necessity to shed blood as a condition for the continued functioning of the cosmos, or the story of Cain and Abel.) Even if one's own blood is shed, one fertilizes the earth, and becomes one with her.

It seems that at this level of regression blood is the equivalent of semen; earth is the equivalent of mother-woman. Semen-egg are the expressions of the male-female polarity, a polarity which becomes central only when man has begun to emerge fully from earth, to the point that woman becomes the object of his desire and love. The shedding of blood ends in death; the shedding of semen in birth. But the goal of the first is, like that of the second, the affirmation of life, even though hardly above the level of animal existence. The killer can become the lover if he becomes fully born, if he casts away his tie to earth, and if he overcomes his narcissism. Yet it cannot be denied that if he is unable to do this, his narcissism and his archaic fixation will entrap him in a way of life which is so close to the way of death that the difference between the bloodthirsty man and the lover of death may become hard to distinguish.

# The Gossamer Art of Making Blockbuster Commercials

*Henry Weil*

*HENRY WEIL is a writer based in New York.*

This is a story about how television commercials are made, and most commercial makers didn't want me to tell it. They were afraid I'd report stories like this one: Last year, a Detroit motor company induced Elizabeth Taylor to chat about diamonds in a commercial celebrating its diamond jubilee. Her fee was $750,000 in equipment which her husband, John Warner, could use on his farm—plus a little spending money for herself (no one will say how much that was).

To present Miss Taylor in the most flattering light, the advertising agency, Kenyon and Eckhardt, hired Hollywood director and former fashion photographer Jerry Schatzberg to shoot the commercial. His fee: $14,000 for two days' work. (A top commercial director seldom earns more than $2,500 per day.) In addition, Kenyon and Eckhardt hired makeup specialist Way Bandy to apply Miss Taylor's *maquillage* at a fee of $5,000. Kenyon and Eckhardt also covered the cost of her overnight stay in New York, when she stopped by for a conference and costume fitting. Her twenty-four hours' worth of hotel room, meals, chauffeur, and incidentals came to roughly $900.

Yet two weeks before shooting was to begin, the motor company, disturbed over a rumor that John Warner might go into politics, canceled the commercial to avoid speculation that Miss Taylor's fee was a political payoff. The actress never collected her $750,000 worth of farm equipment, but—because of money committed, time reserved, and guarantees made—it still cost the motor company $44,000 to call the production off.

Advertisers are always upset when such stories appear in print, because consumers complain that products would cost less if the purchase price didn't have to cover such lavish promotional fees. Corporate stockholders may protest, too, pointing out that dividends would be greater if less cash went into advertising, or they may even argue that sales would be higher if the company spent *more* on advertising.

At any rate, the money poured into advertising today is staggering. No one knows exactly how many hundreds of thousands of businesses buy advertising space on TV, but it's estimated that in 1976 the hundred highest-spending advertisers alone laid out $7.7 billion to arouse our desire to buy. (The most lavish advertiser in America is Procter and Gamble, which doled out $445 million to promote its stable of products.) In 1976, advertising agencies—which suggest to manufacturers ways to advertise their products, then prepare and place their ads, and finally analyze what impact these ads have had on consumers—earned $2.5 billion for their services.

Most of these agency enterprises were, of course, kept strictly under wraps. No one wants to talk about the fierce differences of opinion within agencies, the calculated ways presentations are designed to deemphasize a product's flaws, or what vast sums of money are spent researching consumer opinions. No company wants to air details of how much it pays to hire the famous voices and faces which urge us to buy—a considerable expense for advertisers, a windfall for the "face."

Example: A few years back, when superstar baseball pitcher Vida Blue was still an Oakland Athletic rookie earning somewhere around $12,000 a year, the manufacturer of "ice blue" (get it?) Aqua Velva paid him $25,000 for one day's work shooting a television commercial—more than twice his annual income as a player. Another example: Dick Wilson, once a serious actor, now earns around $10,000 yearly as Charmin's Mr. Whipple, the toilet-paper squeezer. Wilson spends five days a year shooting maybe twenty commercials; then the rest of the year is a vacation. Most of the time, advertisers will opt for actors rather than celebrities because they have fewer problems. If an unknown actor made the commercial and then got hit by a truck, for instance, you could still use the commercial, whereas some famous and popular figure hawking toothpaste a few days after his funeral had been covered by TV news wouldn't do at all. This very circumstance arose last year with Guy Lombardo, who did a commercial for Taylor Wine just before entering a Houston hospital for a heart ailment—from which he never recovered. Taylor Wine had to spend roughly $40,000 to reshoot the commercial, using Count Basie.

Because of all these subtleties, I was politely but firmly turned down by nearly every agency I approached when I went hunting for commercials-in-the-making to profile. Then, at last, I found two brave agencies representing two courageous clients—willing to open their doors to the press. The agencies turned out, interestingly, to be one of the largest and one of the smallest around. Both were planning ad campaigns which were happily shaping up without squabbles in the agencies, the clients' executive suites, or in the production companies hired to film the commericals. Such confident, unified fronts are rare in advertising, and when I luckily contacted these two agencies just when they were most proud of their smooth operations, I was graciously invited to come watch them at work . . .

## The Commercial as a Bright Idea

First, let me introduce the two agencies and their clients: Epstein, Raboy Advertising (ERA), representing Long John Silver's Sea Food Shoppes, and Young and Rubicam (Y&R), representing Dr. Pepper. ERA has only one office with a staff of twenty-six and places roughly $12.5 million in ads annually; its Long John Silver's ads were the first the agency had ever produced for network television. Y&R has fifty-four offices around the world (its New York base takes up most of a midtown skyscraper), and a staff numbering 3,980; it places just under $1 billion worth of ads annually and represents dozens of clients as familiar to any TV watcher as the knobs on her dial—General Foods, Eastern Airlines, Chrysler.

ERA's assignment was to create a set of three commercials (one running sixty seconds, two running thirty) to introduce Long John Silver's Sea Food Shoppes to the nation. There are 1,000 such shops now (up from 285 in 1975), and though ERA ads for the fast-food chain had run on certain local stations, 60 percent of the country had not yet been treated to a pitch on the joys of Silver's seafood.

## Into Production

In the advertising business, the actual work of putting commercials on film is normally done outside the agency. Music is composed, arranged for orchestra and singers, and recorded for sound tracks by one production company. Filming is done by a second, while a third takes care of editing, printing needed on film, and making sound precisely match action. Production houses (mostly in New York and Los Angeles) are asked to submit prices at which they'll work to an agency's specifications, and the lowest submitted price—provided it's from a reputable production house—usually gets the job.

ERA solicited bids in the same manner, but because their campaign was mostly a reworking of an old idea the lowest bids came from music, film, and editing houses which had worked on the *previous* campaign. Naturally, these companies didn't anticipate complications and didn't have to budget for them. Also, because of everyone's familiarity with ERA's concept, the campaign only took one month (from mid-July to mid-August) to produce, from corporate approval to final film prints. Y&R's campaign took longer, from mid-May to mid-October, because while Long John Silver's campaign could be filmed in two days, Dr Peppers' took several weeks. The planning was much more complex.

Once production companies were chosen, agency representatives met in preproduction conferences with craftsmen assigned to put the commercials on film. Everyone at such meetings always works from "story boards," which are cartoon breakdowns, shot by shot, of what will be seen on the screen accompanying the dialogue. All concerned try hard to thrash out in advance all the problems which could arise. Example: At one of ERA's round-table discussions, the film director, Alan Brooks, asked ERA to talk to police at Bethpage, Long Island, where their commercials were to be shot, to ask the cops to hold back traffic during outdoor photography in order to keep truck growls off the sound track. (The cops did.) Later, there was worried discussion about whether the pirates ought to be eating along with the customers. (The pirates didn't.) ERA's creative director, Richard H. Raboy, also asked Brooks to avoid filming such areas as the restaurant's w.c. doors and its posted price list (prices could go up before the campaign was retired).

Many of Y&R's preproduction meetings were less formal. The campaign's Pied Pepper portion was filmed mostly in California and New Orleans, and the vignettes—scheduled to be filmed later in New York—were discussed at random moments while work was going on. And by the time Y&R's creative personnel and production crew returned to New York, many preproduction confusions in the vignettes had been straightened out. Still, I was at one preproduction meeting, three days before New York shooting began, at which everyone suddenly realized that the story boards called for some characters to sing "I'm a Pepper" and to *drink* Dr Pepper at the same time. The song was promptly rewritten—"He's/She's a Pepper"—and assigned to different characters to sing.

Casting was quickly accomplished for both agencies. ERA rehired the four actors who had played pirates before and hired blandly handsome types for the restaurant diners, choosing them from a television tape made at director Brooks's office. One by one, attractive young actors culled from Brooks's file had trooped in, faced a camera, recited their names, and smiled, usually uncomfortably, not knowing what was wanted of them. (Answer: a pleasant, unobtrusive face, not identifiable from other commercials.)

Y&R used a casting agent for its dozen vignettes. Each vignette called for a screenful of different actors, and their auditions were more complicated than ERA's. They were asked to improvise a dance to nonexistent music, sing (anything), and take a sip of Dr Pepper (so they wouldn't look askance—at a crucial moment in the future—as if to say, "Is *that* how Dr Pepper tastes?"). Each audition was conducted in a tiny, white-painted room in front of a small television camera, and many of the actors had to audition a second time as casting was narrowed. All choices were made from television videotapes based on how the actors looked on a nineteen-inch screen.

Casting tends to be a heady experience for advertising people.

Thousands of eager actors in New York and Los Angeles are hungry for the money a commercial pays (they get a "residual" check each time the commercial is shown on television anywhere, though they need to do several commercials a year to make a substantial income); and the imperial process of browsing and choosing among anxious actors is inevitably ego-inflating, even when done by committee, as at both these agencies.

Casting is, simultaneously, rather humiliating for actors who go through dozens of auditions without ever being hired—and without ever knowing *why* they were rejected. The reason is usually their "look"—age, color, degree of prettinesss—about which they can can do little. Most actors effect an attitude—winsome, seductive, Joe College—hoping their style will help land the job. It rarely does. Either they match an agency's preconceptions or they don't. Talent (or lack of it) is usually of secondary importance.

ERA's pirates wore the same buccaneer garb they had worn in earlier commercials, the Y&R production house provided costumes for their vignettes where necessary—an apron for a hot-dog vendor, uniforms for a basketball team. Most actors in both campaigns wore their own clothes (seen and approved first by the production company and sometimes also by the agency).

Another important element in production to be dealt with before shooting begins is the musical sound track, as music for commercials is usually recorded a day or two before shooting begins. Music is, of course, handled as painstakingly as any other production element. One of the three Long John Silver's commercials featured a long song, and another ended with a brief mixed chorus of pirates and customers, both prerecorded so that the actors on location could mouth their words in sync with the music (played for them over a portable loudspeaker while cameras rolled).

The Dr Pepper vignettes were trickier. Singing and speaking voices of principal actors were to be recorded on location, but all orchestral accompaniment, plus the chorus at the end of each vignette, were *prerecorded*. However, the loudspeakers on location had to be turned off whenever live actors' voices were being recorded. Consequently, speaking actors had to be certain that when they burst into "I'm a Pepper" at the end of their brief dialogue, it was at the precise instant that the prerecorded tape *also* broke into appropriate accompaniment. Keeping rhythm and pitch for the music while they were *talking* was a difficult challenge for the performers, and they required many "takes" before catching on.

The recording of music for commercials is done in drab, windowless soundproof studies scattered around Manhattan and Los Angeles. The music is recorded on two-inch-wide reels of tape capable of handling forty separate tracks of music at once, a bit of electronic wizardry that makes it possible for each instrument in a twenty-eight-piece orchestra and each singer in a chorus of twelve to be recorded separately. In the sound tracks for both ERA's and Y&R's commercials, rhythm and brass instruments were recorded first (each instrument playing into a different microphone), followed by special-effects instruments (calliopes, soprano recorders, accordions), followed by strings (for Dr Pepper), followed by a chorus of four singers, who were rerecorded in places to sound like a chorus of eight. Singers and musicians are recorded on separate tracks, and later balanced without their actually being present. This is done as a matter of economy, since musicians and singers are paid by the hour, and it would be quite costly to have them go over material again and again until it is properly balanced. Then, too, the more singers or instruments used in a final sound track, the more money must be paid every period of time the commercial is *shown*. If, after a recording session is over, the sound engineer finds that certain voices and instruments can be eliminated without hurting quality, residuals can also be eliminated.

## Perils and Pleasures of Shooting on Location

At any rate, ERA took its prerecorded tape on location to an immaculate Long John Silver's Sea Food Shoppe at Bethpage, Long Island. (Both campaigns were filmed "on location." Commercials shot inside studios usually look less colorful.) The actors were made up and costumed in Manhattan—starting at the leaden hour of 4:45 a.m.—then driven to Bethpage on a large silver bus. Film crew and agency personnel arrived about 7:00 and immediately began setting up on the restaurant's lawn, assembling cameras, lights, microphones, and recording paraphernalia. Everyone, still sodden with sleep, moved slowly and deliberately. Union rules require food on a film set, and when a large urn of coffee, cartons of orange juice, and two large boxes filled with bagels, Danish, and sliced buttered rolls appeared (paid for by the production company and supplied by an outfit called "Location Catering Service"), a suddenly energetic cast and crew formed a long and hungry line.

By 7:30, the restaurant manager had turned his air conditioning up to an arctic roar in anticipation of the heat from the lights that would soon be brought inside. He also began cooking food for the restaurant's display case for shots in which the actors would be eating. Jerrico, Inc., the parent company of Long John Silver's Food Shoppes, sent an executive to the location, one of whose talents was the ability to cook photogenic food, and eventually she became chief cook. Making photographed food look appetizing is extraordinarily difficult (coleslaw washes out; French fries look greasy), and Jerrico was countng on its artistic chef to work visual wonders.

Outside, the camera was rolled into position (not too far to the left, where garbage cans and a Pizza Hut would be in view, nor to the right, with a busy street and ugly vacant lot to contend with). At 8:00, shooting began.

By then, a small crowd had gathered—children on bicycles,

young mothers with strollers—and traffic had slowed enough for truck-drivers to shout wisecracks over the roar of their engines. Despite the efforts of the police to stop traffic during each take, one gawking driver managed to slam into the back of another car.

A script girl with a stopwatch timed every shot and shouted to the director how close each shot came to its ideal duration: "Half over." "Quarter under." (That's *a quarter of a second!* You have no time to waste in a thirty-second commercial.) After each take, the director would call a comment to the actors. ("You played too much to the camera." Or, "This time you turned too far away; I couldn't see your face." And, after one pirate impulsively picked up a six-year-old actor playing a customer: "I like it. Leave it in.") In all, the director made thirteen takes of the first scene, among which were several that seemed acceptable, and at 8:25 he announced he was satisfied, and it was time to go on to the next scene.

By 10:30, each of the outdoor scenes had been completed and all of the equipment moved inside. Filming indoors was distinctly awkward, as the dining area was small and low-ceilinged, with little room for the camera. For one shot, the camera had to be nailed to a wooden plank balanced between serving counter and a partition; when the director climbed up to look through the camera's viewfinder, his head touched the ceiling.

Getting the camera into position took more than an hour, and before indoor filming could begin, the crew broke for lunch. Union rules require that a hot lunch be served, so again good old Location Catering Service pulled into the restaurant's parking lot in a mobile lunch wagon. The catering service brought chicken, roast beef, rice, mixed vegetables, tossed salad, tuna salad, bread and butter, two kinds of cake, beer, coffee, and soft drinks, and we all picnicked in the parking lot. Meanwhile, a countergirl from the restaurant's staff had been given the job of turning away customers, assuaging them with tickets good for a later free meal. (The outlet lost about $2,000 over two days of shooting—a slow Monday and Tuesday—which the parent company, Jerrico, Inc., later reimbursed.)

The two days of shooting crawled by. Actors would sprawl and doze while the crew set up a shot; then the crew would sprawl and doze while a shot was rehearsed. Professionals all, everyone took the tedium in stride. Shots were spoiled when bites of fish fell off plastic forks, when a spilled Coke inundated a platter of fish, when a bright script girl noticed that an actress, playing a mother, was wearing no wedding ring. (One was immediately borrowed from her "son's" *real* mother.)

The script called for a shot of the actress-mother eating, and before each take a makeup girl touched up the actress's hair and reapplied her lipstick, while a prop girl placed a fresh piece of fish in front of her. After each take, the actress daintily spit her mouthful of fish into a paper napkin. ("Why are you doing that?" asked the six-year-old playing her son. "Because I'm not hungry," she explained sweetly.) Similarly, just before the actor-father was filmed close-up cutting his fish, he was given a manicure.

The director called a halt just before 6:00 that evening, and shooting began again shortly after 7:00 the next morning, continuing the second day again until dinnertime, when director Brooks declared the Long John Silver's filming finished.

Y&R, meanwhile, was shooting the Dr Pepper vignettes over a period of two weeks, with locations ranging from a scruffy Manhattan gym to a Long Island beach, to a shopping mall, to an elegant Manhattan penthouse. Often, locations were chosen a matter of hours before shooting began, from polaroid pictures snapped by the production company's advance scout. Y&R's hours were long: On one horrendous day, shooting began after breakfast and proceeded to four different locations, ending at 3:45 the next morning.

Where ERA's production company would shoot the same scene a dozen times, Y&R's averaged two dozen. Each setup was more delicately considered, carefully composed, and freely manipulated during shooting. After maybe half a dozen shootings of a scene, film director Melvin Sokolsky would suggest a different gesture to an actor, and a new series of "takes" would be made to include the improvement. Once, a *real* Dr Pepper truck was recruited. In keeping with the proposed concept that Dr Pepper has become omnipresent, locations had been "dressed" with Dr Pepper signs and dispensers (laboriously hauled in for shooting and carted away afterward), so when a genuine Dr Pepper truck was spotted making a delivery nearby, the driver was soon talked into parking his truck within camera range. (He was delighted—he'd be on television!—but first he had to clear this brash departure from routine with his boss and his union supervisor, whom he rushed to call from a pay phone in a nearby pizzeria!) Each scene took four or five hours to set up, rehearse, film, and dismantle—much more time than was taken by any ERA shot. There was, however, a *reason* for Y&R's more pronounced meticulousness.

Soft-drink advertising is *always* elaborate and extravagant, because soft drinks (unlike scouring powder, say) solve no inherent *problem* for customers. ("Thirst" is considered too vague for a "problem.") Advertisers must, therefore, try to establish an image which will plant itself in the viewer's memory. Research has shown repeatedly that since consumers don't watch a problem being solved in a soft-drink commercial (a malfunctioning muffler replaced or mounds of photogenic filth removed from some grateful housewife's rug by a spiffy little vacuum cleaner), they forget it almost immediately.

Consequently, Coke, Pepsi, and Dr Pepper are well known within the industry for trying to lick "memory" problems by regularly producing the most lavish and expensive commercials going. Images presented in their commercials are filmed with "maximum elegance" in hopes that viewers will be impressed with the quality of the *ad,* since no known way exists to impress them, on film, with the quality of the *product.* So Y&R's commercials were filmed with utmost care. Dozens of takes were made before the director was willing to abandon a location. It wasn't enough to have acceptable film in the camera—all shots had to be flawless.

## From Camera to Your TV Screen

The footage for each campaign took hours to sit through—and looked surprisingly like home movies. The actors would be seen lolling carelessly, waiting for the cue "Action," after which they cavorted for the camera, then relaxed again—just like a family cutting up in front of Dad's camera. Also, tiny mishaps which seemed insignificant at the shooting were blown up to tremendous importance simply by being framed: a background actor moving so vigorously that he stole the scene from one of the speakers, a suspicious pause before an actress bit into a piece of fish.

At each agency, the film was reviewed by a committee: agency representatives, people from the film production house, and the film editor. Everyone voted on which takes seemed best; then the editor took the film and tried to assemble choice shots into a smooth-flowing commercial.

In the Dr Pepper commercials and in the musical Long John Silver's commercial, shots matching the prerecorded sound fell effortlessly into place. When the sixty-second *spoken* Long John Silver's commercial was edited, however, the film ran four and a half seconds too long. The editor tried to remedy this disaster by trimming away every frame before a pirate began speaking and every frame after he stopped, but he only reduced the time by about two seconds. Finally, one shot filmed for the thirty-second spoken commercial—which was similar to a shot in the sixty-second commercial but used fewer words—was cleverly spliced into the sixty-second commercial. The offending two and a half seconds disappeared.

One Dr Pepper vignette at a cocktail party ("Are you a Pisces?" "No, I'm a Pepper"), which had looked snappy on paper, turned out to be less appealing on film. Though planned as an opening vignette, it was moved into the middle group, necessitating a whole new musical sound track. In all the commercials, new voices were added here and there to cover a few less-than-mellifluous sounds accidentally picked up on location.

Background music was added to the spoken Long John Silver's commercials, and title cards were printed and superimposed by complex photographic processes on the finished film. Finally, several "master prints" were made of each commercial, and the ones with best color balance (not all color prints from a single negative come out the same) were put on videotape cassettes to be used by the networks.

To get a commercial seen on network television, an agency or its client must "buy time." A representative goes to one or more networks, explains how much his client is willing to spend on television advertising, and the network proposes several different "packages" of programs in a variety of time slots. Each package extends over a period of three months, more or less (time slots remain "negotiable" in case the sponsor objects to a show for some reason). Most sales of time are made eight to ten months in advance. A sponsor also may buy time on individual "special" or sports events if he likes, or even initiate a special himself ("Dr Pepper presents . . ."). Agency researchers compile precise figures gauging the "reach" (percent of those viewers the sponsor wants to reach who are actually watching the show) of various shows and specials, so time-buying decisions are based on a crafty analysis of communicating potential.

The cost of producing the Long John Silver's commercials came to just under $73,000, the Dr Pepper commercials nearly $1 million. If you were to produce a movie on a similar cost-per-minute basis, a two-hour film at Long John Silver's rate would cost $4,375,000 (a reasonably generous Hollywood budget) and, at Dr Pepper's rate, would cost more than $18 million (lavish by anyone's book). By Madison Avenue's standards, Long John Silver's per-minute costs were low and Dr Pepper's were high. The *average* commercial cost-per-minute would cover making a movie extravaganza on the order of *Star Wars.*

The commercials produced for both these campaigns were undeniably professional and presentable, yet I felt that neither was especially original. Such personal opinions, however, are beside the point. The *point* is to increase sales of a product. If commercials are brilliantly entertaining, dandy; but if, after seeing them, we still aren't convinced to rush out and buy a certain deodorant, to shampoo with product X, or feed the kids So-and-So's cereal for breakfast, the pitch has failed.

Were Silver's and Pepper's commercials a success? You can answer that question yourself. Did you see them? Did they make you want to eat at Long John Silver's Sea Food Shoppe or drink a Dr Pepper? Now you know.

# In the Land of the One-Eyed Monster The Two-Eyed Parent Rules

*Aimee Dorr*

*AIMEE DORR is Associate Professor of Psychology at the Annenberg School of Communications, University of Southern California.*

Have you ever watched your child—as I have mine—try to kick down the door with his bionic leg one day and then another make a waterwheel like he saw a Zoomguest make? Have you then wondered how you could make television more often work *for* you as a parent rather than against you? As the parent of a nine-year-old and a new baby and a research psychologist interested in the effects of television on children, I have often encountered this question. I would like to try to answer it here. My suggestions come from research studies of children, parents, and television. They apply to all the different kinds of programs children watch on television. Naturally, most of you already have some techniques that work for you and your children, but I hope I can add a few others.

I think there are good reasons to try to make television work for you. There is ample research evidence that children as young as two can and do learn many things from all kinds of television programs. They may often keep this learning to themselves, but under the right circumstances you will have clear evidence that they have learned from television. In many cases you will be delighted with what they have learned, but in many others you will wish that they hadn't learned it or that they understood that they shouldn't perform what they learn. My goal is to help you maximize learning and performance of the things that fit your own values and to minimize learning and performance of everything else.

In addition to helping children get the best possible from their television viewing, many parents want to decrease the total amount of time their children spend with television. This is because they, like many authorities on child development, worry that time given to television is time lost from other important activities such as reading, hobbies or playing outdoors. Even activities such as arguing with brothers and sisters or friends may do more than watching television does to help a child learn how to reason, talk, express emotion, understand someone else's point of view, and resolve disagreements. We do not as yet have much evidence about the effects on children's development of the amount of time they spend with television. All I can do right now is tell you that people are worried about it and suggest some ways in which you—if you too are worried about it—might reduce the time your children spend with television.

> **"What is right for you and your family is what you must choose to do."**

## Know What's On Television

It seems to me that your first goal as a parent concerned about television is to become familiar with all of the programs which your children could watch. Unless you know what is on television, you cannot make decisions about what your children should not watch, what you should watch with them, what they can watch alone, and what you should encourage them to watch. I know it is not always easy or pleasant to watch children's programs in the early morning and late afternoon, but at least once a year—probably in the fall when the new season begins—you should watch at these times and all the others when you would allow your children to view. Armed with the knowledge of what is routinely available on television and how it matches your own values, you can make wiser choices about your children's television viewing. If you take the extra time each week to read *TV Guide* or the television supplement to the Sunday paper, you can spice their regular viewing with the specials which appear so often and can add so much to a child's knowledge.

## Control Viewing

Once you know what is on television and how well it matches your own values, you can begin to regulate your children's viewing. This can be made easier by judicious placement of the television set—or by moving it around as the situation dictates. Young children generally prefer to be with other family members so you can decrease their television viewing time by putting it out of sight in a place where they would be alone when viewing. I put mine on a shelf in an upstairs bedroom closet one year. Older children and adolescents generally prefer to be alone when they watch, so you may be able to decrease their viewing time—as well as keep better track of what they do view—by putting the television someplace where family members are likely to be most of the time.

You may choose to control your children's viewing on the basis of the total number of hours they watch, the specific times they watch, the programs they watch, or with whom they watch. I think there are good reasons for using all of these criteria in deciding on your children's viewing habits. If you believe, as I do, that children learn both from their own activity and from watching television, then as a parent you have the right and the responsibility to control how much and what they watch. If you decide they should not watch television or a particular program, you can make it easier for both of you by explaining why you do not want them to watch, by actively providing them with some engaging thing to do, or by suggesting other programs they may watch. Above all, be absolutely consistent in your regulation. I always remind myself that if it is really important that my child not watch, then it is worth the extra effort it takes to help him accept my decision easily.

Encouraging children to watch programs is usually easier. This can be done by simply suggesting a program, by giving a preview of the appealing parts of it, by turning the television or the program on, or by making a program the only one that can be watched at that time. Your own enthusiasm for television or a particular program will often make a difference to your children. You can show your enthusiasm by talking about programs and by watching with your children. We know that how much children watch and what they watch is related to how much and what their parents watch. So, in all decisions you make about your children's viewing you should remember that what you do yourselves also influences very much what your children do.

## Talk About Television

Discussing television with your children is obviously a good way to find out what they like to watch, what their friends are watching, and what they understand about what they watch. It is also a good way to help them think more realistically about the meaning of the programs they watch. My own research suggests that children who can talk to us more easily about television, particularly about what is real and pretend on television and how they decide this, are generally more able to resist undue influence from entertainment programs. While you are talking about television it is also easy to explain your own values and the reasons for them. This is important because there is good evidence that children from families who clearly communicate their own values generally grow up to be more desirable people. Moreover children who reacted in more mature ways to television in research I have conducted more often told us that their information came from their parents.

There are many magazines and toys which are based upon television programs. You may wish to use some of these, rather like props, when you talk about television with your children. Programs like *Sesame Street* and *The Electric Company* have monthly magazines you can subscribe to. Zoom has an activity book you can buy and activity cards your child can send in for. Many programs have toys based on their characters or actions. Some of them would undoubtedly encourage children to practice the things you don't want them to do, but many others can be easy objects around which to center a productive conversation.

## Watch Television With Your Children

There are many things to be gained from watching television with your children. Looking at their reactions to a program and asking them a few judicious questions are good ways of finding out how they react to a program and what sense they make of it. What they don't understand you can then explain. It is also a good way for you to find out exactly what ideas your children *could* get from the program. If you approve of these ideas, you can encourage them to watch again, and if you don't approve, you can avoid the program in the future.

It also helps to tell your children what you do and do not approve of in the program. Researchers have found that children will take your comments seriously. When you are with them, preschoolers will be more likely to do the things you have approved of and refrain from doing the things you have disapproved of. Somewhat older children will even modify their behavior according to your comments when you are not around to watch them.

If you watch television with your children, you can also see what the program is teaching and help your children to learn it. You can do this by directing their attention to the important parts of the program, by encouraging them to guess the answers to questions asked in the program and to repeat words or actions presented in it, by elaborating on the information presented, or by finding ways for them to use the information in their other activities. We have lots of evidence that children learn more when they actively practice what they see on television, and you can find many ways to help them do this.

Finally, when you take the time to watch and discuss with your children, you demonstrate that their television viewing experiences are important to you. This is a tangible way for you to say to your children that you care about them, that what they watch is important to you, that you want them to be able to understand what they are watching, and that your own values are every bit as important as those presented on television.

## Teach About Television

While you are watching television or talking with your children about television, there are some ideas and skills you should try to teach them. Most of the children, and even many of the parents I have worked with understand very little about how a television program is made and why it is broadcast. You can help your children make more accurate evaluations of television programming by teaching them that nearly all of the programs they see are made up by other people. By the time children are five they are interested in exactly how programs are made. You can demystify television by explaining such things as how cartoons are made, how they make the bionic woman look so strong, how the set for *All in the Family* is constructed, and how television casts learn lines written by other people.

Most children and many adolescents do not understand the economic motivations for the production and broadcasting of most entertainment programming. Many believe that most such programs are designed to teach children things rather than understanding that the goal is usually to attract large audiences which in turn create large advertising revenues for the station. We have found that children and adolescents who do understand these facts about American television *and* use them in thinking about the accuracy of the programs they watch are more likely to resist undue influence from such programs.

You can also help your children develop skills for evaluating television content, in terms of both its accuracy and its congruence with your family's values. Encourage them to compare what they see to what they themselves have experienced. Encourage them to talk to you about the things they question or do not understand. Give them as much information as you can about your understanding of our society. Show them that they can turn to books, magazines, informational programming, and other authorities to get another perspective. These skills will help your children to evaluate television—and every other information source they encounter—more accurately.

## Focus on Family

The suggestions I have made are not all appropriate for children of every age. It seems to me that younger children need more control of their viewing, more viewing with parents, and more teaching. If you begin when your children are very young, you will help establish viewing patterns you approve of and your children are more likely to understand that you have standards for their television viewing just as you do for other aspects of their behavior. As they get older, children need and expect more independence. If your children can talk with you, understand how and why television programs are produced and broadcast, and have the skills for evaluating television content, then you can feel more secure in giving them greater freedom to choose what and when they will watch.

It is hard to learn from American television that there is great variety among the people in our society. This you can teach your children and remember as you decide how to make television work for you as a parent. Each family and each child within that family has its own values and interests, strengths and weaknesses. What is right for you and your children is what you must choose to do. Perhaps you will include many of the things I have suggested, and perhaps you won't. Whatever you decide, remember that television viewing is only one of many experiences your children will have as they grow up. A strong family, guided by sensitive, thoughtful people, is the best insurance that children will grow into kind, intelligent adults and that television viewing will be a positive contribution to this growth.

### *REFERENCES*

Brown, L. *Television: The business behind the box.* New York: Harcourt, Brace and Jovanovich, 1971.

Brown, R. (Ed.) *Children and television.* Beverly Hills California: Sage Publications, 1976.

Frank. J. *Television: How to use it wisely with children.* New York: The Child Study Association of America, 1969. (9 East 89th Street, New York, New York 10028.)

Howe, M. J. A. *Television and children.* Hamden, Connecticut. Linnet Books, 1977.

Kaye, E. *The family guide to children's television.* New York: Pantheon Books, 1974.

Kuhns, W. *Exploring television.* Chicago: Loyola University Press, 1971.

Lesser, G. S. *Children and television: Lessons from Sesame Street.* New York: Random House, 1974.

Liebert, R. M., Neale, J. M., and Davidson, E. S. *The early window: Effects of television on children and youth.* New York: Pergamon Press, 1973.

Littell, J. F. (Ed.). *Coping with the mass media.* Evanston, Illinois: McDougal, Littell, and Company, 1972.

Littell, J. F. (Ed.). *Coping with television.* Evanston, Illinois: McDougal, Littell and Company, 1973.

Melody, W. *Children's television: The economics of exploitation.* New Haven: Yale University Press, 1973.

Schrank, J. *TV action book.* Evanston, Illinois: McDougal, Littell and Company, 1974.

Shayon, R. L. *The crowd catchers: Introducing television.* New York: Saturday Review Press, 1973.

*Teacher's guides to television.* P. O. Box 564, Lenox Hill Station, New York, New York 10021.

Valdes, J., and Crow, J. *The media works.* Dayton, Ohio: Pflaum/Standard, 1973.

Winn, M. *The plug-in drug: Television, children, and the family.* New York: Viking Press, 1977.

# Television and The Child Under Six

*Dorothy H. Cohen*

*The late DR. DOROTHY H. COHEN was a member of the senior faculty, Graduate Programs, Bank Street College of Education. She wrote* The Learning Child, Kindergarten and Early Schooling *as well as other books and articles.*

*EDITOR'S NOTE: Dorothy Cohen died as this book was in press. Her article, commissioned in the late winter, is one of her last pieces of completed writing.*

The elasticity of the human species has been its outstanding characteristic, giving humankind its capacity both to adapt to an existing environment and to create environments to suit developing needs. With the intense, dramatic and seemingly endless creations of new, man-made environments since the end of World War II, thinkers like Rene DuBois and Jerome Bruner, among others, have raised the question of whether the species is indeed endlessly adaptable (or whether it is the environment that is now altering human beings.)

The question has special relevance to the development of the young, who are particularly vulnerable in any species. Although biological patterns follow certain directions according to genetic endowment both for the species and for individuals, such endowment takes form and shape only in interaction with environmental elements, and when used. Thus there is no "unfolding" without environmental support; similarly, the absence of environmental support (or its inadequacy), leads to no growth, or poor growth at best.

Thus the question whether dramatic changes in the environment can affect development is not an unreasonable question when applied to young children. Nor is the direction of that question toward the effect of television as a major environmental element in the lives of children an argument against other factors of equal importance, so much as a focus on one important aspect of environmental change and influence.

Long before the advent of any organized study of children, apparent agreement emerged across the centuries from literature, philosophy and sermons admonishing parents as to the nature of young children's behavior. Whether this behavior was subject to positive or negative appraisal, young children were always observed to be endlessly active, endlessly curious, and endlessly absorbed in non-adult kinds of interest. In the last century, carefully delineated long and short range studies of infants, toddlers, and young children have confirmed the activity and curiosity that spring forth without teaching. This more sophisticated knowledge blueprints such activity as a force for growth and life that impels the young organism to participate in activities that support its development. For example, at the basic survival level there is the cry of hunger, the demand for relief from pain and the appeals for loving. At a later level, motor patterns are learned simultaneously with discoveries about soft and hard, wet and dry, hot and cold, under and over, two hands or one, push or pull, shapes, smells, tastes.

## Sensory Learning

All during early childhood there is sensory outreach for feedback which results in a construction of reality which grows out of an individual's action and experience. The insatiable search for feedback around which to construct an image of the world (which must in time share some agreement with others) has, until our era, encountered clearly human and clearly non-human environments which young children had to disentangle. Never before was that task confused by long hours of contact with a nonhuman environment that simulates humanness as it does in our time. And therein lies the problem of television and the young child.

Piaget[1] has uncovered for us much we may have forgotten about our own early experiences, but which takes on a déjà vu quality when we read his work. He describes how the young child endows all things with life—"animism"—and interprets all things as made by people and for the express purpose of serving people—"artificialism." From the psychoanalytic backlog of experience we learn of the difficulty young children have separating fantasy from reality, a not too dissimilar finding on the emotional level from Piaget's on the cognitive.

The inevitability of early childhood activity, curiosity, animism, artificialism, confusion between fantasy and reality and the need for interaction are all part of a stage of development. Resolution of the confusions and conflicts incurred in the processes indicated are tied in with language development as well as with physical, social and intellectual growth.

Basic to the young child's functioning is the egocentrism that sees the world and its events as revolving around the child himself. "It's raining outside," says the three-year old, "because I am thirsty." Egocentric views of the world do not disappear lightly or easily. But in early childhood, largely through children's contact with each other, and especially through their play with each other, they learn gradually that there is another view than their subjective one. It is this experience with people, for which there is no verbal or vicarious substitute that works, that stretches a child's mind toward larger horizons. Piaget insists that it is interaction with others in early childhood that leads to the more objective thinking associated with cognitive maturity.

The egocentric child learns a great deal more, however, than a base for objective thinking. From the social-emotional perspective, Erikson[2] speaks of autonomy and initiative as emerging characteristics of personality in early childhood. By autonomy, Erikson means the internal sense of "I am the captain of my ship", a feeling the toddler may somewhat arrogantly misinterpret as "I am the captain of the whole world", but which for the preschool period and ever afterwards is the basis for independent, self-propelled behavior. As the child comes closer to four, five and six, the urge to undertake projects of independent character becomes part of his/her functioning. Children develop initiative, a type of functioning which demands sufficient independence from parental or other adult direction to let imagination and originality play their part in the child's actions and consequence learning. To many an adult this is a period of mischief, but to the children themselves their activity springs from innocence tied to an experimental urge that will not be denied.

## Play

Other important developments also take place during early childhood. One of these, which seems to emerge quite spontaneously across cultures, is the children's capacity to involve themselves in activity which is symbolic in nature, i.e. play. Apparently, unless tied to work too early, all young children play; a portion of their play includes symbolic activity.

A look at the biological underpinnings for the importance of play in childhood reveals a growing consensus among biologists that play among young primates which include monkeys, apes, and humans, is preparation for adult life. It is Harlow's[3] contention, for example, that play provides the behavioral mechanisms by which activities for adult functioning can be imitated and perfected. In fact, Harlow's well-known studies of monkeys led him to conclude that "no play makes for a socially very disturbed monkey."

The behavioral mechanisms that Harlow refers to include, among human children, the capacity for symbolization. This capacity, given the opportunity, develops in an orderly course. A recent study of play by Virginia Stern of Bank Street College, measured this trend among children aged three to five.[4] As though in pursuit of an instinctual impetus to fulfill a biological inheritance, the increase from solitary to group play went from 47% to 71% during ages three, four and five.

**"The active child learns to be unnaturally passive."**

It is, therefore, probably not happenstance that the capacity for symbolization, which is unique in its complex form among human beings—and without which we could not learn to read and handle numbers—is both a first requisite capacity for play, and increases as an important characteristic and ability during play. It is worth repeating here a long-ago comment by Susan Isaacs that the more complex the animal along the evolutionary scale, the longer is the childhood and the longer its period of play. To this observation, as we go on to consider the phenomenon of television, we might add a well-known comment by Piaget:

> It appears that many educators, believing themselves to be applying my psychological principles, limit themselves to showing the objects without having the children manipulate them, or, worse still, simply present audio-visual representations of objects (pictures, films, and so on) in the erroneous belief that the mere fact of perceiving the objects and their transformations will be equivalent to direct action of the learner in the experience. The latter is a grave error, since action is only instructive when it involves the spontaneous participation of the child himself, with all the tentative gropings and apparent waste of time that such involvement implies. It is absolutely necessary that learners have at their disposal concrete material experiences (and not merely pictures) and that they form their own hypotheses and verify them (or not verify them) themselves through their own active manipulations. The observed activities of others, including those of the teachers, are not formative of new organizations in the child.[5]

Language is an early symbolic activity, learned in babyhood as a child listens to and interacts with the adults around. Children understand language before they can speak—they absorb syntax long before they can explain it. Yet this apparently inherent trait among humans will not develop without interaction with other humans. Listening only, without response, does not cause fully developed speech to follow. Being listened to, as one uses speech, is an equally important part of the process of learning to use language. Using it in interaction with others is thus highly significant.

## Perception

A last major learning that occurs in early childhood is the development of perception—the ability to draw meaning out of the environment. This calls for the ability to see differences and similarities, to make comparisons, to recognize what will not change and what is subject to change. What is more, a child also learns to recognize stability within change, so that, for example, a parent remains a parent whether in pajamas or evening dress, whether at work or at home.

The sum total of the many concrete and direct sensory experiences normally encountered by a healthy young child over the years from infancy to five or six, leads to a sense of rootedness in the world of the human and non-human. The child gains strength out of awareness of this reality and this, together with the sense of inner rootedness, gives him/her impetus to go on to further learning. More of what is learned will in time be less dependent on the sensory, will be abstract as well as concrete, as the children grow older.

In what ways may the present heavy involvement of young children with the television set interfere with the growth and development indicated? A look at a fairly typical course of experience with television sets the stage for such analysis.

Children's encounters with television may start in infancy when the set is on as a steady accompaniment to the adult's activities around the house. At six months, a child may be sitting next to a parent or housekeeper who is taking a well-earned rest watching a program. The child's eyes will be drawn to the light and remain fixed on it, a practice that Dr. Ed Gording, psychometrist, believes will cause the baby to cease to use its eyes for properly searching out stimuli in a variety of directions. By 18 months, children go to the knob in imitation of their parents, and by 2½ large numbers are "hooked" into programs. By age four, television content is a major source of social discussion at nursery school, and nursery and kindergarten teachers at conferences and meetings speak with concern of the negative effect of television on children's play.

It must be granted that before the advent of television, there were many children whose lives were not optimal and whose development reflected inadequate environmental support. Unfortunately, the effect of television on normal development is likely to be felt by many more children, since hardly a home is without a set and television takes up most young children's time to an average of 23-25 hours per week. This reflects a marked increase since a study in 1959 showed that young children who had sets watched for three-fourths hour a day. The amount of time spent viewing is obviously an important variable in how effective television actually can be on development. Variables too are the more difficult-to-access vulnerability of some children and the protective ability of others to turn away. But because the amount of time for most children is a given, and because many experienced teachers of young children as well as pediatricians, hospital nurses, and psychologists are agreed that some new kinds of behavior are emerging in children, it is possible to look at what television demands of children, or imposes upon them, and perhaps accounts for such behavior.

The active child learns to be unnaturally passive. The talking child learns to listen and expects not to have someone to whom to respond. (Even programs that call for response cannot have interaction.)

The child struggling to assess the relative meanings of animate and inanimate, of man-made and natural, of fantasy and reality, is not only not aided in the struggle, but further confused or convinced that imagery *is* reality. What happens to the still existent "old" reality in the child's mind? What is the basis for "the construction of reality" of which Piaget speaks, if imagery replaces concrete reality?

As children take on the formulas of programs for their source of play, what happens to their own autonomy and initiative; to imagination and originality?

As they are inundated with too much of some sensory modes (sound and sight) and lose totally on the screen the other sensory modes of touch, smell and taste, what happens to the clarification of the world by a sensory-guided child?

As the limited, yet exaggerated world of the television screen with its speed of images and strength of sound becomes an indigenous part of a young child's world, the capacity for perceptual development through feedback of the senses becomes questionable, at the least.

Investment of self in exploration and experimentation so that one can arrive at conclusions, and one can learn to reason as a result of comparison and awareness of evidence, are not usually considered properties of early childhood growth. Yet they are, and are known to be such to any careful observer. Television interferes with the child's search for reality through his/her own efforts, and offers instead a hodge-podge of noise and fast, incomprehensible but seductive images that may well be creating the new type of child seen by many teachers—the child at nursery school who is frenetic, runs around playing out a script with peers who also know the script, which is limited to jumping off heights, shooting, running, playing dead, and jumping up to climb heights again. It is play which gets nowhere. Many of these children will not get interested in nursery school materials, and unless running around aimlessly, have "nothing to do."

Too many children in first grade will not exert effort to learn; expect to be entertained but not to be a participant and exert effort; have real difficulties of perception; are bored easily because they have few inner resources on which to draw; have sophisticated "knowlege" they do not understand; and ignorance of the simplest realities of the natural world.

Television has become a way of life, no longer a source of entertainment. As a way of life, it permeates every phase of our lives. Do young children need protection from this new environment? The contradictions between their normal needs and the way television functions, would certainly seem to suggest that it is wisest to come to the screen and its images quite a bit later than children now do.

Basic growth into thinking, feeling and active people can only occur out of interaction with the human and non-human environment, not by passive observation of even the most instructive imagery. To deny this right to children is to court the danger that, like so many already seen, they will neither think, feel, nor act on the basis of both. The outer limits of elasticity of human adaptation may be stretched too far by the loss to young children of the kind of experiences that start them off well able to adapt even to television and its imagery when they are older.

### *REFERENCES*

1. Piaget, Jean, *The Child's Conception of the World,* Totowa, N. J.: Littlefield, Adams & Co., 1960
2. Erikson, Erik, *Childhood and Society* (Chapter 7), New York: W. W. Norton & Co., 1950.
3. Suomi, Stephen J. and Harlow, Harry S., "Monkeys at Play," *Natural History.* New York: The Museum of Natural History, December 1971.
4. Stern, V., Bragdon, N., Gordon, A., *Cognitive Aspects of Young Children's Play.* Final Report. New York: Research Division, Bank Street College of Education, 1976.
5. In Milton Schwebel, ed., *Introduction to Piaget in the Classroom.* New York: Basic Books, 1973.

# Children as a Special Audience

*Peggy Charren and*
*Natalie Rothstein*

*PEGGY CHARREN is president of Action for Children's Television, based in Newtonville, Mass.*

*NATALIE ROTHSTEIN is director of special projects for ACT and a freelance writer.*

It is 10:00 Saturday morning. Do you know where your children are?

Of course you do. They are flopped in front of the set, watching television along with 16 million other American children. They are watching cartoons and commercials—often without being able to tell the difference—while being entertained, amused and manipulated.

As guardians—whether parents, educators, or child professionals—we know how special children are. Because of their inexperience and vulnerability, we know how much they need our protection and guidance. Concerned as we are for their health, welfare and development, we teach them how to cross streets safely; we don't let them out alone at night and we guide their choices while we try to be responsive to their needs.

It is time for all of us to extend this concern for the special vulnerabilities of children to the hours they spend with television. Television is the single most powerful selling medium that exists in our society. It is accessible: 97 percent of all American homes have it; it is easy: just switch it on and you've got instant entertainment; and, it creates immediate celebrity. While only half of the world's adults can identify their national leaders, Fred Flintstone is recognized by 90 percent of American three year olds. The term "hero" seems to take on new meaning when applied to the ubiquitous presence of bumbling Fred and other such TV characters as Tony the Tiger or (I dream of) Jeannie. Whereas once a hero might be expected to display some portion of talent, courage or intelligence, it now seems that all one needs is just to be on TV. Familiarity-superstardom-role models, patterns and attitudes of behavior for children to admire and emulate.

The role models that children see portrayed on television say many things to young people. For the most part, our children are getting the impression that the world is peopled primarily by white American males, ages 18-35. Where are the mirrored self-images of little girls or members of racial minorities? What does their absence say to our children?

And what of the other messages transmitted to them? Children's commercials, cartoons and reruns, and much adult programming that children see, carries the sock 'em, bang 'em, shoot 'em-up message of violence. Though there is less of it, it still carries the impact of a one-two punch. You hit and your problem is solved.

As parents we should be aware of the inescapable influence of television and the need to assert some control over our children's exposure to it. We can limit the number of hours our children watch television and guide them in the selection of what to watch. We can encourage them to look at some worthwhile, creative programs, many of which appear on Public Broadcasting stations. And we can watch with our children whenever possible, talking to them about what they've seen and how they feel about it.

But parents can't do it alone. The broadcast industry must share the responsibility for children's television. In return for the use of the public airwaves, broadcasters are required to serve the public interest. We should encourage broadcasters to remember that meeting the needs of children is part of the "public interest." Only public pressure will ensure concrete improvements in children's television.

Children are a special audience with special needs. Their need to be entertained, instructed and delighted is in conflict with the broadcaster's economic imperative. To the industry, children are regarded as a market to be delivered to potential advertisers. Programs that have the highest ratings (largest audiences) can command the highest price for advertising time. Commercial television is a ratings game which is won by the broadcaster who can attract the largest number of young viewers.

Through television, we are allowing the equivalent of a door-to-door salesman into our homes each day, a salesman whose pitches are beamed to our children. Research has shown that a child often lacks the cognitive ability necessary to understand the commercial significance of the advertising message. A child may believe an entire commercial, parrot its words, and remember the name of the product without appreciating the motive of the advertiser.

Products are pitched to the 2- to 11-year-old "market" in ads that help sell billions of dollars worth of goods, at the same time bringing in nearly $600 million a year to the television industry. Every day our children are being "sweet-talked" as they watch the heavily sugared food ads that account for 60 percent of all children's television advertising. With 90 percent of all American children suffering from tooth decay, the question of whose interests are best served by these commercials is a question with a ready answer.

What resistance do our children

have to the lures of the advertising message? For, if as adults we can succumb to the hard sell and even to the subtle pressures of the soft sell, what hope can children have?

Children, the most impressionable of audiences, are being courted and cultivated for their buying power, an illusionary power to say the least. What buying power are we talking about? The 50¢ a week that Cindy or Charlie earns or receives isn't going to buy too many Snicker Bars or Farrah Fawcett dolls. No, it's mom and dad whose money is needed to support the habit that TV advertising works to create. A habit in search of an appetite and in need of constant feeding.

The young child, without sophistication and experience, is not equipped to make informed consumer decisions. The ensuing adversary relationship that can develop between parent and child has been cited by child development experts. Child psychologist, Professor Robert Liebert of the State University of New York at Stony Brook has stated, "mistrust results when legitimate authority figures—such as parents—are implicitly silenced or discredited, as they are if they pit their meager persuasion techniques against the might of television advertising directed at their young children."*

What can be done? Colossal though the impact of television may be, there are things that we can do. And it is not necessary to be a giant killer in order to gain some control over the giant. The practice in ancient times of killing the messenger if the message was unacceptable is hardly what we advocate.

The need for parental involvement and awareness cannot be overstated. Beyond the individual supervision of our children's viewing habits, however, there are avenues of organized activity. Action can be initiated through local church, school and community groups, developing programs and discussion which focus on children's television. Parents can write letters—both compliments and complaints—about programs and commercials to local stations, networks and advertisers. And concerned parents can join with other concerned adults as members of Action for Children's Television.

ACT, begun by a handful of suburban Boston parents ten years ago, is a nationwide consumer organization with more than 12,000 members. Since its inception, improving children's television has been the goal of ACT. As an advocate, not a censor, ACT has always focused on education and legal activism to change broadcast practices. Because of this advocacy, several things have been accomplished.

The selling of vitamins on children's television has been eliminated. Promoted like candy, the visual message to kids to ingest brightly colored capsules and pills, was clearly dangerous.

There is now 40 percent less advertising on children's weekend television. And violence in children's programming has diminished.

But one of the most significant events in the history of children's television is currently underway, one that was first initiated by ACT petitions. Investigations are being conducted into the issue of children's television by both the Federal Trade Commission and the Federal Communications Commission. While the FCC is directing an inquiry, the FTC is conducting a rulemaking procedure. Among the considerations of the FTC investigation, are the following proposed regulations:

—a ban on all television advertising directed at children under 8;

—a ban on television advertising to children between ages 8 and 12 of the highly sugared foods most likely to promote dental decay;

—corrective advertising (nutritional disclosures) to balance commercials for other sugared foods.

These restrictions are endorsed by major health, consumer and public interest groups. Comments that parents and other individuals have filed with the Commission testify to public support for regulations that would offer greater protection to children. Concerned parents should continue to make their views known by writing to the Federal Trade Commission, Washington, D. C.

While activitism and concern have brought improvement to children's television, there is still much to be done. By being concerned and getting involved, we can work to see that the public airwaves operate in the public interest. And children—a large segment of that public—will have their special interests better served when they are better protected.

---

* Statement made during testimony before the Federal Trade Commission on January 16, 1979.

# How TV Changes Children

*Claire Safran*

*CLAIRE SAFRAN is an education writer with Redbook.*

• Susie, who once hung back, playing in corners by herself, now joins in with the other three- and four-year-olds of her nursery class.

• Young Mark has stopped playing at "blowing things up." It is two weeks since his last fist fight, and he doesn't tease and poke at the family pets as often as before.

• Mary, who was getting Ds and Fs on her third-grade spelling tests, came home the other day with B plus.

These are some of the changes in some of the children who took part in a recent *Redbook* experiment. Young children, according to most surveys and polls, watch television an average of two to three hours a day. What effect, *Redbook's* editors wondered, does this amount of TV watching have on them? Some of what the children see is good, even educational. Much of what they see has been called "bubble gum for the mind." Can this spoil a child's appetite for better kinds of nourishment? Can it make her sick?

Looking for answers, we asked a group of parents in New Milford, Connecticut, to put their children on a "television diet," cutting them back to no more than an hour a day for a four-week test period. The differences in the children's behavior and in the quality of family life in just that short a time are something that we think other parents should know about.

Like most diets, ours wasn't easy. Asked to change a habit, some of the children began by being cranky, fidgety, downright nervous—almost as if they were having "withdrawal pangs," as adults do when they cut back on food or cigarettes.

There were parental pangs too. In one family, the wife attends graduate school in the early evenings and the husband prepares dinner and looks after their three-year-old. He'd been using television as an electronic cookie, something to offer when the child was fussy or bored or underfoot. Now, he invited the youngster to join in the cooking. "It is not an unmitigated joy," he reports, "to have a three-year-old help you fix lasagna."

For other parents it was like losing a treasured baby sitter. "I used to wake up on weekends to the sounds of cartoons blaring," says one mother. "I could turn over without a qualm. Now I wake up to toy trucks going up and down the stairs and the sound of the refrigerator door opening and closing. There is nothing like the vision of a quart of milk spilling all over the floor to get me out of bed—fast."

Some of the effects of television on young children have been widely studied and documented. On the plus side are the indications that TV broadens a child's vocabulary and increases her awareness of the world around her. On the minus side, however, studies of video violence have followed one upon another, as inevitably and almost as frequently as those messages from the sponsor. Admittedly much of what the studies have said about the effects of television's mayhem and gore has been theoretical, or else focused on that small minority of children who are especially suggestible. Yet these investigations have been persuasive enough to compel the three major networks to institute the "family hour," banning the worst kinds of violence until after 9 p.m., when little children should be, but aren't always, in bed.

There are a few easily recognized symptoms to warn parents that their child may be spending too much time with that electronic baby sitter or is staring at the wrong programs. Nightmares. Aggressive, overheated play. A preoccupation with "bad people," guns, fire, death. Even the wail of an ambulance siren sometimes attest to an overdose of TV. Almost every neighborhood has its own horror story: A young boy drapes a towel around his shoulders to simulate a cape and leaps out of a window "like Batman." After Evel Knievel performed one of his motorcycle jumps on television, there was an epidemic of broken young bones as, across the country, children tried to duplicate the feat on their bicycles.

What, though, of the symptoms that aren't so dramatic or obvious? What of the normal, healthy, happy child? Does television—even when it's nonviolent—have a hidden impact on her too?

Evidence on that is beginning to trickle in. Among her files, psychologist and author Eda LeShan has the case history of a 16-month-old baby, a perfectly healthy infant except that he slept fitfully and was crankier and more restless than is typical. The child's mother was in the habit of keeping the television set on most of the day, not really looking at it, using it "just for company." Though there seemed no obvious connection, she decided to turn it off. Very quickly the child's behavior changed. He slept better; he was less easily distracted; he developed a new ability to concentrate on his own child's play. "Most of television is too much noise, too much stimulation, too much syncopation, for a young child," says Eda LeShan. If it can have so dramatic an effect on a young child who isn't even "watching," what does it do to an actual viewer?

Across the country, nursery-school and kindergarten teachers have been noticing a changing pattern of behavior in children, and they are attributing it to television. These teachers report an increase in violent and frightening elements in youthful play—a phenomenon that Dr. Benjamin Spock commented on, ascribing it to the same source, in his *Redbook* column of July 1970. The usual games go ignored as children—especially the boys but often the girls as well—swoop and roar, playing at being a superhero, anyone from Batman to the Bionic Woman. A schoolteacher friend tells me of a five-year-old pupil whom she's never seen walk normally; his every movement is an exaggerated, slow-motion imitation of the Six-Million-Dollar Man. Many veteran teachers report a definite sense that children's play is not as rich, imaginative and spontaneous as it has been in the past.

Along with that, teachers see an increase in passive behavior, more shyness, more withdrawal. They complain that children ask fewer questions and volunteer fewer answers. The child who spends too much time with the sit-down, one-way experience of television may come to settle for that and not reach out for more-active, direct, two-way involvement with other children or with creative toys and materials.

## School Policy

Last year Eleanor Brussels, principal of the Horace Mann School for Nursery Years, a highly regarded private school in New York City, decided to do something about this. She sent a letter to parents explaining her concern about television and urging them to monitor the programs their children watched and to limit them to an hour a day. Most of the parents cooperated, and within three weeks there were some remarkable changes. The children seemed calmer and more relaxed at school. Instead of acting out *Superman* or *The Flintstones,* they began to play out their own minidramas and to reinvent the running games that children have played for generations, games that grow out of a child's own imagination. They were less easily distracted and could work creatively either alone or as a group—something only a few had been able to before.

We talked with Ms. Brussels and her staff of teachers about their experience in limiting television. Would the same thing happen with other children? What would the children do during the hours that once were filled by TV? What would a mother do if she couldn't rely on her handy-dandy combination pacifier and baby sitter?

We decided to repeat the experiment, this time following up on it so that we could know what happened at home as well as in the classroom. In the quiet suburb of New Milford, Connecticut, we worked with the Family Resource Center, which includes nursery, kindergarten and day-care facilities, and with the First Congregational Church Co-op Nursery. About 15 mothers agreed to join the television study, monitoring the shows their children watched and keeping diaries on what happened as they limited the viewing. Some of the teachers also contributed their observations, and when the test period was over, parents, teachers and editors met for an exchange of impressions and feelings.

Here, then, are the results, concentrating on just a few of the families whose experiences seem typical of the others and changing their names for the sake of privacy.

If there were a prize for the most-changed child, it would go to Susie Richards, a bright, usually well behaved four-year-old. Before the *Redbook* study she was a passive child, a bit too quiet, very much a loner. During her mornings at nursery school she fluttered on the edge of things. Despite her teacher's urgings, Susie refused to join the other children or to work by herself with clay, paints—anything that took real involvement.

During the first week Susie had "fits of temper" over not being allowed to watch *Batman.* At nursery school she was "moodier than usual, sitting and staring at the ground." During the second week, though, she began to ask her mother to invite a playmate home for the afternoons, something she'd rarely wanted before. Her teacher noticed that she played more with the other children and seemed "happier and more talkative." By the end of four weeks, Susie seemed a different girl. She was no longer just an onlooker. Now she was a participant, playing creatively by herself or joining willingly in a dodge-ball game or a group project. To both her mother and her teacher she seemed a happier child.

Did Susie change just because she was watching less television? There always have been shy or withdrawn children, of course, and many of them become less retiring in the natural course of time. Could that be what had happened to Susie?

During our study Susie kept asking when "the test" would be over. After four weeks her mother allowed her to resume her normal television habits. A couple of weeks after that, Susie's mother and teacher talked again. "Susie was doing so well," the teacher said, "But now she's off by herself alone again." When the fantasies and adventures of television were taken away from her, Susie had felt a real need to reach out, to be involved, to find her own adventures. When they were returned, she withdrew again. There seems to be a definite cause-and-effect relationship between Susie's behavior and her television watching, and her mother has put her back on a "television diet."

Some of the preschoolers watched only a bit more than an hour a day even before our study, and on a sunny day the set might never be turned on at all. Yet many children watched more than their parents realized, perhaps because they did it in bits and pieces. Often children use television as "background noise." (One boy came home from school, turned on the set and practiced his clarinet.) Sometimes it's the parents who are watching while the child plays off to one side, seemingly paying no attention to the set, glancing up only occasionally, catching only the commercials and one or two other especially noisy moments. Psychologist and author Bruno Bettelheim thinks those times should be counted too. "If the television is on," he says, "it changes the character and the quality of a child's play."

Studies by the A.C. Nielsen Company, the television ratings and research firm, show that boys and girls between the ages of two and five watch an average of 23¼ hours of television a week, with the figure going down slightly in the summer months, when *The Flintstones* have to compete with outdoor play, and up a bit in winter. For some reason not yet explained, very intelligent children tend to watch more than others, and suburban children, despite all those back yards, average about an hour a week more than do city children. Wherever they live, children who remain "typical" viewers from the age of three through 17 will end up spending more time at the television set than at school.

There is a widely quoted figure that has young children averaging 54 hours a week in front of the set, but the Nielsen people say that's an old misquote that keeps coming back to haunt them. Perhaps this is another reason many parents mistakenly believe their children watch television "less than average."

Mark's mother was one of those parents. "We're really the wrong family for this experiment," she wrote. "I think Mark watches less television than most children." Yet before the study Mark, a boisterous, energetic three-year-old, ate breakfast with *Captain Kangaroo;* in the afternoons he watched creaky reruns of reruns—*The Partridge Family* or *The Brady Bunch, Batman* or *Star Trek*—and occasionally, at his mother's urging, *Sesame Street.* After supper he'd sit, half dozing, in front of a game show. Weekend mornings, while his parents slept, he was up early, gorging on peanut-butter sandwiches and cartoons.

It seemed harmless enough, two or three hours a day, a mix of educational, action and family programming that is typical of what his age group looks at. But all in all, Mark fitted neatly into the preschool picture drawn up by the Nielsen Company.

**"I am thrilled about how much we all talk to one another now."**

## Effects

During the first week of the *Redbook* experiment, Mark's mother wrote, "The biggest problem we have with our son is that he is very rough and mean with our pets, a young puppy and a kitten. His punishments have ranged from spankings to no dessert, from sitting in the corner to going to bed early, but nothing has helped so far."

Like many of the children, Mark was angry, even "bitter," at the new limits on TV watching. At nursery school his teacher found him "even more aggressive and overactive than usual." Mark's parents vetoed any violent shows, but the decision of when and how to use his TV hour—all in one chunk or in segments—was left up to him, "to give him some sense of control over his own life." Mark saved his hour for after supper, but that led to a new problem. Without a quiet time in front of the set before supper, Mark was keyed up and came "exploding" to the table. One night he had a crying fit and needed to be held on his mother's lap throughout the meal.

Television is a stimulant, but for some children it's also a sedative, something to help them unwind after the day's excitement. Eventually Mark's mother found that he could calm down by playing records or working on puzzles. Other mothers shifted the bath time around for children who needed something pacifying before supper or before bed.

Almost all the mothers in our study read aloud to their children, but some of them preferred not to do that at bedtime, when stories tend to stretch on and on. "And it's a waker-upper," wrote one woman. "TV puts my daughter to sleep but a book gets her involved. She responds; she asks questions." That, of course, is the value and the joy of books. Most of television goes by so quickly that a young child has no time to think about what's happening or to respond to it. The action moves at a helter-skelter pace, a dazzle of sounds and images, and that's one of the major criticisms that some educators and psychologists have even of the *Sesame Street* program.

By the third week, Mark was calmer at nursery school. He settled down more easily and listened to directions more readily. At home the situation with the family pets steadily improved.

Most mothers and fathers know almost by instinct how to say "no" to touching a hot stove or crossing a busy street. Perhaps because it's a newer phenomenon, one their own parents weren't sure how to handle, they're less certain how to enforce discipline about TV. Yet after a week or so of the test, most of the children stopped nagging about it. "They no longer ask to watch television," one mother wrote, "because they know what the answer will be."

Many of the parents in our study are aware that the habit is theirs as much as the child's. A father, one of two who kept diaries in this study, admits, "When we're trying to get everybody off to work or school in the morning, or when I hear the kids squabbling in the evening, I keep wanting to yell, 'Okay, go watch some television.' I used to say that oftener than I realized. I have to remind myself that if they were watching TV, they'd still be squabbling—over what to watch."

As parents cut back on television they looked around for something to take its place. Some invited playmates to visit oftener. Others became freer with some of the "messier" things they once rationed—

paints, paste, clay. Coloring books enjoyed a revival. The children were encouraged to play outside oftener. "I never knew that Susie was a close neighbor," reports her teacher. "Now I see her bicycling up and down the street all the time."

"I didn't have TV till I was about ten years old," writes Beth Thomas, now the mother of two children. "What did I do without it? You know, it's hard to remember." For the first week or so of our experiment, she put on a dazzling series of entertainments for her children.

First they tried baking. "Not too good," Beth reports. "My eight-year-old is the one I'm trying to keep away from food too." The next afternoon she bought marching records. The following day she and the children made papier-mâché animals. They followed that with paper-bag puppets, designing them one day, putting on a play with them the next.

"I really enjoyed getting more involved with the kids," says Beth Thomas, "but I also missed having some time and privacy for myself." The children seemed to expect a new treat every day. By the second week they were bringing their friends home "to see what Mom is up to now." At that point Beth resigned as full-time social director. She told them to find something to do on their own. They complained. When she offered a few suggestions, they turned them down one by one. Eventually, though, they themselves came up with something to do.

Like Beth, a number of mothers over-reacted, trying to fill every spare moment for their children and then feeling annoyed at having to do it. Yet most of the children, some sooner and others later, did find ways to entertain themselves. "That's the best part of this experiment," reports Beth. "My children have imagination, and it's wonderful to see them using it."

Beth's four-year-old didn't have much of a television habit, though he liked to be in the room with his sister while she watched. Eight-year-old Mary used to do her home work in front of the set and then watch until bedtime. With the set turned off, her homework is now done on time. "And done better," reports her mother. "Her grades have gone way up."

Television, of course, wasn't teaching Mary to misspell; it wasn't giving her the wrong multiplication answers. Without too much TV, though, she is better able to concentrate at school and in the new-found quiet at home. She feels better about herself, more in control of her own habits. When her friends chat about the latest episode of *The Bionic Woman,* she is unimpressed. She's teaching her younger brother to read; she's building a puppet theater; she doesn't need to watch "that old thing all the time."

The change in Mary is measurable, at least in her grades—up from Ds and Fs to a B plus. In many of the other children the differences are subtler. But they do seem changed—not necessarily because television, except for its violence, is by nature pernicious and evil, but perhaps because whatever has taken its place is more valuable. Some of television can certainly be called "educational," but a one-way communication is not the best way for children to learn and grow. They do that through involvement—with parents, with other children, with material they can manipulate.

There are a string of small benefits. A mother who's begun to go bicycling with her children during the former TV hour is delighted with her slimmer thighs. Instead of Saturday-morning cartoons, some children actually clean their rooms. One mother talks of the new freedom that comes with breaking any habit. "If we're out shopping, we no longer rush home in time for *Batman.* If there's something interesting along the way, we now can stop." Many parents are pleased that their children no longer *vroom* about, playing at being superheroes. When her youngsters turned their bedroom into a mock spaceship, one mother asked, "Are you playing *Star Trek?*" They shook their heads. "No," they said, "This is ours. This is better." And it was.

During the third week, another mother wrote, "No TV, but our very own Wonderama. As usual, our three-year-old was first awake and first up. He came downstairs, let the dog out and went into the cold garage to get the library books we'd picked up the day before. He spent about twenty minutes reading them. The he hung up his sisters' coats. Wow! I thought. I wonder how many reruns of *that* are possible."

Along with their children, many of the grownups reduced their own viewing. "We used to sit there saying how awful the program was," a young wife relates. "Now we read aloud to each other." Some couples are playing board games during the evening; almost all find they are talking more.

Communication between all members of the family is livelier. "I am thrilled about how much we all talk to one another now, " a mother writes. "To be honest, we *have* to. The TV sounds are gone, and silence seems very strange with all of us at home. The TV doesn't delay those things the kids are avoiding—chores, homework, baths, bedtime. Oh, those things are still avoided, but somehow I find a half hour of casual conversation and wandering about is easier and nicer than the same time in front of the tube."

The parents who took part in *Redbook's* experiment were loving and concerned mothers and fathers to begin with. For the most part their children were happy and healthy. With less television in their lives, the family members are a bit more involved with one another now; they seem a shade happier.

A couple of months after the test period ended, we checked back with some of the families. Now that they're no longer counting the video minutes to record them in diaries, some parents have added a bit to their children's viewing time. But they are still monitoring the shows carefully, filtering out the violence. Almost all are confident they will never slip back into the old patterns, the automatic click of the "on" button, the mindless watching.

These families have broken the habit. They have looked closely at their children, at the television they were watching—and back again. For any parent, that can be an eyeful.

# Learning and TV Violence

*Alberta E. Siegel*

*ALBERTA E. SEIGEL is Professor of Psychology in the Department of Psychiatry and Behavioral Sciences at Stanford University. She was a member of the Surgeon General's Scientific Advisory Committee on Television and Social Behavior.*

Technological change alters the fabric of our social life. Often this social alteration is a side effect of the technological innovation, not its intended purpose. As an example, new modes of transportation—the automobile and the jet airplane—have enlarged our socially significant peer groups, enabled us to remove our homes considerable distances from our places of work, enabled our children to attend schools and colleges far distant from their homes and perhaps remote from their parents' alma maters, and so forth. Similarly impressive effects on our social life could be cited as stemming from changes in the technology of warfare, of healing, of food manufacture, of construction. The ways human beings relate to each other, in families, at work, in communities, are altered by technology.

My concern here is with yet another set of technological innovations: modern means of communication. At the same time that psychiatric theorists are concentrating on communication between people as a central issue in both normal and psychopathological development, electronic engineers have devised new technologies for communication. I believe that these have profound implications for our social system, including our socialization of the young.

The higher primates communicate primarily by visual signals: gestures, postures, expressions. The visual display is supplemented by auditory signals: grunts, calls, murmurs, smacks, howls. Usually these two forms of signaling are to convey emotional meanings, and usually they are redundant.

Man uses these same communication modalities. In the human infant, in fact, they constitute his total communicative repertoire. In older children and adults, visual and auditory signaling continue to be important, but speech is the central communication modality, the most sophisticated and the most differentiated. Washburn estimates that man has had speech for the last 40,000 years. It is during the years that man has had speech that most of the distinctively human social institutions have emerged.

Written language is much newer. Although early man may have communicated with his fellows by pictures, the earliest evidence of a written language meeting minimal linguistic criteria appears only about 5000 years ago, with the Sumerian lexigraphic systems.

Written communications were not mass communications until the development of printing. Forms of printing existed in China and Korea at least 2000 years ago, but the ideographic scripts of the Oriental languages did not lend themselves to mechanization. The same limitation did not constrain the European alphabet of the 15th century, with its 23 letters, so that when printing and moveable type were developed in the West five centuries ago, they found ready use.

With the advent of printing, nonverbal communication and speech and writing were supplemented by books, newspapers, journals and even comic books and penny dreadfuls. Our modern systems of education—with their textbooks, libraries, lectures, assigned reading—are based on a combination of face-to-face speech and printed communications, as is our system of scientific discourse, based on professional meetings plus printed journals.

In the last 100 years, a panoply of new communications media has been developed. We may expect them to alter our social arrangements just as profoundly as printing has done over the past five centuries. In fact, the alterations may be more fundamental, for these new media reach all humans, whereas the print media reached only those who were literate. Until very recently the literate have been only a small minority of the human population.

The phonograph was developed in the final quarter of the 19th century, and its commercial heyday was the initial decades of our own century. Commercial radio broadcasting followed in the 1920's. Here were mass media of communication which did not require the audience to be able to read. Small children as well as adults could be reached. And the technique spanned both time and space: a radio listener in Oregon could hear the President speak in Washington, a housewife in Nebraska could hear a gramophone record preserving Caruso's performance years earlier in a New York concert hall.

Motion pictures were developed during the same era. Film appealed to the eye whereas the phonograph and radio were beamed to the ear.

Reprinted from: AGGRESSION. Res. Publ. A.R.N.M.D., Vol 52

Film technology was more complex, so that whereas radios and gramophones were home equipment, the movies were shown in nickelodeons during the first decade of our century and in vaudeville houses and cinema theaters thereafter. The audience paid for each sitting. No special training was required to enable a child to comprehend film material, and by the 1920's most American children spent most of their cash on movie tickets.

When sound was synchronized with visual image, it seeemd that the pinnacle had been reached in the development of techniques to communicate with the illiterate. With the talkies, eye and ear received an integrated message. Al Jolson's *The Jazz Singer* was screened late in 1927, and by 1930 the silent movies were a thing of the past. Children and adults flocked to American movie theaters in the 1930's as do children and adults today in underdeveloped nations, where movies are still the most advanced medium of communication. When color was introduced (the first commercially successful color film, *Gone With The Wind,* appeared in 1939) it seemed that man had bridged both time and space with a totally effective communications technology, one attuned to the primate's keenest senses—vision and audition—and capable of transmitting gestural, spoken and written communication to these senses, with a musical soundtrack thrown in at no extra charge.

From the vantage point of 1972, we see this past as merely prologue. We recognize how movies were circumscribed by the walls of the movie theatre, their impact blunted by the barrier of the box office. We remain at home with today's medium of eye and ear; we pay no admission fee to watch *Hawaii Five-O* or *Mod Squad.* And the TV receiver brings us events the moment they occur. No longer must we wait until the film is processed, printed and transported to the movie house. As we watch the political conventions and football games live on TV, we wonder how we could have been so patient with those quaint newsreels and their week-old "news."

## History

The first regular program of entertainment offered on TV was in London in 1936. Three years later, regularly scheduled telecasting was initiated by NBC in the United States. World War II interrupted the commercial development of this new device, but when wartime manufacturing restrictions were lifted the TV broadcasting industry grew very rapidly. By 1949, there were one million TV receivers in use in the United States. Two years later that figure had increased ten-fold. Well before 1960, there were 50 million receivers in use here, a number close to the saturation point. Today, over 60 million homes in the United States receive television and more than one-third of U.S. families own more than one TV set. All but about 4 percent of American homes have TV, more than have refrigerators or indoor plumbing. The Kerner Commission report called television "the universal appliance of the ghetto." In 1971, Americans shelled out $3.8 billion for new radios and TV sets. In the same year, advertisers spent $3.5 billion to support radio and television broadcasting.

The other media remain onstage in America, but frequently, like aging actors, their assigned role is that of the supporting player, enhancing the glamour of the new young star, television. Among magazines, for example, the one with the largest readership today is *TV Guide.* Here we see the 500-year-old art of printing being used to guide the literate in their attention to a nonprint medium. Newspapers remain an important medium, and the TV page is one of the best read of their features. A survey was conducted among 300 students at the University of Oregon shortly after Governor Wallace was shot. All had heard the news, but only 3 percent had learned it from a newspaper. About 60 percent had first learned of the event from radio or TV, whereas more than 33 percent did so from another person. The aristocrats of the print media are books. Today, many books become best sellers when their authors are featured on TV talk shows. Movies continue to be important, of course, but when Hollywood produces a film today it is with an eye to the TV audience, for profits come from selling the Hollywood products to the TV stations for airing to the home viewer.

## Habits

In the average American home, the TV set is turned on over 6 hours a day. No one person in the home is watching throughout that period, but most Americans watch TV at least 2 hours a day.

American children become TV watchers at a very early age, just about as soon as they can toddle over to the set and turn the knob. In interviews with mothers of preschool children in Los Angeles, Lyle and Hoffman[1] learned that the majority reported their children were singing TV jingles by age 2, and over 90 percent had joined this youthful chorus by age 3. Of these mothers, many from poor and welfare families, 87 percent reported that their preschool children asked for food items they had seen on TV, and 91 percent said the preschoolers asked for toys seen on TV. The preschoolers themselves were interviewed individually. Although they could not express themselves very well, the 3-year-olds made it clear that they watch TV regularly on a daily basis. Viewing was particularly heavy during afternoons and on Saturday mornings, but the majority also indicated that they watched on weekday mornings and in the evenings. Many of the 3-year-olds could state what their favorite program was. Almost 9 out of 10 of them could identify Fred Flintstone from a photograph. Seven out of ten could identify Big Bird (from *Sesame Street*). Among the 4- and 5-year-olds, recognition of various TV characters was much higher.

It is not only Los Angeles toddlers who watch a lot of TV. Stein and Friedrich[2] studied 97 4- and 5-year-olds in a university community in central Pennsylvania. Detailed interview with their mothers revealed that these preschoolers watched TV an average of 30 hours a week. This means they are in front of the set well over one-third of their waking hours.

In Washington, D. C., Murray[3] observed in the homes of 27 black male children, ages 5 and 6. All were TV watchers. Viewing hours per week were highly variable, ranging from 5 to 42, but on average these young boys spent half of an adult's work week—21 hours—watching TV. About half of the programs they watched were adult shows: situation comedies and action dramas.

One might think that children would have fewer hours to watch TV at home after they begin attending school. The facts show the opposite. School-age youngsters watch more TV than do preschoolers. Lyle and Hoffman[4] studied TV-watching hours over the week among well over 1000 first graders, sixth graders and tenth graders. The weekly averages were: 23 hours, 31 hours and 28 hours. These results are similar to those of several other studies, showing a buildup of viewing time during the school years, then a drop in viewing time during adolescence.

Time does not permit my reviewing the extensive data on adult viewing. Suffice to say that teenagers and college students are light watchers, viewing becomes heavier during young adult years when more time is spent at home and heavy viewing is characteristic of the middle-aged and even more characteristic of the elderly.

To the child psychologist, what is of interest is that the average American child spends more time watching TV than he spends in school, for TV engages his attention 7 days a week, 52 weeks a year, and TV recruits his attendance several years before he begins attending public school.

I am offering this capsule history of the mass media and this commentary on how the newest medium is especially attuned to the receptive sensibilities of the human young in order to dramatize my point that in attempting to evaluate the far-reaching effects of TV we are faced with the same difficulties that face those attempting to evaluate any technological revolution.

The worst mistake is to take the new technology at its face value. From the vantage point of 1972, what can we say about the effects of the automobile on human social organization? No exhaustive list of those effects could be offered, but surely they already exceed anything dreamed of by the social scientists of Mr. Ford's era, and we are far from seeing their end—only now is mass use of autos taking hold in the European nations, and the technological revolution has not even started in Asia except in Japan. Those who produce and market a new technology have in mind a particular purpose. In the case of the car, it was transportation. But to think about the car's effects only within the domain of transportation is simply to miss the boat.

With American television, there are two avowed purposes: entertainment and advertising. Social scientists cannot ignore these purposes, but it would be folly to limit our attention to them.

The nervous system of the immature primate is adapted for learning. This learning occurs not only through direct reinforcement of acts, but also, and I think more importantly, through observation. The child does not need to be rewarded for learning to occur; reward bears on the probability that he will perform what he has learned and maybe on the probability that he will attend to certain stimuli from which he might learn. Attention and observation are sufficient for new learning to be stored by the child. Practice and imitation perfect his acquisitions from observation.

To my mind, the most brilliant research on television's effects on young children is that pioneered by my colleague at Stanford, Professor Albert Bandura, in the early 1960's. He demonstrated that preschool-age children imitate the aggressive behavior they observe on film and TV. In his experiments, he devised several TV sequences in which adults performed novel aggressive acts. Young children who observed these sequences subsequently performed the same aggressive behaviors. Bandura's findings have now been replicated by many psychologists, and their limits are currently being explored. When the *Report* to the Surgeon General was being prepared by his Scientific Advisory Committee on Television and Social Behavior, we were able to list about 20 references to scientific papers on children's imitation of TV aggression.[5] The fact that the television industry appreciates the implications of these findings is reflected in their successful efforts to blackball Professor Bandura for membership on the Advisory Committee.

Although the media men would have us think that their electronic wonder is simply a method of entertaining the public and selling soap to them, as behavioral scientists we ought to be thinking of television as a teacher, a source of information, a form of cultural transmission, an agency for socializing the young, a technique of displaying behavior that children will observe and imitate.

The learning need not be antisocial. Stein and Friedrich[2] conducted a careful study in which children in a nursery school had one of three possible TV "diets" in school: aggressive, neutral or prosocial. The prosocial series were episodes from *Mister Rogers' Neighborhood,* with themes of cooperation, sharing, sympathy, affection and friendship, understanding the feeling of others, verbalizing one's own feelings, delay of gratification, persistence and competence at tasks, learning to accept rules, control of aggression, adaptive coping with frustration. The children's behavior in play in the nursery school was observed before, during and after their four week exposure to these three diets. Among lower socioeconomic children, those who had the prosocial diet on TV increased markedly in prosocial interpersonal behavior in the nursery school: there were no changes in such behavior for lower socioeconomic children having other TV diets. This finding awaits replication, but it is in accord with certain laboratory experiments on the modeling of prosocial behavior in suggesting that young children can learn prosocial as well as aggressive behavior from watching TV.

## Content

What do American children watch on TV? This has been widely studied, and the findings are fairly consistent.

Only the very youngest children spend the preponderance of their viewing time with so called "children's" shows. (These have the derogatory label of "kid vid" in industry circles.) Four-year-olds have well established program preferences. Cartoons are their favorites, *e.g. The Flintstones*. Situation comedies also rank high: *The Courtship of Eddie's Father, Here's Lucy, Bewitched.* In some communities, *Sesame Street* and *Mister Rogers' Neighborhood* attract some of the very young viewers; in others, these noncommercial shows are not available or not widely watched. Middle-class children are among the most loyal viewers of the public television shows designed for the disadvantaged.

By the time they are in first grade, children have family situation comedies as their favorites. More adult shows and fewer children's shows appear on thir lists of TV programs regularly watched.

When a child reaches sixth grade, cartoon shows and "kiddie" shows no longer attract him at all. He watches family situation comedies, other situation comedies and "hip adventure" shows like *Mod Squad, It Takes A Thief* and *Star Trek.*

The preferences of tenth graders are dominated by adventure programs—those already mentioned, *Dark Shadows, Highway Patrol,* and so forth. Music and dramatic shows also rank high.

Adolescents and adults are more favorably disposed to very violent shows than are schoolage children. Men watch more violence than women. Blacks watch more violence that do whites. The poor and less educated watch more violence than do white collar workers and the college educated. Violence viewing is especially high among adult males who were high school dropouts.

Young children see a great deal of violence on commercial TV not because they seek it out or prefer it, but rather because commercial programming is so saturated with violence and children watch so much television. Also significant is the cartoon-maker's regular recourse to aggression as a method of conflict-resolution, inasmuch as cartoons are favored by young children.

As a child psychologist with a special interest in the social psychology of childhood, I am impressed by two facts about television. The first is that it is a superb technique for communicating with children. It reaches the home, the child's ecological niche. It brings vivid visual and auditory images to children, requiring no ability to read or write. That is, it speaks their language. Young children learn primarily by observation and imitation. TV presents them with behavior to observe and imitate, in a form which they can assimilate. The second fact which impresses me is that there is simply no social institution to govern television as a socializing agency for America's next generation: no board of TV education, no professional school to train the television socializers and child care workers, no textbook commission to screen, evaluate and upgrade the contents of the TV curriculum.

## Effects

The work of the Surgeon General's Scientific Advisory Committee on Television and Social Behavior was initiated because of concern among some U.S. Senators about TV violence and its possible social consequences. The Committee was organized by U.S. Surgeon General William H. Stewart in 1969. Funds and organizational support were made available through the National Institutes of Mental Health (NIMH), under the experienced and very capable leadership of its Assistant Director for Behavioral Sciences, Eli Rubinstein. Twenty-three research projects were supported by NIMH contracts after the proposals had received approval under review procedures modeled on the peer review system which has been so effective in assuring quality control of extramural research. These investigations, in various parts of the nation, concentrated on the violent and aggressive content of entertainment materials presented on commercial TV; this is the material which has been the focus of public and Congressional concern. Although adults were studied in some of the investigation, children were the subjects in most. The research is reported in five volumes of articles spanning almost 2500 pages. In these volumes, the authors of the research speak in their own words. The Advisory Committee's words appear in a separate volume, offering a cautious interpretation of the new studies as well as some consideration of earlier research on the topic.[5] These materials appeared in print early in 1972, and at that time they were the subject of four days of hearings by the Senate Subcommittee on Communications under the Chairmanship of John O. Pastore.[6] A capsule summary of the research findings was published simultaneously in a scholarly journal[7] by NIMH staff members who had earlier published a very useful annotated bibliography on the entire previous research literature.[8] We are now seeing scholarly reviews of the whole effort[9] and a book is in press which will offer an independent assessment of the research sponsored by the Program.[10]

Over the years since the advent of TV, many investigators have analyzed the content of TV drama, repeatedly demonstrating the preponderance of violent themes on American mass entertainment. George Gerbner is among the most original, thoughtful and persuasive of these investigators. He has surveyed the contents of TV dramatic shows on prime time in 1967, 1968 and 1969, making detailed content analyses of 281 programs appearing over 182 hours in these three years.[11] His materials reveal that about 8 out of every 10 TV programs contain violence. Incidents of violence occur at the rate of about 8 per hour during prime time. Saturday morning cartoons have the heaviest saturation of violence on all television, with one violent episode every two minutes.

The incidence of violence in cartoons beamed to child viewers increased from three times the level in programming for adults in 1967 to six times that level in 1969.

The world of TV drama is a mythical world with its own social rules, norms and mores. It is peopled by persons who may look and talk like our neighbors and friends, so that children and foreign viewers may confuse them with America's citizens, but their lives follow a script which in some ways mirrors American social realities and in some ways diverges radically from those realities. In TV drama, law enforcement agents are involved in only one-tenth of all violent episodes. In those episodes in which they are involved, law enforcement agents are themselves violent more often than not. Weapons, most commonly guns, are used in about one-half of all violent episodes on TV. TV violence is eerily vacuous. In one-half of all the violent episodes chronicled by Gerbner's painstaking observers, no painful effect of any kind was discernible, whereas in the other one-half it was often difficult for the observers to agree whether any pain at all had been shown, or indeed any consequence. In only one-third of the violent episodes does the viewer see either death or injury as a result. The long term disabling effects of violence, so familiar to the physician and the nurse, are almost never evident to the observer of TV drama. Nor does he see the emotional and social effects of violence. He is spared any exposure to the bereaved: the children who are left fatherless, the widow left to mourn her husband's loss and to attempt creating a life for herself and his family without him. This "sanitizing" of violence is but one way in which TV drama differs from life as we know it. There are other ways. In TV dramas, the leading character is typically male, American, middle or upper class, unmarried and in the prime of life. Of all the TV dramatic characters we see on prime time, only one-half are gainfully employed, and their work is usually not central to the dramatic portrayal. Those who do not have visible means of support are typically in high status jobs. Gerbner suggests that his fascinating ethnography of the TV culture, in some ways so similar to American culture and in some ways so different, is to be understood as a description of a mythology which functions to convey values and norms about power and influence to the TV viewers who fall under its spell.

Gerbner's social anthropological view of the social implications of the contents of TV entertainment is not congenial to many psychiatrists. Instead, psychiatrists have commonly spoken of TV watching as an essentially harmless opportunity for emotional catharsis of unacceptable impulses, including aggressive and sexual impulses. Freud followed Aristotle in his view that vicarious participation in dramatic themes allows an emotional purging in the audience. Most psychiatrists follow Freud and Aristotle, and whereas one must admire their fastidious good taste in selecting such intellectual leadership, one must also note that neither of these men based his work on observing children and that both were discussing serious drama presented by living actors and were not in a position to comment on entertainment in the modern commercial electronic media.

Investigators of TV's effects have found it difficult to get a handle on the catharsis notion in its psychoanalytic form. When it is formulated in behavorial terms, it becomes amenable to experimental test. The results of the many experiments on catharsis in the past two decades are not terribly clear-cut and are hardly satisfying to the research purist. But at least one can say that the evidence does not favor the emotional catharsis theory. Where it has been possible to isolate emotional effects of viewing violence on film or TV, usually it is an increase in aggressiveness which has been found. The catharsis notion would predict a decrease. In our 1972 report, the Surgeon General's Advisory Committee reported that the weight of the evidence is against the catharsis notion (p. 65 to 67). I have commented elsewhere[12] on the one major investigation which has yielded findings dissonant with the dominant trend.

Many psychiatrists have not stayed in close touch with the relevant research, and their thinking has been relatively uninfluenced by the results that emerged in the late 1950's and the 1960's. Unfortunately, the television industry's leadership has exploited statements by psychiatrists expressing the catharsis notion, inasmuch as these statements seem to provide expert opinion in support of the industry's position that TV violence has no appreciable antisocial consequences. A notable and admirable exception to the general trend of psychiatric opinion has been the views of Dr. Frederic Wertham.[13] The respect which the public extends to psychiatric opinion and the concern which parents and other citizens feel about TV violence is reflected in the wide and sympathetic hearing which Dr. Wertham has gained for his incisive statements.

In their preoccupation with the significance of early experience in the family in shaping the level and the expression of a person's aggressiveness, many psychiatrists have seemed to imply that other influences are negligible. As a child psychologist, I quite agree that the family's influence is central, but I disagree that we can afford to minimize or dismiss the effects of other influences in a child's life, especially those that reach into the family home and intrude on family life.[14] [15] I have been delighted to see efforts toward increased communication between psychiatrists and behavioral scientists, and I am hopeful that the public pronouncements of members of the psychiatric profession will increasingly reflect knowledge of contemporary research findings as well as personal clinical observations and psychodynamic theory.

Among the seven distinguished social scientists who were blackballed by the TV industry from membership on the Surgeon General's Committee, only one was a psychiatrist, Dr. Leon Eisenberg, Professor at Harvard and editor of the leading journal of child psychiatry. It is hardly an accident that the blackball was reserved for a specialist in child behavior, a scholar who is notable in his profession for his cordial familiarity with social scientific research and his ability to synthesize and integrate clinical and research findings, for his thoughtful dissent from tired psychoanalytic cliches and for his personal qualities of leadership, articulateness and social vision. It is especially to be regretted that the Committee was denied the benefit of his participation.

In brief, then, the new research confirmed that children and adults watch a lot of TV, that TV entertainment is dominated by themes of violence, that young children can and do imitate the aggression they observe on TV and that there is no convincing evidence of any cathartic or purging effect of TV viewing. What else did we learn?

Both short term experiments, studies and long-term correlational field studies yielded evidence that watching TV violence instigates or arouses aggressive emotions and behavior. This finding holds for preschoolers, school-age children and teen-agers.

There is one longitudinal study of TV viewing, conducted in upstate New York.[16] Among boys, those who preferred aggressive TV material when studied in the third grade were rated higher in aggressiveness by their peers 10 years later. This correlation holds up after the obvious statistical correctives are applied.

We found that both children and their parents think of TV as providing information on how to behave, how to speak, how to dress, what to eat, and so forth. As a social institution, commercial television provides a definition of what's new, what's important and who's who. Because the contents of TV are widely thought of as defining "what's happening," the violent content of TV is a special cause for concern.

Research on the effects of movies on children has been conducted since the late 1920's, and TV research got underway in the 1950's. Why have results been so slow in coming, and why must our statements be so cautious even though we base them on literally hundreds of investigations? In the first place, we have no animal model for media use. Only human beings serve as our subjects. This limits the interventions we may envisage. Second, aggressive tendencies and proclivities are not easy to measure. Aggression, like homosexuality or drug use, is a taboo behavior, and the difficulties in studying taboo behaviors accurately are well known. Many of our studies rely on self-report or parental report, and there is convincing research evidence that both forms of reports are frequently distorted. Third, the research designs of social psychology are best suited to the study of explosive events. In the before and after experiment conducted over a short period of time, we can study dramatic single events which have a strong impact. But TV violence may have effects more like corrosion than explosion. The effects may be subtle and continuous, chronic rather than acute, insidious rather than blatantly overt. I suspect they are. If so, short-term experiments will not reveal the full measure of the effects.

We are left with evidence which gives some cause for concern, plus common sense which tells us that any activity occupying so many hours in a person's life must have lasting significance for him. Both the evidence and common sense converge to suggest that TV's continuous preoccupation with stereotyped violent conflict and its resolution through violence can hardly be constructive and healthful for the child viewer.

### *REFERENCES*


1. LYLE, J., HOFFMAN, H. R.: Explorations in patterns of television viewing by preschoolage children, in Television and Social Behavior, Reports and Papers, Vol. IV. Television in Day-to-Day Life: Patterns of Use, edited by Rubinstein, E. A., Comstock, G. A., Murray, J. P. Washington, D. C., U. S. Government Printing Office, 1972, p. 257.
2. STEIN, A. H., FRIEDRICH, L. K.: Television content and young children's behavior, in Television and Social Behavior, Reports and Papers, Vol. II. Television and Social Learning, edited by Murray, J. P., Rubinstein, E. A., Comstock, G. A. Washington, D. C., U. S. Government Printing Office, 1972, p. 202.
3. MURRAY, J. P.: Television in inner-city homes: Viewing behavior of young boys, in Television and Social Behavior, Reports and Papers, Vol. IV. Television in Day-to-Day Life: Patterns of Use, edited by Rubinstein, E. A., Comstock, G. A., Murray, J. P. Washington, D. C., U. S. Government Printing Office, 1972, p. 345.
4. LYLE, J., HOFFMAN, H. R.: Children's use of television and other media, in Television and Social Behavior, Reports and Papers, Vol. IV. Television in Day-to-Day Life: Patterns of Use, edited by Rubinstein, E. A., Comstock, G. A., Murray, J. P. Washington, D. C., U. S. Government Printing Office, 1972, p. 129.
5. Surgeon General's Scientific Advisory Committee on Television and Social Behavior: Television and Growing Up: The Impact of Televised Violence. Washington, D. C., U. S. Government Printing Office, 1972, p. 159.
6. Committee on Commerce, Subcommittee on Communications, United States Senate, Ninety-Second Congress, Second Session. Hearing on the Surgeon General's report by the Scientific Advisory Committee on Television and Social Behavior. March 21-24, 1972. Serial Number 92-52. Washington, D. C., U. S. Government Printing Office.
7. ATKIN, C. K., MURRAY, J. P., NAYMAN, O. B.: The Surgeon General's research program on television and social behavior. J. Broadcasting 16:21, 1972.
8. ATKIN, C. K., MURRAY, J. P., NAYMAN, O. B.: Television and Social Behavior: An Annotated Bibliography of Research Focusing on Television's Impact on Children. Public Health Service Publication No. 2099 Bethesda, Md., National Institute of Mental Health, 1971.
9. BOGART, L: Warning: The Surgeon General has determined that TV violence is moderately dangerous to your child's mental health. Pub. Opinion Q 36: 491, 1972-73.
10. LIEBERT, R. M., NEALE, J. M., DAVIDSON, E. S.: The Early Window: Effects of Television on Children and Youth. New York, Pergamon Press, 1973.
11. GERBNER, G.: Violence in television drama: Trends and symbolic functions, in Television and Social Behavior, Reports and Papers, Vol. I. Media Content and Control, edited by Comstock, G. A., Rubinstein, E. A. Washington, D. C., U. S. Government Printing Office, 1972, p. 28.
12. SIEGEL, A. E.: Can we await a consen-

sus? A review of S. Feshbach and R. D. Singer's Television and Aggression, Jossey-Bass, 1971. Contemp Phychol 18:60, 1973.

13. WERTHAM, F.: A Sign for Cain: An Exploration of Human Violence, New York, Macmillan, 1966.
14. SIEGEL, A. E., Statement before the Senate Subcommittee on Communications of the Committee on Commerce of the United States Senate, Ninety-Second Congress, Second Session, March 21, 1972. Hearings. Washington, D. C., U. S. Government Printing Office, Serial No. 92-52, p. 62.
15. SIEGEL, A. E.: Educating females and males to be alive and well in century twenty-one, in Changing Education: Alternatives from Educational Research, edited by Wittrock, M. Englewood Cliffs, N. J., Prentice-Hall, 1973.
16. LEFKOWITZ, M. M., ERON, L. D., WALDER, L. O., HUESMANN, L. R.: Television violence and child aggression: A follow-up study, in Television and Social Behavior, Reports and Papers, Vol. III. Television and Adolescent Aggressiveness, edited by Comstock, G. A., Rubinstein, E. A., U. S. Government Printing Office, 1972, p. 35.

## Positive Uses of Television
*Gloria Kirshner*

*GLORIA KIRSHNER, editor of Teachers Guide to Television, has taught (from kindergarten in East Harlem to a class in Behavioral Science Findings and TV at Columbia), served as educational consultant, script writer, and educational supervisor for network television children's series, and edited over a thousand teachers guides to television programs on all four networks. She is the author of "From Instinct to Intelligence. How Animals Learn," and articles in "Elementary English" and other publications.*

### Make a Difference

The figures are familiar. Fifty-six million families are watching television on a Monday night. And among them, after the 8-9 "Family Hour" is over, and the youngest are supposedly safely tucked in bed, there will still be 11 million children, ages 2-11, watching from 9-9:30.

From 9:30 to 10:P.M., the number wil be 9.7 million. From 10 to 10:30 p.m., there will be 6.8 million children below 12 in the audience. From 10:30 to 11 P.M., 5.6 million children below the age of 12 will still be watching.

"Television has become a flickering blue parent, occupying more of the waking hours of American children than any other single influence—including both parents and schools," states psychologist Kenneth Keniston, chairman of the Carnegie Council on Children.

Parents who would prop up the sleeping, soft-skinned infant, alone with a baby bottle, understanding the importance of a mother's presence or a father's voice, of eyes that smiled into a newborn child's or wiped away its tears, sat the toddler in front of the TV set—alone. And left it there. It was still viewing, alone, when the baby boy felt for fuzz on his upper lip, and the baby girl felt her nipples and longed to feel a breast.

"But how old do you have to be?" Rose Goldsen asks, "to notice that all that lovely stuff going on right there on the television screen—all those people and animations and puppets who are so entertaining, such fun to watch—are not paying attention to you? How old do you have to be to learn that you can wet your pants or fall off a chair and break your arm, but the people on the television screen will go right on doing what they're doing. The voice will keep right on coming out of the loudspeaker. The music will keep right on playing. You? You can cry your heart out. Nobody will reach a hand out of the television set to pick you up. Nobody on the screen will even notice you!"

A term popular in education technology is feedback. It originally meant the dialog that went on between machines, a thermostat and a furnace, for example, working together to keep a room comfortable. Dr. Goldsen urges "setting up feedback loops appropriate to our human nature—people noticing one another and responding by adjusting something in themselves to the needs they notice in someone else."

When parents participate in a child's television experience, providing the human feedback a television machine cannot, will their children respond differently to this all-pervasive influence in their lives?

Recent findings suggest that they will, no matter what the child's age.

"It seems to me that selective turning *on* is just as important as selective turning *off,*" family counselor Eda LeShan points out to parents of grade school children. " . . . No program can seriously damage a child in any way if a concerned and loving adult shares the experience and uses it as an opportunity to talk about feelings."

"If parents show respect for, or at least interest in, their teenagers' favorite programs," Louise Bates Ames, Co-Director of the Gesell Institute for Child Development suggests, "useful and interesting discussion can often be stimulated. There is nothing more important to 16-year-olds than to have somebody *listen to* and *respect* their opinions, whether agreeing with them or not. Situations and values shown on the television screen can be used as topics or lively and productive discussion."

"Me and My TV," a research report on the role of popular television in developing verbal skills and bringing together adolescents and adults, cites this experience of Vera, a young black mother.

When her estranged husband died, and her five-year-old child had many questions about death, Vera asked her young son to watch "Sunshine" with her. The program was a popular made-for-TV movie about the death of the mother in a family. Vera used the program to answer her son's questions. The young mother reported that the shared television experience made it easier for her and her son to share their emotions.

At Teachers Guides to Television we began in late 1976 our first Parent Participation TV Workshops, designed to explore using television to open communication between parents and children, working with an NBC series and schools in various cities.

Parents were invited to come to school at 4:00 P.M. to watch the NBC *Special Treat* broadcast with their children. Teachers then led a cross-generation discussion, demonstrating for parents how to use the television experience at home with their children, sharing ideas and emotions, encouraging the child's growth and development.

In Charlottesville, teachers of fourth and fifth graders used the programs, "Big Henry and the Polka Dot Kid," to lead children and parents in a discussion of decision-making. Had young Luke and his Uncle Henry made their choices for emotional or practical reasons?

The parents were astonished by the perception the children brought to the television drama, and said they had had no idea that the children took in as much or were as intuitive as their nine- and ten-year-olds seemed to be.

Charlottesville parents were unanimous that discussions made possible by a television broadcast would help them with their home relationships. "Parents grow to realize," Charlottesville Superintendent of Schools William Ellona, says, "that selected TV programs present a tremendous opportunity to increase communication between different age levels. TV can be used as an entree to discussion of issues, values, ethics, and other topics that might be difficult to bring up unless there was some reason to do so."

In New York City, where Cathedral High School girls come to school from all over the city, only two parents showed up for the first TV Workshop.

The high school girls saw the cross-generation discussions as a way to educate their parents. So the teenagers went to work. They put up posters, sent invitations, made telephone calls, and organized a prize drawing and a tour of NBC—a resource in their own backyard neither they nor their parents had ever seen.

Over 40 adults asked to participate in the second Workshop! Some mothers brought younger children. One girl, whose mother was divorced, asked her local pastor to come and he did. Another girl invited four social workers.

Following the viewing of "*Big* Henry", the groups gathered around tables in the school library. The talk ranged from the propriety of chewing gum to a 12-year-old girl's demand to see an X-rated movie in New York City, all stimulated by a TV drama about a 10-year-old boy and his uncle in the Northwoods.

"The television medium," said a mother, "is the best way to induce discussion. The situations hit home many times in many different ways."

Drama can reach below the surface to the deepest human emotions—emotions parents and children can share and use to open communication with each other.

**"It need not be added that the adult sets the model for self-control. . ."**

## Use TV

"We shape our buildings," Winston Churchill said, "and our buildings in turn, shape us."

We shape our children's experiences, and their experiences will, in turn, shape them.

It is a parent's responsibility to guide a young child's choice of programming. The younger the child is, the easier it will be to fulfill that responsibility, *as long as the parent does not use TV as a babysitter.*

That is easier said than done.

"We can be better guides to television watching," Eda LeShan suggests, "when we ask ourselves such questions as: What do I want my children to believe in? Can I teach them to trust other people? How do I want them to cope with the problems they will face? How do I want them to judge "success" in life? How do I want them to feel about other people? Such searching questions, and the answers we supply to them—will help us to decide what television programs are welcome in our home."

If we fail to make the decision we must remember Sartre's warning, "Not to choose is to choose."

But any parent is entitled to quiet moments alone. Mothers *do* need a chance to rest. Fathers *do* need time to relax. And if the parent uses TV properly, even the youngest child can often watch television alone without developing donkey ears and a tail because he or she has been to "the land of Boobies."

Television viewing can provide even the very young child with a great deal of mental stimulation.

The human baby begins to learn on the first day it is born. Through sound, sight, touch, smell and taste, information is pouring into its brain. The tiny brain is a far more complex information processor than any manufactured computer. The baby must organize the information received by the sense. What does this face mean? What does that sound mean?

But what happens if no one is feeding information into the baby's brain? What if there is only a white ceiling to stare at, and a bare wall beside the crib? What if no one bothers with the baby all day? Day after day.

Doctors Burton White and Richard Held at the Tewksbury State Hospital in Massachusetts wondered if the richness of environment could affect the speed of a baby's learning.

Would a baby learn faster if there was more to look at? Babies at the hospital usually started reaching toward toys with their fists closed when they were 65 days old. They usually started grasping an object when they were 145 days old.

The doctors experimented with placing gayly colored objects over the babies' cribs. The nurses spent more time holding and handling the babies each day. The experimenters watched to see what would happen.

The babies who had the more exciting visual stimuli and more attention reached out for the toys ten days earlier than the other babies did! They were able to grasp objects as much as two months earlier! You *could* change how fast a baby's mind grew.

There are no tests, yet, of how television stimulates infants' minds, but we know the results that *Sesame Street* and *The Electric Company* have achieved. And the St. Mary Center for Learning successfully used popular television to develop verbal skills among teenagers and adults.

*Even if a child watches alone, the participating parent can find opportunities to talk about what has been seen on the TV screen, and to share feelings.*

A toddler unable to talk will feel warmed and encouraged by a mother passing by during a commercial who points to the cat on the TV screen and gently persuades the child to say cat. The nursery age child will love a quick rhyming game—"cat, fat, sat, hat, flat, that"—and an affectionate "pat"—and take one step closer to reading readiness.The sophisticated ten-year-old is ready to compare and classify cats, lions, tigers, and other mammals—and, pleased with a parent's pride, to take one step closer to abstract thinking.

Dr. Benjamin Bloom and other researchers have found a very high correlation between school achievement and what happens in the home. Today's researchers have been able to identify the five parent behaviors that make it possible for children to achieve (or not achieve) in school. Below are each of the five "successful parent" behavior patterns with brief examples of how parents can provide "the human feedback loop" that will make TV a tool for encouraging their child's mental and emotional growth. Parents will be able to think of dozens of others—for there are many ways in which if the adult provides only a few moments of encouragement, television opens a child's windows on the world. It is up to the parent not to permit a child to be a passive viewer. For that passive viewer is learning something—if nothing more than to be a spectator of life instead of an active participant. It is only active participation in his or her own growth that will permit a child to reach full potential. Here are the scientists' findings about the behavior of parents whose children were able to achieve in school.

1. The adult sets a model of the language expected from the child. E.g., if the toddler points to, but does not name a desired object, the adult helps the child to use words before providing the object. If an incorrect word is used the adult corrects the word.

With TV: The adult helps the child to name the objects seen on TV (the cat in the commercial), or to explore the meanings of words and phrases learned from programs or commercials.

2. The adult continually conveys a desire for the child to achieve and points with pride to the child's achievement.

With TV: The adult encourages the child to tell what has been seen on a program the adult could not watch or to draw or paint a picture for the adult depicting the program, and the child's feelings, or to "act out" a scene. The adult praises the child for successful efforts.

3. The adult talks about interesting ideas with the child, encourages the child's curiosity, and encourages the child to *think* and *plan.*

With TV: The adult talks about the ideas presented on TV programs they have watched together and helps the child to follow-up with trips to the neighborhood library, or fire or police station, or museum or zoo, and encourages other projects that grow out of the child's TV interests—build a tepee, make a police badge, put together a peep-show, plan a TV party.

4. The adult supports the child and provides help whenever the child encounters a learning difficulty.

With TV: The adult provides needed word meanings (the nightly news?) or explains difficult visuals (weather maps?) within the limits of the child's ability to comprehend.

5. The adult structures the child's life with schedules of time and place, thus providing a sense of control of self vs. acting on impulse.

With TV: There is a time to eat and a time to sleep, a time to study, and a time to watch TV; there is a place to eat, a place to sleep, a place for studying, and a place for watching TV.

It need not be added that the adult sets the model for self-control in this regard.

As the child grows older, television plays an ever-increasing role in emotional as well as intellectual development, and the child turns to the role models found on TV in a search for the answer to that overwhelming question of adolescence, "Who am I?"

The National Commission on the Causes and Prevention of Violence reported " . . . as the child grows older, he learns increasingly more from what he observes in the behavior of others. His own behavior is shaped by the observation of the success and failures and the rewards and punishments meted out to those around him. . . .

" . . . many adolescents consciously rely on mass media models in learning to play real-life roles. . . . This is especially true of those adolescents who are not well integrated into family and school life. . . . In the absence of family, peer, and school relationships, television becomes the most compatible substitute for real life experience."

The participating parent who uses television to keep open the lines of communication between child and adult can help the adolescent understand how we enlarge ourselves by our experiences—our inner experiences as well as our outer—and how television drama, as it reaches below the surface to our inmost emotions, can help us learn from the inner experiences of others.

By identifying with the characters met in television dramas, by feeling their dilemmas as his or her own, the adolescent comes to understand the human predicament, to experience "the unresolved choice between two directions of action," that Jacob Bronowski reminded us will be found in every work of literature.

Was there ever a more crucial time for a parent to be present?

Sharing the experience, open and accepting (it is a "TV drama" and not "real life"), the parent can help the young person discover the consequences of choice. Families find it easy to communicate calmly about what they have seen on television, to provide practice in decision making when the experience is a vicarious one and their young are not personally involved. The TV experience presents no threats. It does present tremendous opportunities for young persons to try on roles, guided by their own and their parents' questions: Do I like him? Is he trustworthy? Courageous? Cruel? Do I wish to be like that?

"Working with a group of kids I can always identify those whose parents have taught them decision making," Dr. Barbara Varonhorst has said, "They know their own values and they can resist group pressures." Here again, television, properly used, becomes the parents' tool.

Writing of her experiment in using popular television programs in the classroom with both teenagers and adults at the St. Mary Center for Learning, Ann Christian Heintz cites the class's reaction to a *Maude* episode in which Maude had to choose between husband and political career.

"There were many places for students to take a value stand," she writes. "Was Maude putting her ambition for political office above her love for her husband? Was Walter reasonable in his refusal to accept Maude's career? What are the criteria for decisions on this issue?"

The class went from discussion to this TV drama to discussion of their own lives.

"This shift from the stimulus of a TV show to a penetrating analysis of real life situations convinces me that television can provide a focus for value clarification that is matched in no other medium," she concludes.

The St. Mary teacher reports still another television experience, equally suited to family viewing, that led immediately to values clarification—the documentary drama on "Judge Horton and the Scottsboro Boys.

"In this true story, seven young blacks are accused of rape in 1931, convicted, and then re-tried. It was an obvious case of injustice and racism. The students were incensed.

"In the retrial a prominent white Southern district judge being considered for governor, risks his entire career in order to give the boys a fair trial. After the trial he was forced to retire, his public life ended. In this case the concepts of justice and integrity had a much greater hold on the students than anything else we could have presented. . . .

"Again there was a breakthrough. The students were willing to think about the meaning of integrity and the consequences of decisions. In this case it was easy to discuss the way that other people's integrity and courage is the basis for some happiness that we enjoy. What will we pass on to the next generation?"

## Teach Values

The discoveries of Dr. Lawrence Kohlberg and his colleagues suggest some of the potential in a parent's use of television for character development. Dr. Kohlberg has developed a six-stage sequence through which a child passes before reaching the stage where he or she is capable of truly principled moral judgment—a stage that few children—and few adults—ever attain.

What is exciting to parents about Kohlberg's approach is his discovery that it is possible to raise children's level of moral understanding by exposing them to moral judgments that are one level above their own. This exposure must occur in a situation where genuine moral conflict, uncertainty, and disagreement about a moral dilemma have been aroused. The arousal of conflict in the child's mind is crucial. An adult's "right answers," or appeals to abstractions far above the child's level, are doomed to fail. The child must be involved, must seek to resolve the conflict.

What better device for involving the child than television, which allows us to share the experiences of others—the dilemma and integrity of a Judge Horton, for example—and make it our own? Television can help us to observe options, see the consequences of choice, and discover how our lives will hold us responsible for our decisions.

"The parent who spends time with his or her children," Ann Ryan points out, "communicates to them a feeling that they are people worth spending time with. The parent who shares his or her thoughts with them, in a constructive and caring fashion, imparts to them a belief in themselves and a desire to respond in kind."

Parents can share the television viewing experiences to help their children discover themselves, or their values, to find meaning in their lives and hope for their future. It can help them think about life purposes, or careers, discover how to handle conflict, explore the moral responsibility of citizenship, evaluate the meaning of religion, or of love—all of these bedrocks of life.

What is important is that parents who share these experiences and their children's responses—who "adjust something in themselves to the deeds they notice in someone else"—are using television to lead their children to think and to feel—and so to grow into their humanity.

# Television As A Moral Teacher

*Robert M. Liebert*
*and Rita Wicks Poulos*

*ROBERT M. LIEBERT and RITA WICKES POULOS are Professors of Psychology at the State University of New York at Stony Brook and Old Westbury, respectively.*

Until relatively recently development psychologists paid scant heed to the role of television in socialization. Then, beginning in the early 1960s, there was an enormous urge of interest in the effects of this medium upon the young (e.g., Bandura, 1963; Maccoby, 1964). A look at television's spectacular commercial rise may partially explain the shift. In 1948 less than 1 percent of all American homes had television sets; by 1972 virtually ever dwelling, even the humblest, had a working receiver. In fact there seems little doubt that more children now have television available than have adequate heating or indoor plumbing.

The spectacular rise of the medium would not justify its mention in this volume, though, unless its content were pertinent to morality and moral development. A. H. Stein (1972), in a thoughtful review for the National Society for the Study of Education, addressed this issue in a summary of recent analyses of television content:

Aggression and illegal actions are often portrayed as successful and morally justified. . . . Law enforcement officers and other heroes use violence as frequently as villains and often break laws and moral codes as well. In both adult's and children's programs, these socially disapproved methods of attaining goals are more often successful than socially approved methods. . . . While criminal and illegal activities frequently escape punishment, goodness alone is rarely sufficient to achieve success. . . . The fundamental philosophy manifest in most current television programming is that the end justifies the means, and the successful means are often immoral, illegal, or violent. . .

Does this enormous diet of morally relevant television affect the child in more than transient or trivial ways? This question has been asked about all the pictorial media—movies, comic books, and television—at one time or another. It is undeniable that television has emerged as a significant agent of socialization. One team of psychiatrists has even stated bluntly that "it is a matter of fact and concern that television has increasingly replaced parents as a definitive adult voice and national shaper of views" (Heller & Polsky, 1971.).

But what specific role does—and can—television play in the development of moral values and behavior? While few people doubt that television viewing has some influence, its impact is not easy to analyze. The child is continually interacting with his environment—soaking up, weighing, and judging all kinds of information from an array of sources. How can the effects of television be separated from a multitude of other influences? How can its impact be followed over time when there is so much interference? How can the differential effects of so great a variety of programs be sorted out? It is no wonder that our ability to analyze television's effects is commonly described as primitive, and that many researchers have been discouraged completely from the task.

On closer inspection, though, the problem of understanding this medium is no more complicated than that of analyzing the effects of parental nurturance, academic encouragement, or socioeconomic factors. Recent studies of the effects of television are no more primitive, and probably somewhat more advanced, than investigations of most other influences on a child's development. And the solution to the dilemmas raised above has been the same for television research as for other areas: We do not ask for or attempt definitive superstudies or a single methodology, but rely on converging and complementary theory and research approaches. Only in this way can we hope to produce information relevant to the long-term effects of television and the complex interaction of forces that doubtless come into play in the real world.

In trying to make sense out of TV's effect on children, we need a basic theoretical model, a superordinate framework to account for the various data. Our view focuses on observational learning as the critical process underlying television as an instrument of socialization.

*Observational learning* is, first of all, a vicarious process in which the behavior of children (or adults) changes as a function of exposure to the actions of others. These other individuals may be viewed directly *(live modeling)* or indirectly *(symbolic modeling);* in the latter case the model's actions are displayed through media such as books, newsprint, motion pictures, and television. Thus, a range of situations far too vast to be experienced directly by the average person is made

*"Television As a Moral Teacher" is reprinted from* Moral Development and Behavior: Theory, Research, and Social Issues, *Thomas Lickona, Editor, copyright © 1976 by Holt, Rinehart and Winston. Reprinted by permission.*

available. Investigations over the last fifteen years have firmly established that observational learning is a basic means by which the human acquires and modifies his behavior, standards, values, and attitudes (for relevant reviews, see Bandura, 1969a; Flanders, 1968; Zimmerman & Rosenthal, 1974).

## Observational Learning

How exactly does observational learning occur? It is useful to conceptualize the process as involving three major stages (cf. Liebert, 1972, 1973): *exposure, acquisition,* and *acceptance.*

For observational learning to occur the observer must first be exposed to the specific acts or *modeling cues* of the exemplar. Note, however, that *exposure* to a particular behavioral example can occur without necessarily leading to learning or retention. A child may simply fail to attend to what is being shown, or he may fail to process and store this information effectively. Yet it is only if the second stage, *acquisition,* also occurs that modeling cues can have a further effect.

The third and final stage of observational learning is *acceptance.* Does a child who has been exposed to and has acquired the modeling cues now accept this information as a guide for his own subsequent actions? Acceptance effects can be grouped into two broad categories: observer actions that are *imitative,* in which case the viewer's behavior becomes more like that of the model than it would otherwise be, and actions that are *counterimitative,* in which case the viewer's behavior becomes less like that of the model. Counterimitative behavior is illustrated by the child who, upon seeing a friend bitten by a dog, becomes less likely to approach the animal than he might have been previously. In this situation the viewer accepts the exemplar's actions as a guide for what action to avoid.

Modeling cues may also serve to bring about acceptance of a more general class of behaviors of which the cues are perceived as being only an instance. For example, the child who sees her parent donate money to a variety of charities may subsequently be more willing to share her toys with other children or to divide a piece of chocolate cake with her little brother. In like manner, a youngster who observes a variety of peers punished for handing in homework late or talking back to the teacher may avoid other school "transgressions" which he has not seen modeled. In the former example, *disinhibition* has occurred; in the latter example *inhibition.*

## Observational Learning and TV Viewing

We assume that the foregoing analysis, derived from a variety of studies on live and symbolic modeling, is equally applicable to television. Beyond potentially accounting for television effects, this analysis can also guide and refine the nature of the inquiry itself. Note, for example, that the emphasis on exposure to specific modeling cues directs our attention to the television content seen by individual children; overall exposure to television per se, as a medium, would tell us little about what any particular youngster might have been exposed to, learned and accepted from television.

An important aspect of the application of the three-stage analysis to television is derived from a distinction traceable to Bandura and Walters (1963). While only a small percentage of viewers may show immediate acceptance effects, others who have been exposed to the content may also have acquired some, or all, of what they have seen. If so, they will be able to activate their knowledge when a more favorable occasion arises, thereby manifesting delayed acceptance.

An experiment by Bandura (1965a) provides a clear demonstration. Children were shown one of three films which began with identical modeling cues: an adult female model behaving aggressively toward a Bobo doll. The films differed in their endings, however, as the model's actions had various consequences: in the first, she was rewarded with food treats and praise; in the second, she was castigated verbally for her untoward action; and, in the third, she had nothing done to her (no consequences). An immediate test of imitative aggression in the playroom revealed that children who had watched the model being punished produced fewer imitative aggressive responses than those in the other two groups. Did the punishment-viewing youngsters fail to *acquire* the model's aggressive repertoire or did they simply fail to adopt her hostile course of action because of the unacceptable consequences it would bring?

In the experiment, children were offered attractive rewards if they would demonstrate all they could remember of what the model had done. Now the differences between the groups disappeared entirely; those who had witnessed the punished model had indeed absorbed her exemplary hostilities and were quite able to perform them when the circumstances were altered to make their performance acceptable and profitable.

Bandura and Walters (1963) thus argued that although measures of acceptance necessarily indicate some of what the observer has acquired, they do not always indicate all of the knowledge gained. Acquisition can be evaluated using any procedure that demonstrates learning from modeling cues. The child can be asked to choose between alternative behaviors (one of which has been modeled), to describe what a model did, or to perform on a test that permits extrapolation from the exemplary performance. Without broad assessment, though, caution must be exercised in judging the limits of acquisition.

The previous sections have described the theoretical perspective we have used to organize and understand television as a possible moral teacher. Employing this framework, we can turn now to an examination of the medium itself.

## Exposure to Television

Television obviously cannot be an important teacher of moral lessons unless children are exposed to it. In fact, virtually every American home has at least one television set and soon a majority will have at least two. It seems incongruous now that warnings were once frequently voiced that TV would be just another fad because people could not use it, as they did radio, while they went about their everyday work and rest.

What has happened, instead, is that other activities have been sacrificed to the television set. In the United States television has affected the lives of the entire family; by the mid-1960s 55 percent of American families had changed their sleeping habits for television, 55 percent had altered their mealtimes, and 78 percent had begun to use television as an "electronic babysitter" (N. Johnson, 1967). Visiting others and entertaining at home also have reportedly decreased (e.g., Cunningham & Walsh, 1958), and the use of other media such as radio and movies has been adversely affected (e.g., Baxter, 1961).

These kinds of changes have been demonstrated wherever television exists. Robinson (1972), for example, compared the daily activities of TV owners and nonowners in fifteen locations in eleven countries by having his subjects fill out diaries concerning all of their activities throughout a full, twenty-four-hour day. Such time budgets showed decreases in radio listening, book reading, and movie viewing. Sleep, social gatherings away from home, other leisure activities (correspondence and knitting), conversation, and some types of household care also decreased. From his results Robinson (1972) concluded that "at least in the temporal sense, television appears to have had a greater influence on the structure of daily life than any other innovation in this century" (p. 428).

But what about more specific viewing habits? How much television do children actually watch? It is estimated that in America the average set is on for 6 hours and 18 minutes daily *(Broadcasting Yearbook,* 1971). This does *not* mean that it is being watched for all of these hours. Still, the usual figure for children's viewing is between 2 and 3 hours daily. A recent study, for example, looked at the viewing habits of children in a small town in southern California (Lyle & Hoffman, 1972). Interviews, viewing records, and TV diaries for over 250 first-graders, about 800 sixth-graders, and 500 tenth-graders revealed that most children watched every day for at least 2 hours, and significant numbers (especially of younger children) were exposed to twice that amount of viewing.

Lyle and Hoffman's findings concerning the time of day when children watched is of additional interest. Although first-grade viewing peaked at 8:00 p.m., 10 to 15 percent of these youngsters were still watching at 9:30 p.m. About half the older children were watching at 9:00 p.m. and over 25 percent were still watching at 10:30 p.m. It would appear, then, that children view a great deal of entertainment that is sometimes justified as being shown only at "adult hours." The reality is seen most clearly in the report of other investigators that "on one Monday during the period covered, over five million children under the age of 12 . . . were still watching between 10:30 and 11:00 p.m." (McIntyre & Teevan, 1972).

So television seems to have become part and parcel of daily life. It now serves children, for better or worse, as a constant informer, a faithful teacher, a window on the world. Exactly what is seen through this window? As far as we know there have been no attempts to characterize all the potential lessons of a sample of television shows at once. Instead, certain aspects of broadcast entertainment have been surveyed and counted according to the interests of the investigators or the times. It is perhaps not surprising that violence and aggression have been carefully and repeatedly surveyed. Some attention has been paid to the presence of national, ethnic, and social stereotypes which may convey undesirable prejudices. On the other hand, there have been few attempts to identify and systematically study the positive lessons of television entertainment, even though there is little doubt that these can be found. Our subsequent discussion therefore continues by first considering within the context of observational learning the effects of televised aggression and violence. Other relevant, and often related, content areas such as rule breaking and prejudicial stereotypes will be discussed more briefly; and, finally, consideration will be given to positive moral lessons that television might provide. Despite this division, it is important to remember that the basic process of observational learning underlies effects in all of these domains. *What* moral lessons television teaches depends simply on what is shown.

## Aggression and Violence

A great deal has been written about violence and aggression on television: research, critique, speculation, and defense. In discussing this area, it is first necessary to define terms. By *aggression* we mean any action that is harmful to others, such behavior often, but not always, contains a component of hostility or antisocial means of goal seeking. While this definition is admittedly broad, it excludes behavior that is simply of high magnitude or energetic. A child (or a television character) can be bold and assertive without being aggressive as defined here.

### Violence in TV Entertainment

Since television's beginning, violence and aggression have been part of American TV fare, but they have gradually and consistently increased. The National Association of Educational Broadcasters, reporting to a Senate subcommittee on juvenile delinquency, noted a 15 percent increase in violent incidents in television entertainment from 1951 to 1953. In 1954 about 17 percent of prime time was given to violence-saturated adventure programs; by 1961 this figure had risen to 60 percent. Two-thirds of the violence in 1964 was aired before 9:00 p.m., indicating that it was available even to children not among the late viewers (Liebert, Neale, & Davidson, 1973). In 1968 it was estimated that "the average child between the ages of 5 and 15 watches the violent destruction of more than 13,400 persons on TV" (Sabin, 1972).

George Gerbner (1972, 1973), Dean of the Annenberg School of Communications, has provided one of the most compelling estimates of the amount of violence appearing during the five-year period from 1967 to 1972. Gerbner's trained observers recorded the number of violent episodes on prime-time and Saturday morning cartoons during one week in October which was representative of each year's programming.

The major results of this research are startling. In 1969, for example, eight in ten plays contained violence, with five violent episodes per play. Further, the most violent programs were cartoons designed exclusively for children. The average cartoon hour in 1967, according to the Gerbner studies, contained more than three times as many violent episodes as the average adult dramatic hour. By 1969 there was a violent episode at least every two minutes in all Saturday morning cartoon programming, including the least violent and also commercial time. The average cartoon had nearly twelve times the violence rate of the average movie hour. This same pattern of high-level violence has held for several years.

## Acquisition of Televised Violence and Aggression

As Bandura (1965a) has shown, a single exposure to novel aggressive actions portrayed on a television screen is often sufficient for children to learn how to be exact "carbon copies" of their exemplars, precisely imitating complex sequences of verbal and physical aggression. What is more, behavior learned in this way is often retained for long periods of time; after a single viewing many children can reproduce what they have seen six to eight months later (Hicks, 1965, 1968).

Recall of physical acts portrayed in specially prepared film sequences must, however, be distinguished from the acquisition of the somewhat more subtle themes and relationships characterizing television stories. There is a growing body of evidence that young children respond to television dramas quite differently from adults, both cognitively and affectively, and thus learn quite different things than their elders from the same content. Children, for example, may inaccurately perceive the underlying plot even if it is a simple one; or they may fragment the overall content into discrete and unrelated segments (cf. Collins, 1970).

In view of these findings, it is reasonable to ask whether children perceive the motives and consequences of a television character's aggression as accurately as they perceive the aggressive actions themselves. The issue is an extremely important one in light of the industry's claim that televised violence is usually negatively sanctioned, inasmuch as "bad guys" are not rewarded for antisocial behavior. Unfortunately, though, the evidence shows rather clearly that commercial entertainment does *not* communicate the negative effects of "bad" violence to young viewers, whatever the broadcaster's intent.

Leifer and Roberts (1972) have attempted to identify age differences in the comprehension of motives and consequences surrounding aggressive acts shown on actual television programs. Almost 300 children, ranging in age from kindergarteners to twelfth-graders, served in their study. Striking age differences were found. Kindergarteners answered only about one-third of the questions correctly, third-graders about one-half, and twelfth-graders about 95 percent. Clearly, then, younger subjects did not learn or retain much about the motives and consequences involved in the story lines. Consistent with these findings is the work of A. H. Stein and Friedrich (1972), which has showed that preschoolers could remember some details of the programs they had viewed, but their recall was far from perfect. Stein and Friedrich's data suggest that "non-action" detail observed only once was not easily remembered.

In contrast to the consequences reaped by villains, televised violence by heroes is not punished. Rather, it is justified by moral pronouncements about the rectitude of aggressing for "good" reasons. We might also ask, then, about what children acquire from exposure to televised heroes who pronounce virtuous statements as they break laws and thrash their opponents. In a series of experiments Bryan and his associates (e.g., Bryan and Walbek, 1970a, 1970b) have documented that a child's moral statements bear little relationship to his actual behavior. "[These] investigations," write Bryan and Schwartz (1971) in a thoughtful review, "suggest the possibility that the aggressive hero who verbalizes socially sanctioned norms may well be teaching the observer how to be brutal and what to verbalize."

## Children's Acceptance of TV Aggression

Findings described in the previous sections show that children in our society are exposed to a substantial amount of violent televised material, and that repeated exposure to such symbolic modeling can teach specific responses and convey more general impressions about society at large and how people deal with each other. The remaining, and perhaps most important, question is whether children actually accept what they see in television's symbolic format as a guide for their own attitudes and actions in the moral sphere.

*Correlational field studies.* To obtain a broad view of the possibility that television violence contributes to aggressiveness, correlational field studies are particularly useful. In general, these investigations involve determining the relationship between the amount of TV violence viewed in the home and the degree to which observers engage in, or otherwise express approval and acceptance of, aggressive acts. While it is widely recognized that the method cannot definitively establish causality, it does provide suggestive evidence while maximizing the naturalness of the events of interest.

Many correlational studies have been conducted. They converge in showing that for youngsters from a variety of backgrounds and ages, television violence viewing is related positively to many aspects of aggressive behavior, from petty meanness to delinquency.

Numerous older studies (for example, Bandura, Ross, & Ross, 1963a, 1963b; Hicks, 1965; Rosekrans & Hartup, 1967) have shown that aggressive acts seen on a television screen will be copied spontaneously by young children when they subsequently interact with inanimate victims such as toys or plastic Bobo dolls. More recently, this basic finding has been amplified by a series of studies showing that preschool and young elementary school children who have seen televised aggression will respond similarly to real people—for example, by hitting a human clown with a mallet, sometimes forcefully (Hanratty, Liebert, Morris, & Fernandez, 1969; Hanratty, O'Neal, & Sulzer, 1972; Savitsky, Rogers, Izard, & Liebert, 1971). Even more striking, such effects have now been shown for peer aggression. Liebert and Baron (1972) exposed boys and girls aged 5 to 6 and 8 to 9 years to either a sports sequence (neutral content) or a violent sequence taken from the television program *The Untouchables*. When given an opportunity to help or hurt another child who was presumably seated in the next room, the youngsters who had seen the violent film chose to hurt the other child for a longer period of time than those who had viewed the neutral program.

Other studies have also shown such effects in the more natural situation of children's play. A. H. Stein and Friedrich (1972) showed violent cartoons or neutral or prosocial fare to 3- to 5-year-old children over a period of four weeks (twenty minutes a day, three times a week). Children who had been above the median in interpersonal aggressiveness on a pretest became significantly more aggressive against peers on the playground and in the nursery school if they had watched the aggressive cartoons. Steuer, Applefield, and Smith (1971), in another study of preschoolers, found that children who watched aggressive cartoons for eleven days became significantly more aggressive toward others than a control group, and in some cases showed increases in overt aggression (e.g., kicking and pushing) of 200 to 300 percent. Ellis and Sekyra (1972) showed first-grade children either an aggressive or neutral animated cartoon, both selected from the film library of a local TV station. The aggressive film featured hitting, tackling, and kicking within the context of a football contest; the neutral film depicted a musical variety show in which the characters sang and danced. Youngsters who had viewed the aggressive cartoon engaged in significantly more aggressive behavior when they returned to their classrooms than those in either the neutral or no-TV control groups. This finding is of particular importance because it was demonstrated in a classroom setting, and the stimulus material was regularly broadcast TV films of the cartoon variety so popular in children's programming.

*Longitudinal data.* Experimentation has shown that the impact of controlled violence viewing endures for several weeks, even when the subjects' home viewing diets are uncontrolled (A. H. Stein & Friedrich, 1972). But a more direct basis for suggesting that television violence does have a cumulative, adverse effect on the young has been provided by a longitudinal correlational study which spanned a ten-year period (Lefkowitz, Eron, Walder, & Huesmann, 1972).

Lefkowitz and his associates obtained data from more than 400 youngsters at 9 years of age and again at 19 years. The measure included peer ratings of aggression, self-reports of aggression, self-reports of various aspects of television viewing, and information on family background and parental practices. The results of this longitudinal study disclosed that, for boys, exposure to television violence at age 9 was significantly linked to aggressive behavior ten years later, at age 19. In fact, of the great variety of other socialization and family background factors measured, viewing of television violence was the best single predictor of aggressive behavior in late adolescence. Complex statistical analyses by the investigators and others (Kenny, 1972; Neale, 1972) have revealed that this relationship is most likely a causal one, not a spurious association. What is more, careful inspection of the evidence shows that television violence effects were not limited to a small number of boys who were already highly aggressive at age 9, but affected the entire spectrum of youngsters with an impact that was socially as well as statistically significant (Huesmann, Eron, Lefkowitz, & Walder, 1973.).

## Moral Lessons of Television

The previous section focused almost entirely on the possibility that exposure to televised violence would increase young viewers' overt aggressiveness. Clearly, this is one result that has been repeatedly demonstrated by the research. Other effects of television viewing, related in various degrees to the violence issue, may be equally or even more important in the child's moral development.

### Breaking Rules

We have already noted that one correlate of violence on television is that both "bad guys" and "good guys" use violence to break established rules. Robert Lewis Shayon (1966) gives this example:

> . . . it is difficult to recall an instance in which the wraps have been taken off popular entertainment's international morality so callously as in the case of *Mission: Impossible*. . . . It tends to legitimatize unilateral force for solving international problems. . . . It pretends that individual Americans are morally impeccable when they break the laws of a foreign nation. . . . The fact that they are not punished endorses the propriety of their acts. . . . Finally, it argues to viewers that a nation may enjoy a double standard—one for domestic, the other for exterior relationships . . . television writers too long have taken the easy way out, solving their problems by dubious moral stan-

dards. In a world at the door of satellite communications, it is time to introduce some international dimensions of ethical sensitivity.

What are the effects of televised rule breaking on young viewers? An extensive literature discloses that observational learning can play a potent role in shaping responses to situations in which breaking an established rule will bring immediate gratifications or benefits to the transgressor. It has been repeatedly shown that children exposed to an exemplar who breaks an established rule will themselves break the rule more often in the absence of adult surveillance than will those who have no such example (M. K. Allen & Liebert, 1969; Bandura & Kupers, 1964; A. H. Stein, 1967). On the other hand, it has been equally well established that live exposure to exemplars who adhere to established rules increases the likelihood that children will also adhere to rules even when they are highly tempted and seem to be able to transgress the rules without being detected (J. H. Hill & Liebert, 1968; McMains & Liebert, 1968; Rosenkoetter, 1973).

Investigators have recently determined that the above pattern of findings also applies to the effects of televised models. Exposure to a deviant television model induces more rule violation in elementary school youngsters than exposure to no example (Walters & Parke, 1964; Wolf, 1973; Wolf & Cheyne, 1972) or exposure to a conforming model (G. M. Stein & Bryan, 1972; Wolf, 1972). Wolf and Cheyne's comparison (1972) of live and televised rule deviations is a particularly important one because a one-month follow-up was included. They found that the effect of televised example was generally as potent as that of live example; both of these types of observational experiences were more important influences on the child's rule breaking than live verbal statements presented by an adult. Impressively, youngsters exposed to a deviant televised example in simulated television programs lasting less than three minutes were more likely to break the rule when tested one month later than were youngsters in the no-television control group. This demonstration that a brief television exposure can instill a remarkably stable tendency to transgress provides strong refutation of the argument that television examples exert only highly transitory influences on moral behavior.

## Blunting Sensitivity to Violence

Whether exposure to television violence leads young viewers to accept aggression as a mode of behavior is not limited to the direct performance of aggressive acts. An equally important outcome is that children's sensitivities may be blunted to aggressive and violent actions performed by others.

Support for this possibility has emerged from a correlational study by Cline, Croft, and Corrier (1972). These investigators divided a sample of 5- to 12-year-old boys into high and low TV users on the basis of interviews with the parents and the children themselves. All youngsters were then exposed to a film containing both violent and nonviolent segments while a variety of physiological measures of arousal (GSR, blood volume, pulse amplitude) were obtained. High users of TV showed less emotional reactivity to the violent scenes than low users. This finding suggests that television violence had indeed blunted the sensitivity of the former group.

As in other areas discussed in this chapter, experimental research is beginning to converge with correlational evidence. In one study (Rabinovitch, McLean, Markham, & Talbott, 1972), sixth-grade children watched either a violent program (a *Peter Gunn* episode) or a neutral one (from the series *Green Acres*). Then, to determine the youngsters' sensitivity to violence perpetrated by others, each child was simultaneously exposed to a pair of slides (one to each eye) using a stereoscopic projector which presented the images so quickly that only one could be seen. In each pair one slide was violent and the other was not. In all other respects the two slides were similar. For example, in one slide pair, the first slide showed a man hitting another person with a book, while the second slide showed a man pointing out something in a book to someone. After each pair was presented, the child was asked to describe in writing what he had seen. Those who had previously watched the violent TV show were less likely to report seeing violence in the slides than those who had watched the neutral program; this suggests that the former group had at least temporarily been desensitized to violent actions by others. Extending these results further, Drabman and Thomas (1974) have recently shown that children exposed to an aggressive television program were subsequently slower to intervene when observing a fight between other youngsters than were children who had not seen violent television fare.

## Stereotypes

Beyond acquiring information about the perpetration and justification of violence, children may also acquire stereotypes about aggressor and victim that will extend beyond the sphere of aggression itself. In this regard it is significant that the white American male is the leading character in half of all television programs. Usually young, middle-class, and unmarried, he is frequently an aggressor. Women, regardless of ethnic or racial background, make up about one-fourth of television characters. Two of every three women are married or engaged; employed women are more likely to be depicted as villains than are housewives.

Similar stereotypes can be found in regard to ethnic or national background. Minority groups are underrepresented and certain national groups are more likely than white Anglo-Saxon Protestants to be lawbreakers. One study has reported that Italians-Americans were represented as lawbreakers in over half the times they appeared on TV (Smythe, 1954). Until recently blacks were hardly portrayed, and then only in minor roles or as lovable clowns. Now blacks appear as stars, but the black experience in America is portrayed no more realistically than before. The black hero is a middle-class white with a dark complexion—a dauntless policeman or an ardent professional supporter of the status quo.

In representing these stereotypes, TV does not portray the world as it is. Still, it may convey information, accurate or not, to children about what is expected of others and themselves:

Several times a day, seven days a week, the dramatic pattern defines situations and cultivates premises about power, people, and issues. Just as casting the dramatic population has a meaning of its own, assigning "typical" roles and fates to "typical" groups of characters provides an inescapable calculus of chances and risks for different kinds of people.

In real life, of course, people are not simply good or bad, and estimates of an individual's moral stance can rarely be made from sex, occupational, or ethnic characteristics. Facial scars, voluptuous moustaches, and black or white hats are useless predictors of a person's decency or worth. Yet these very flat, simplistic categorizations are those that television most often presents. Rather than helping youngsters to outgrow simplistic stereotypes, TV seems to foster them (cf. DeFleur and DeFleur, 1967).

Given the stereotypic portrayals that dominated television in the late 1950s, it is perhaps not surprising that one of the earliest studies showed that boys who were high users of pictorial media (television, movies, and comic books) were more likely to invoke social stereotypes than were low users, even after the researchers statistically controlled for such third-variable explanations as social class and IQ (Bailyn, 1959). Similarly, an early study conducted by Siegel (1958) illustrated the same process experimentally. She asked second-grade children to listen to a radio program in which taxicab drivers were aggressive or to a similar radio play in which the drivers did not act aggressively. When questioned later about the likely behavior of cab drivers, children exposed to the more aggressive story were more likely to believe that drivers in their own town would act aggressively. This effect occurred after a single exposure, and may be only a weak reflection of the impact of repeatedly presented stereotypes through the more captivating audiovisual format of television.

## Reduction in Cooperative Behavior

Televised violence may also teach children by implication that people do not cooperate with each other, that negotiation and working together are not the usual (or effective) ways to settle differences or divide resources. If this is a fact, TV's influence should "spread" to other areas of behavior. The manner in which such spreading may occur has been demonstrated by Hapkiewicz and Roden (1971). In this study children saw either an aggressive cartoon, a nonaggressive cartoon, or no cartoon. Shortly thereafter they had an opportunity, in pairs and in the presence of an adult, to look at a movie through a "peephole" so that only one member of the pair could look at a time. Those who had seen the aggressive cartoon were dramatically less likely to share the peephole. Specifically, those in the nonaggressive cartoon group shared an average of 34 times during the observation period. Among the boys in the aggressive cartoon group there was an average of less than 9 instances of sharing during the same period—almost a fourfold difference.

## Implications

At the simplest level one implication of the research reviewed above is clear: Television *is* a moral teacher, and a powerful one. Contemporary television entertainment is saturated with violence and related antisocial behavior and lessons, which have a clear and, by most standards, adverse effect on young viewers' moral development and behavior.

But what are the further implications of these facts? Some might be tempted to damn the medium entirely, urging parents and educators to take private and public steps to reduce television usage by the young. We feel, however, that beyond the questionable practicality of such a plan—all indications are that massive exposure of children to television is here to stay—a blanket condemnation misses the more fundamental lesson of the research: Television does not stimulate antisocial behavior, *certain types of television content stimulate antisocial behavior.* And if certain content can teach undesirable moral lessons, there is a good reason to believe that other content might be equally potent in teaching positive, prosocial, moral lessons instead.

Loosely paralleling the exploration of aggressive and antisocial acts through observational learning, an equally impressive body of literature shows that desirable, prosocial forms of behavior can be acquired and shaped through exactly the same processes. It has been demonstrated repeatedly, for example, that exposure to live exemplars can increase children's sharing and stimulate cooperative behavior (Bryan & London, 1970; D. L. Krebs, 1970a; Poulos & Liebert, 1972).

Furthermore, as was the case for observational learning and aggression, the accumulated findings suggest a continuity between live modeling and television formats. Elliot and Vasta (1970), for example, have shown that sharing can be stimulated by exposure to a televised model, and that sharing may generalize to forms other than that demonstrated. Children who viewed a peer model give away some of his candy on television were more willing to share their money than children who observed no model. An extensive series of investigations by Bryan and his associates (e.g., Bryan, 1971; Bryan & Walbek, 1970a, 1970b) has provided further evidence of the potency of televised behavioral examples of generosity on young observers. Subjects were given the opportunity to play and win gift certificates. They then viewed a televised model who, upon winning certificates in the same manner, either kept all of them or gave some away to a charity. Subjects who viewed the charitable model were more likely to share than those who viewed the selfish exemplar. It is noteworthy that Bryan has tested hundreds of children in several geographical locations, and has consistently found this result. And generosity is not the only prosocial behavior influenced by televised models. Friendly, cooperative acts can be transmitted via a television format (Fechter, 1971; Fryrear & Thelen, 1969), as can adherence to a rule (G. M. Stein & Bryan, 1972; Wolf & Cheyne, 1972).

Nor are we limited in this realm to laboratory demonstrations. As noted earlier, A. H. Stein and Friedrich (1972) exposed 3- to 5-year-olds to one of three conditions. Children saw either aggressive programming (Batman and Superman cartoons), prosocial programming (episodes taken from *Mister Rogers Neighborhood*), or neutral fare (scenes such as children working on a farm). Effects were measured by observing the youngsters' naturally occurring behavior for two weeks prior to the viewing period, four weeks during the viewing, and two weeks after the viewing. Prosocial behavior was categorized into two classes: *self-control,* including rule obedience, tolerance of delay, and persistence; *interpersonal prosocial behavior,* including cooperation, nurturance, and verbalization of feelings. The results generally showed that self-control was increased by the prosocial programs and decreased by the aggressive ones as compared to the neutral programming; for youngsters in lower socioeconomic ranges, exposure to the prosocial shows increased positive interpersonal behavior as well.

Experimental demonstrations of what television *could* do, of course, will have little social significance until broadcasters know how to produce such programs and are convinced that these shows will enjoy popularity among young viewers in the competitive world of commercial television. It is here that we feel concerned social scientists can play one of their most significant roles in the future. In our own work we are proceeding in several ways.

One of our efforts involves an attempt to construct a code that can evaluate regular television programs for prosocial content. Such a code is necessary in the selection of programs for research purposes, but could also provide parents with a tool for assessing current programming—thus enabling them to meaningfully monitor their children's viewing. Feedback to the industry, in turn, could be an effective way to influence broadcasters' decisions about the kinds of programs they will air.

Our second thrust involves an effort to demonstrate further the positive impact of commercial programs high in prosocial content. From data recently collected there is no doubt that such an effort will be fruitful. In conjunction with this work we are also investigating children's attention to, and liking for, specific aspects of a variety of programs. This information can be fed into the production of prosocial shows which will be compelling in their lessons and enjoyable in their format and style of presentation.

Finally, as the culmination of our other efforts, we are collaborating in the production of children's programs. The first of these projects, now under way, involves miniature, commercial-message-length stories depicting nonaggressive, prosocial solutions to conflict situations. In this work, the full battery of knowledge regarding observational learning, vicarious consequences, and modeling techniques is being employed; equally important, children's understanding and reactions are being tested throughout production, a procedure rarely, if ever, employed in making standard fare for the young.

# What's Right with Sight-and-Sound Journalism

*Eric Sevareid*

*ERIC SEVAREID has been a CBS journalist through four decades as a member of the news team assembled by Edward R. Murrow in 1939. He has had long experience as a war correspondent political reporter and essayist.*

A kind of adversary relationship between print journalism and electronic journalism exists and has existed for many years. As someone who has toiled in both vineyards, I am troubled by much of the criticism I read. Innumerable newspaper critics seem to insist that broadcast journalism be like *their* journalism and measured by their standards. It cannot be. The two are more complementary than competitive, but they *are* different.

The journalism of sight and sound is the only truly new form of journalism to come along. It is a *mass* medium, a universal medium; as the American public-education system is the world's first effort to teach everyone, so far as that is possible. It has serious built-in limitations as well as advantages, compared with print. Broadcast news operates in linear time, newspapers in lateral space. This means that a newspaper or magazine reader can be his own editor in a vital sense. He can glance over it and decide what to read, what to pass by. The TV viewer is a restless prisoner, obliged to sit through what does not interest him to get to what may interest him. While it is being shown, a bus accident at Fourth and Main has as much impact, seems as important, as an outbreak of a big war. We can do little about this, little about the viewer's unconscious resentments.

Everybody watches television to some degree, including most of those who pretend they don't. Felix Frankfurter was right; he said, "there is no highbrow in any lowbrow, but there is a fair amount of lowbrow in every highbrow." Television is a combination mostly of lowbrow and middlebrow, but there is more highbrow offered than highbrows will admit or even seek to know about. They will make plans, go to trouble and expense, when they buy a book or reserve a seat in the theater. They will not study the week's offerings of music or drama or serious documentation in the radio and TV-program pages of their newspaper and then schedule themselves to be present. They want to come home, eat dinner, twist the dial, and find something agreeable ready, accommodating to *their* schedule.

For TV, the demand-supply equation is monstrously distorted. After a few years' experience with it in Louisville, Mark Ethridge said that, "television is a voracious monster that consumes Shakespeare, talent, and money at a voracious rate." As a station manager once said to a critic, "Hell, there isn't even enough mediocrity to go around."

TV programming consumes eighteen to twenty-four hours a day, 365 days of the year. No other medium of information or entertainment ever tried anything like that. How many good new plays appear in the theaters of this country each year? How many fine new motion pictures? Add it all together and perhaps you could fill twenty evenings out of the 365. As for music, including the finest music, it is there for a twist of the dial on any radio set in any big city of the land. It was radio, in fact, that created the audiences for music, good and bad, as nothing ever had before it.

Every new development in mass communications has been opposed by intellectuals of a certain stripe. I am sure that Gutenberg was denounced by the elite of his time—his device would spread dangerous ideas among the God-fearing, obedient masses. The typewriter was denounced by intellectuals of the more elfin variety—its clacking would drive away the muses. The first motion pictures were denounced—they would destroy the legitimate theater. Then the sound motion picture was denounced—it would destroy the true art of the film, which was pantomime.

To such critics, of course, television is destroying everything.

It is destroying conversation, they tell us. Nonsense. Non-conversing families were always that way. TV has, in fact, stimulated billions of conversations that otherwise would not have occurred.

It is destroying the habit of reading, they say. This is nonsense. Book sales in this country during the lifetime of general television have greatly increased and well beyond the increase in the population. At the end of a program with Hugo Black, we announced that if viewers wanted one of those little red copies of the Constitution such as he had held in his hand, they had only to write to us. We received about 150,000 requests—mostly, I suspect, from people who didn't know the Constitution was actually down on paper, who thought it was written in the skies or on a bronze tablet somewhere. After my first TV conversation with Eric Hoffer, his books sold out in nearly every bookstore in America—the next day.

TV is debasing the use of the English language, they tell us. My friend Alistair Cooke, for one. Nonsense. Until radio and then TV, tens of millions of people living in sharecropper cabins, in small villages on the plains and in the mountains, and in the great city slums, had never heard good English diction in their lives. If anything, this medium has improved the general level of diction.

The print-electronic adversary relationship is a one-way street. Print scrutinizes, analyzes, criticizes us every day; we do not return the favor. We have tried now and then, particularly in radio days with "CBS Views the Press," but not enough. On a network basis it's almost impossible because we have no real national newspapers—papers read everywhere—to criticize for the benefit of the national audience. Our greatest failure is in not criticizing ourselves, at least through the mechanism of viewers' rebuttals. Here and there, now and then, we have done it. It should have been a regular part of TV from the beginning. The Achilles heel of TV is that people can't talk back to that little box. If they had been able to, over the years, perhaps the gas of resentment could have escaped from the boiler in a normal way; it took Agnew with his hatchet to explode it, some years ago. The obstacle has not been policy; it has been the practical problem of programming inflexibilities—we don't have the fifteen-minute program anymore, for example. If we could extend the evening news program to an hour, as we have wished to do for years, we could do many things, including a rebuttal period from viewers. It is not the supposedly huckster-minded monopolistic networks that prevent this; it is the local affiliates. It was tough enough to get the half-hour news; apparently it's impossible to get an hour version.

I have seen innumerable sociological and psychological studies of TV programming and its effects. I have never seen a study of the quality—and the effect—of professional TV criticism in the printed press.

TV critics in the papers tell us, day in and day out, what is wrong with us. Let me return the favor by suggesting that they stop trying to be Renaissance men. They function as critics of everything on the screen—drama, soap operas, science programs, musical shows, sociological documentaries, our political coverage—the works. Let the papers assign their science writers to our science programs, their political writers to our political coverage, their drama critics to the TV dramas.

Let me suggest also that they add a second measurement to their critiques. It is proper that they judge works of fiction—dramas, for example—entirely on the basis of what they see on the little screen because in that area the producers, writers, and performers have total control of the material. If the result is wrong, *they* are wrong. News and documentaries are something else—especially live events, like a political convention. Here we do not have total control of the material or anything like it. On these occasions, it seems to me, the newspaper critic must also be a reporter; he must, if he can, go behind the scenes and find out why we do certain things and do not do other things; there is usually a reason. In the early days such critics as John Crosby and Marya Mannes would do that. A few—Unger on the *Christian Science Monitor,* for example—still do that, but very few.

Let me suggest to their publishers that a little less hypocrisy would become them. Don't publish lofty editorials and critiques berating the culturally low common denominator of TV entertainment programming and then feature on the cover of your weekly TV supplements, most weeks of the year, the latest TV rock star or gang-buster character. Or be honest enough to admit that you do this, that you play to mass tastes for the same reason the networks do—because it is profitable. Don't lecture the networks for the excess of violence—and it *is* excessive—on the screen and then publish huge ads for the most violent motion pictures in town, ads for the most pornographic films and plays, as broadcasting does not.

Now, at this point the reader must be thinking what's this fellow beefing about? He's had an unusually long ride on the crest of the wave; he's highly paid. He's generally accepted as an honest practitioner of his trade. All true. I have indeed been far more blessed than cursed in my own lifework.

But I am saying what I am saying here—I am finally violating Ed Murrow's old precept that one never, but never, replies to critics—because it has seemed to me that someone must. Because the criticism exchange between print and broadcasting is a one-way street. Because a mythology is being slowly, steadily, set in concrete.

Why this intense preoccupation of the print press with the broadcast press and its personae? Three reasons at least: broadcasting, inescapably, is the most personal form of journalism ever, so there is a premium on personalities. The networks are the only true national news organs we have. And, third, competition between them and between local stations is intense, as real as it used to be between newspapers.

So the searchlight of scrutiny penetrates to our innards. Today we can scarcely make a normal organizational move without considering the press reaction. Networks, and even some stations, cannot reassign a reporter or anchorman, suspend anyone, discharge anyone, without a severe monitoring in the newspapers. Papers, magazines, wire services, don't have to live with that and would very much resent it if they did.

We live with myths, some going far back but now revived. The myth that William L. Shirer was fired by Ed Murrow and fired because he was politically too liberal. He wasn't fired at all; but even so good a historian as Barbara Tuchman fell for that one. The myth that Ed Murrow was forced out of CBS. At that point in his great career, President Kennedy's offer to Ed to join the government, at cabinet level, was probably the best thing that could have happened to him. The myth that Fred Friendly resigned over an issue of high principle involving some public-service air time. There are other such myths, and a new generation of writers are perpetuating them in their books, which are read and believed by a new generation of students and practitioners of journalism.

We have had the experience of people leaving, freely or under pressure, then playing their case in the papers to a fare-thee-well; they are believed because of that preconceived image of the networks in the minds of the writers. What does a big corporation do? Slug it out with the complainant, point by point, in the papers? Can it speak out at all when the real issue is the personal character and behavior of the complainant, which has been the truth in a few other cases? It can't. So it takes another beating in the press.

There is the myth that the CBS News Division—I am talking about CBS, of course, because it is the place I know and because it is the network most written about these days—has been somehow shoved out to the periphery of the parent corporation, becoming more and more isolated. What has happened is that is has achieved, and been allowed, more and more autonomy because it is fundamentally different from any other corporate branch. Therefore it is more and more independent.

There is the myth that the corporation is gradually de-emphasizing news and public affairs. In the last sixteen years since CBS News became a separate division, its budget has increased 600 percent, its personnel more than 100 percent. It does *not* make money for the company; it is a loss leader, year after year. I would guess it spends more money to cover the news than any other news organization in the world today. This is done because network news servicing has become a public trust and need.

There is the myth that since the pioneering, groundbreaking TV programs of Murrow and Friendly, CBS News has been less daring, done fewer programs of a hard-hitting kind. The Murrow programs are immortal in this business because they were the *first*. Since then we have dealt, forthrightly, with every conceivable controversial issue one can think of—drugs, homosexuality, government corruption, business corruption, TV commercials, gun control, pesticides, tax frauds, military waste, abortion, the secrets of the Vietnam War—everything. What shortage has occurred has been on the side of the materials, not on the side of our willingness to tackle them.

In case I had missed something myself, I have recently inquired of other CBS News veterans if they can recall a single case of a proposed news story of a documentary that was killed by executives of the parent organization. Not one comes to anyone's mind. Some programs have been anathema to the top executive level, but they were not stopped. Some have caused severe heartburn at that level when they went on the air. Never has there been a case of people at that level saying to the News Division, "Don't ever do anything like that again."

**"Today we can scarcely make a normal organizational move without considering the press reaction."**

For thirteen years I have done commentary—personal opinion inescapably involved—most nights of the week on the evening news. In that time exactly three scripts of mine were killed because of their substance by CBS News executives. Each one by a different executive, and none of them ever did it again. Three—out of more than 2,000 scripts. How many newspaper editorialists or columnists, how many magazine writers, have had their copy so respected by their editors?

There is the perennial myth that sponsors influence, positively or negatively, what we put on the air. They play no role whatever. No public-affairs program has ever been canceled because of sponsor objection. Years ago, they played indirect roles. When I started doing a six P.M. radio program, nearly thirty years ago, Ed Murrow, then a vice-president, felt it necessary to take me to lunch with executives of the Metropolitan Insurance Company, the sponsors. About fourteen years ago, when I was doing the Sunday-night TV news, a representative of the advertising agency handling the commercials would appear in the studio, though he never tried to change anything. Today one never sees a sponsor or an agency man, on the premises or off.

There is the myth, which seems to be one of the flawed premises of so successful a reporter as David Halberstam, that increased corporate profitability has meant a diminished emphasis on news and public affairs. The reverse, of course, is the truth.

There is the new myth, creeping into print as writers rewrite one another, that an ogre sits at the remote top of CBS Incorporated, discouraging idealistic talents down the line, keeping the news people nervous, if not cowardly. His name is William S. Paley, and the thesis seems to be that the tremendous growth of CBS News in size and effectiveness, its unmatched record of innovation and boldness in dealing with public issues, its repeated wars with the most powerful figures of government and business, have all taken place over these forty years or more in spite of this man's reluctance or downright opposition. The reverse is much closer to the truth. *Only* with a man of his stripe could all this have been done. Think what it is to sit up there all those years, whipped by gales of pressures from every public cause group, politicians, Presidents, newspapers, congressional committees, the FCC, affiliates, stockholders, and employees, individually and organized. To sit up there under unrelenting pressures of an intensity, a massiveness, rarely endured by any print publisher and still keep the apparatus free and independent and steady on its long course. After all, in this country networking might at its inception have become an appendage and apparatus of government; it might have gone completely Hollywood. It did neither. It grimly held to every freedom the law allows, and it fights for more. This has not been accomplished by weak or frightened men at the top.

I am no appointed spokesman for William Paley. We are not intimates. I owe him nothing; I have earned my keep. I have not got rich. I have had my differences with him, once or twice acutely. I have been a thorn in his side a number of times. But we had our differences out, and never once was his treatment of me less than candid and honorable. He is now in the evening of his career; I am now pretty much the graybeard of CBS News. I must soon go gently into that good night of retirement. But I shan't go so gently that I shall not say what I think of the mythologists who now surround us, what I think of these ignorant assaults on Paley. It would be cowardly of me not to say that many of these critics are simply wrong—wrong in their attitude, wrong in their premises, repeatedly wrong in their facts.

We are not the worst people in the land, we who work as journalists. Our product in print or on the air is a lot better, more educated and more responsible than it was when I began, some forty-five years ago, as a cub reporter. This has been the best generation of all in which to have lived as a journalist in this country. We are no longer starvelings, and we sit above the salt. We have affected our times.

It has been a particular stroke of fortune to have been a journalist in Washington these years. There has not been a center of world news to compare with this capital city since ancient Rome. We have done the job better, I think, than our predecessors, and our successors will do it better than we. I see remarkable young talents all around.

That's the way it should be. I will watch them come on, maybe with a little envy, but with few regrets for the past. For myself, I wouldn't have spent my working life much differently had I been able to.

# The Gulag of Television Reviewing

*Karl E. Meyer*

*KARL MEYER is Television Critic for* Saturday Review.

A glance at the calendar alerts me to the surprising fact that I have just begun my fourth year in this berth. At the risk of appearing self-serving, I would like to say a few words about the state of television reviewing. It is not good. Questions of quality aside, the quantity is derisory. I know of no other form of expression in which there is a greater disparity between popular availability and critical attention—in the case of television, familiarity seems to breed contemptuous neglect.

An off-Broadway play, seen at most by thousands of people, is taken more seriously than a TV mini-series that will be viewed by tens of millions. The feeblest movie, perpetrated by the most inconsequential director, will attract more critical notice than a first-rate television documentary that addresses a subject of obvious importance. This is true even though most television programs are available for advance screenings, affording the critic an enviable chance to inform a national audience about an event that can be seen at home, at no charge.

Television remains scarred by its ugly birthmark. It is still condescendingly perceived, as it was in the formative 1950s, as the boob tube and the idiot box. In a sense, television has taken over the cultural Gulag once occupied by the movies; it was a rash highbrow, in the 1920s, who would brave titters by arguing (as did Gilbert Seldes in his pioneering *The Seven Lively Arts*) that Charlie Chapin and D. W. Griffiths were "serious" artists.

Nowadays, every national magazine employs a film critic, and few jobs are as coveted (one gets the impression that the life's ambition of every red-blooded American youth is to become a cinema critic). No movie ad is complete without its garland of approving comments by the likes of Kael and Canby, Sarris and Simon, Schickel and Kroll, Haskell and Schlesinger, Crist and Cocks. In the other established arts—drama, music, dance, painting, and sculpture—there are critics with the same authority, if not glamour.

By contrast, television criticism is conspicuous by its absence. To be sure, a few major newspapers, such as *The New York Times* and the *Washington Post,* feature daily reviews, often written with verve and bite. But aside from the *Saturday Review* (an unwelcome distinction), I cannot name a single general-interest national magazine that publishes a TV column in most issues.

The *New Yorker's* Michael Arlen has few peers as a reflective essayist on television, but he appears too infrequently. Over at *New York,* which faithfully covers the other arts, there has never been sustained coverage of television. *Esquire* deals with TV only when its regular columnists, like Richard Reeves, feel so inclined. The *New Republic* publishes James Lardner's television column from time to time, but on just that basis—from time to time. *New Times* has the usual film critic, but no television counterpart. Neither the *Atlantic* nor *Harper's* cares to dignify television with continuous attention. Ditto the *Nation.* However, after ignoring TV for years, the *Village Voice,* to its credit, now features the hyperfluent James Wolcott.

Without such specialized journals as *American Film* (edited by Hollis Alpert, a former *SR* critic, and published by the American Film Institute), the situation would be close to desperate. Suspecting that even *Time* and *Newsweek* also keep their television departments on starvation rations, I undertook a column-by-column census of back-of-the-book coverage in 13 consecutive issues of *Time,* with results tabulated above. According to my count, *Time* found space to review 62 books, 28 movies, 11 plays, 8 art exhibitions, and just 3 television programs. If one is to credit *Time,* the only TV programs that mattered in nearly

TIME'S CULTURAL COVERAGE

| Number of columns devoted to: | June (4 issues) | July (5 issues) | August (4 issues) | Total (13 issues) |
|---|---|---|---|---|
| BOOKS | 32 | 41 | 29 | 102 |
| CINEMA | 10½ | 25½ | 16 | 52 |
| MUSIC | 11 | 21 | 10½ | 42½ |
| ART | 15 | 12 | 12 | 39 |
| THEATER | 12 | 8 | 5 | 25 |
| SHOW BUSINESS | 10 | 6 | 11 | 27 |
| TELEVISION | 2 | 4 | 0 | 6 |

three months were ABC's fine documentary, *Youth Terror: the View from Behind the Gun;* PBS's *The Norman Conquests,* a British import; and a retrospective on public television of Edward R. Murrow's *Person to Person* interviews. Though the newsweeklies devote considerable attention to TV profits, salaries, and ratings, they rarely find any TV program deserving of critical comment.

With the new fall season upon us, we can expect that the newsweeklies will churn out the predictable cover story on the prime-time puerility of network programming. And the same magazines will continue to lavish space on third-rate movies while routinely ignoring anything superior on television. Small wonder that so many of television's talented people regard the medium as a preparation for a really serious career—like making mediocre movies for a coterie audience. If it is a boob tube, we are getting only what we deserve.

# RATINGS

## The Numbers Behind Programming Decisions

With millions of dollars at stake when programming decisions are made, *the ratings* provide the great magical numbers for the television industry. Ratings determine which programs stay on the air, which are cancelled, where a program will be scheduled in the broadcast day, and the amount of money advertisers will pay to get their commercials into that same time slot.

Ratings are also watched closely by the persons who are planning and developing new programs. In their attempts to create winners, they often closely imitate the top-rated programs of today so ratings play a large role in future programming.

Ratings, like political polls, are based on relatively small samples of people. Despite that, they are considered to quite accurately represent the viewing patterns of the mass audience of more than 73 million U. S. television households.

It is important to note, however that the ratings indicate choices made by viewers *from the programming that is available.* That available programming is often very similar on the various channels. Given our tendency to watch television no matter what is on, a choice can as easily be a vote for what one dislikes least as for what a viewer would like most to see.

Ratings deal in the big numbers. Someone's favorite program may not even show up on the ratings of what a station broadcasts. As far as the ratings go the program does not exist, a fact that can well mean the program soon will not exist for the viewer, either.

The major rating companies are Nielsen and Arbitron. They compete in local markets, but Nielsen ratings are the yardstick used at network level.

The Nielsen Television Index (NTI) measures the audience for network TV and is drawn from a sample of 1170 households selected on the basis of census information. Each household usually participates for a five-year period, the TV's connected to a device that records whether or not a set is on, what channel it is turned to when on, how long it is on each channel and all channel switching. The device, called Storage Instantaneous Audimeter (SIA), of course is collecting information from the TV, not from the viewers. It doesn't know if anyone is watching the screen or not. Actually, research using cameras hidden in TV screens discovered some interesting activities in front of the TV that didn't involve watching the screen.

A separate panel of households participates in Nielsen's National Audience Composition (NAC) sampling. In those homes a Recordimeter records the amount of time the set is on, and members of the household keep a written log indicating which person viewed and what they viewed.

The Nielsen Station Index (NSI) is a local market service. Households in the three largest metropolitan areas, New York City, Chicago and Los Angeles are equipped with SIA devices. The electronically collected information is supplemented by information from viewer diaries. However, it is the SIA *electronic* information which determines the overnight ratings.

In all other metropolitan area markets, Nielsen Station Index information is gathered from diary information gathered three to eight times a year depending on the area's population. Those who keep the diaries are selected from a sample of residential phone listings. Each household keeps a diary for one week.

Arbitron offers services similar to Nielsen, using a Household Collector in Los Angeles, and New York. Diaries are used for Arbitron's Television Local Market Reports.

The so-called "Sweeps", ratings issued three times a year by Arbitron and Nielsen, represent some 200 individual TV markets. Each sweep covers a four-week period and the information collected is used in determining rates for national commercials. Since the ratings have a direct connection to station income, the Sweeps are very subject to manipulation with high audience appeal programs, such as top-rated movies and highly promoted specials, often taking the place of programs that have less appeal. For the viewers this can lead to a feast and famine situation. During the four-week Sweep there seems to be a lot of great programming available, then the sweep period ends and it is three months before the screen offers so many appealing choices again.

Both Arbitron and Nielsen report rating information in many ways.

*Average Audience*—the program's audience during the average minute, which indicates audience size during any one average commercial.

*Total Audience*—the number of households tuned to at least six minutes of a program. It indicates how many households may be reached by at least one commercial message during a telecast.

*Share of Audience*—a ratio of a telecast's audience as compared to the total number of households using TV at that time. A program getting a 30 percent share was the program on the screen in 30 percent of all households where TV was in use.

*Rating*—the percentage of households tuned to a program as compared to all the TV households in the nation or in a market. A national rating of 20 means that 20 percent of U. S. TV homes (73 million in 1978) are estimated to be watching that program, or about 15 million homes.

Nielsen and Arbitron also provide extensive projections of the make-up of the audience, called demographics. The person categories are:

Child—2 to 11

Teens—12 to 17

Men—18 to 34, 18 to 49, 25 to 49, 25 to 54, and 18 and up.

Women—15 to 24, 18 to 34, 25 to 54, 18 and up.

These demographic figures provide a comprehensive picture for who is watching what kinds of programming at what times of the day. It becomes very valuable information for advertisers who want to aim their sales message at specific groups. Sometimes the demographic figures show a series with a satisfactory overall rating is failing to reach the audience which is most responsive in its buying habits to the sales messages. That can mean cancellation of a program, despite its satisfactory rating.

Arbitron and Nielsen both offer a variety of explanatory material at no charge. Write to:

ARBITRON TELEVISION
1350 Avenue of the Americas
New York, N. Y. 10019

A. C. NIELSEN COMPANY
Media Research Services Group
Nielsen Plaza
Northbrook, IL 60062

# A Letter to the President of ABC Television Network

*Sister Elizabeth Thoman*

*Written by Sister Elizabeth Thoman, CHM, Director of the Sisters Communication Service, after the airing of the ABC Sunday Night Movie,* Most Wanted. *The film's storyline includes the raping and murdering of nuns.*

Mr. Frederick Pierce
President, ABC Television Network
1330 Avenue of the Americas
New York, New York 10019

March 31, 1976

Dear Mr. Pierce:

The National Sisters Communications Service, a national liaison and resource office in communications for 140,000 American Catholic sisters, would like to express our great disappointment and even anger about the March 21 ABC Sunday night movie, "Most Wanted," produced by Quinn-Martin Productions.

We feel, first of all, that the movie not only stereotyped sisters as naive, innocent and faceless but by creating a story in which the sisters are rape victims, over-sensationalized rape as a sexual crime rather than an agressive act of violence against a human person. We would ask, for instance, whether the "Most Wanted" police brain trust force that is central to the plot, would have been created if the victims were elderly women or minority women or mothers of 2-year-olds or telephone operators or any other "group" of women besides nuns? The fact that the city councilman and the press bring pressure on the police department because "Brides of Christ" (an extremely out-of-date and pre-Vatican II theological concept, by the way) had been violated and murdered leads the viewer to believe that it is somehow a worse crime to rape a nun than to rape any other woman. We reject that concept totally and consider it a storyline that insults all women, for whom rape is not just a physical danger but a violent attack on one's very personhood.

In addition to the basic storyline concept, we must also fault "Most Wanted" for passing up a marvelous opportunity to make a *positive* contribution to women viewers, using the story to provide information about rape defense. The scene in which this could have been done easily is when the detective (Robert Stack) comes to the "Convent" to speak to a group of sisters about defending themselves against the rapist-murderer. The scene is set, the opportunity is available but it failed in several ways.

1) The detective is not only patronizing and protective but, in actuality, gives absolutely *no clear information* about defense, other than to scream or to hit with a heavy book. It would seem more logical that the communities of sisters would have organized workshops in self defense given by women's groups or *at least* a woman police officer, instead of just sitting demurely in straight-backed chairs listening to a fatherly *male* detective's words of caution.

2) The portrayal of the sisters as naive and nervously laughing like school girls leads to a false impression that sisters are not mature adult women, but some kind of repressed neuter gender. In addition, this scene, by zooming in on the sisters, eyes cast down at the suggestion that they might have to submit to the rapist to avoid being killed, reinforces the concept of rape as a sexual experience, *which it is not,* rather than a physical assault, which it is. Furthermore the detective's lines, by not being specific, border on innuendo, thus titillating the late-night viewing audience, who, in turn, laugh at the nuns' naiveté.

A third aspect of the film continues this undertone of sex vs. naive innocence. To have the sister, who was raped but not killed, upset over "the potential life she might be carrying" was simply not believable. Rape is an assault and a nun would be treated medically as any other woman victim so that there would be no worry of pregnancy as a result of the rape. We regret that there was no information given in this scene, or anywhere in the film, about rape crisis clinics or other self-help programs developed by women for rape victims. No, instead, the detective continues his fatherly role, even dispensing moral advice to the distraught sister. Later, in order to trap the rapist, he indicates that she has been sent "on retreat" alone. Both of these scenes leave the impression that the sister is totally alone in dealing with the reality of her assault. There is no indication of the loving supportive *community* of women who would sustain her psychologically and emotionally if the event had happened in real life. The absence of such a community in the film leads to a further distortion of the life of women religious as cold, disciplined and lonely.

Finally we want to say something about the dress of the supposed nuns in the film. For the most part they wore contemporary, but subdued and tailored dress with a short veil. One of the first signs of renewal in the Church was to separate a sister's mode of dress from her ministry or her work as a woman in the Church. Consequently, many sisters today do not wear a veil or any kind of distinctive dress. But in the film, the wearing of the veil seemed essential to the story—how else would the rapist know whom to rape? May we suggest that if the story had actually happened, the existence of such an attacker would be cause enough for the sisters to stop wearing their veils in public and to dress as inconspicuously as possible. Even in days of the traditional habit, sisters did not hesitate to abandon it when wearing it might have endangered their lives.

In addition to these aspects of the story involving sisters, we would like to mention the poor character development of the woman on the brain trust team, the clnical psychologist. At the beginning she is identified as a top-notch clinical psychologist and an experienced police woman as well. She is hired because she's the best. But as the story develops, she is portrayed as vacillating, having to "check things out" with another psychologist—a male—because she's "not sure." The other male members of the brain trust do not defer to others in this way, why should she? We find her development as a character as stereotyped and offensive as the portrayal of the sister-characters.

As you can gather from these observations, we found the plot, the characters and the whole development of "Most Wanted" not just inadequate or stereotyped but totally unbelievable and demeaning to women, especially women religious. We find it regrettable that ABC, which has had the reputation of being somewhat sensitive to women and women's issues, sold out to sexual sensationalism by buying and airing "Most Wanted."

As for the future, may we just remind you that the National Sisters Communications Service, staffed by professional sister-communicators, is anxious to help the media industry portray the roles of women in the Church accurately and fairly. Our Los Angeles location makes us easily accessible as consultants to producers such as Quinn-Martin and even to writers in the early script stages. We would, of course, much prefer having input on a production as early as possible, rather than simply writing letters of complaint after the fact. In addition, we feel that our values are in solidarity with other women who find themselves imaged in stereotypes and therefore would be pleased to serve as a sounding board for questions of taste, values or image concerning the portrayal of women, including minorities, the elderly and others that are too often stereotyped and voiceless.

For your information, copies of this letter are being sent to the following:

*Broadcasting and Film Commission, National Council of Churches
*United States Catholic Conference, Department of Communications
*National Organization of Women, Media Committee
*Media Committee of the Presidential Commission on the Observance of International Women's Year

as well as to the sisters' organizations in our own constituency:

*Leadership Conference of Women Religious
*National Sisters Vocation Conference
*Sister Formation Conference
*National Assembly of Women Religious
*National Coalition of American Nuns

We would appreciate your sending copies of this letter to the appropriate ABC offices that would find this letter informative or helpful.

Thank you.

Sincerely,
Sister Elizabeth Thoman, CHM
Director

P.S. We call special attention to the recently issued "Entertainment Programming and Advertising Checklist" developed by the Media Committee of the Presidential Commission on the Observance of International Women's Year:

#4—Is the exploitative "Woman as Victim" theme the main entertainment value of your piece? Is she the hapless object of brutalizing forces? Does she make things worse by making panicky choices? Would the piece work just as well if a man were in her shoes?

#5—If a rape is shown, is it dealt with as a basically sexual experience, which it is not, or as a physical assault, which it is?

# The Role of Mass Communication in Society

*William F. Fore*

*WILLIAM F. FORE is Assistant General Secretary for Communication, National Council of Churches. He is a member of the Advisory Council of National Organization, Corporation for Public Broadcasting. Mr. Fore has a Ph.D. in education and media from Columbia University, a D.D. from Yale Divinity School.*

As society gets more complex, theology gets more complex. It's like the theologian and the astronomer who happened to sit together on the airplane. After introducing themselves and discovering each other's occupation, the astronomer said, "I've always thought that theology coud be summed up in the fatherhood of God and the brotherhood of man." "And I've always thought," said the theologian, "that what we know in astronomy could be summed up in 'Twinkle, twinkle little star.' "

The situation is not simple, either in society or in religion. Yet it is clear that our social engineering simply has not kept pace with our technological engineering. Everywhere we are faced with *things* that are out of control: air pollution, armaments, automobiles—I invite you to run through the complete alphabet of Frankenstein horrors we have created. This failure of social engineering is at root a *religious* problem, because we have failed to examine our religious heritage and our sense of the holy in a systematic way, and to relate it to our lives in a complex social environment.

To do this we need first to look at the nature of society and what it needs in order to function; second, to understand the role of communication in society and how it meets society's needs; and third, to analyze the underlying myths, values and assumptions which energize our society, and how they relate to Christian myths, values and assumptions.

## What Every Society Needs

What every society needs and must have if it is to remain a society is *commonality*—common interests, common language, common traditions, common institutions, common values, common ends. And there must be a set of common assumptions—assumptions about who we are, who has the power, what we can and cannot be, what we can and cannot do.

But these underlying assumptions are hidden. They come to light only when we begin to ask such questions as: What are those things we never have to ask about? What are those things we know are not only true but are simply *there*? What are those things that are given to us in the way things "are"? One reason advanced geometry is important is that it makes students consider worlds quite different from the world they assume to be "true," worlds in which parallel lines meet, in which the shortest distance between two points is a curved line. Science fiction and *Mad* magazine and foreign languages get at the social world the same way, by questioning the given, the assumed reality.

But society resists this probing, this questioning of what *is*. Society needs stability, and stability depends upon commonality, uniformity, conformity. Thus every society propagandizes and every society censors. Jacques Ellul, in his book on propaganda, describes it as an all pervasive aspect of communication in society which is not an arbitrary creation by the people in power, but is something which grows out of the need of the whole group in order to sustain the group.[1] It uses all the media of communication, but it is most effective when it reaches a person "alone in the mass," that is, when a person is cut off from group participation.[2] It tends to cut a person off from outside points of reference,[3] such as, for example, transcendent religious reference.

The society also employs active censorship against those communications which tend to threaten the common values, actions and assumptions. The censorship may be legal, as with pornography. It may be political, as with the press silence on American involvement in Cambodia. It is most likely to be economic, as in the case of TV's exclusion of minority points of view because they would tend to reduce profits.

Propaganda and censorship are not something visited on the people by evil manipulators. They are an inevitable process that gives most people—that is, the society—what they want and need very badly: stability, cohesion, and common purpose.

## What Media Provide Society

The way the society does this is primarily through mass media. Virtually every communication contributes to the creation of commonality. Every activity—games, work, play, sex, study, eating, resting—and every medium—verbal, non-verbal, signs, symbols, architecture, paintings, books, memos, letters, maps, and so on—all are mediators of the culture. But only in the last seventy-five years have there developed the mass media of communications—the telegraph, the large volume newspaper, the wireless telegraph, radio and television—all of which are primarily *social* inventions, because they fundamentally changed the speed, the extent and nature of the process by which a society maintained commonality, and thus changed the nature of society itself.

And just as society as a whole must propagandize and censor to maintain a reasonable commonality of experience, so also the mass media select and distort what they mediate, for two reasons. First, because it is their *nature,* and second, because the society needs them to create the common world of which all can be a part. For example, television is indeed a window on the world. But a window by its very nature selects out only a small piece of reality. And though its glass seems transparent, it actually shuts out heat and cold, noise and smells, and as in the tinted glass in today's buses and airports, it may even totally change the color of everything "out there." Thus with TV: it acts as a filter, selecting things out, extracting unpleasant (and pleasant) elements, coloring others, and making a whole world "out there" seem real to us when it is in fact nothing more than bright phosphors dancing on a piece of glass in a real room which looks quite dull by comparison.

Rudolf Arnheim, the author of *Visual Thinking,* says that a child who enters school today faces "a 12 to 20 year apprenticeship in alienation." He points out that as soon as a child learns to name something, the child begins to separate himself from it, and before long learns to handle words and concepts, but at the risk of becoming estranged from the objects talked about. The child learns to manipulate a world of words and numbers, but does not learn to experience the real world. Instead, the child has been conditioned to live in our culture. Exposure to television for hours every day simply further separates the child from the world of reality, or rather creates for the child a *new* reality.

For example, according to *The New York Times,* a commanding officer of a U. S. Army base in Germany attributed the high divorce rate of servicemen stationed there to the absence of English language television. "When they go home at night," he said, "there's nothing to do but to talk to each other, and what they see and hear they don't like."

And when television and reality conflict, TV often has the greater power. Recently two communities lying within the Salt Lake City broadcasting area, but in another time zone, petitioned the Department of Commerce for rezoning. They wanted clock time to conform to broadcast time.

Abraham Moles, Director of the Social Psychology Institute at Strasbourg, points out that while television has been a cultural life-buoy for farmers, lonely people, and the impoverished, at the same time it has been a pressure toward the banal and the constricting for those already experiencing a communication-rich life. But in both cases, as the individual is exposed to more and more TV, he or she is a little bit less able to differentiate between the fictional universe and the real world. Thus by its very nature television, and all mass media, filter and change the reality they mediate.[5]

## Myths, Symbols, Images and Fantasy

The mass media are also expected to provide commonality for the society. The tools which the media use are myth, symbol, images and fantasy. The myths are most important, because they stand behind everything else. In essence, myths tell us who we are, what we have done, and what we can do. Myths deal with power—who has power, who does not have power, where we are in the power scale. Myths deal with value—what is of value and what is not. Myths deal with morality—what is right and permissible, what is not right and is forbidden.

The myths of our society thus constitute a kind of religious framework, both in the sense that they provide us with a belief and value system, and also in the sense that they express the things which we uncritically assume as *given* in our lives. The myths express not the rules that are written down in our laws and our Bibles but the *rules behind the rules* that are written down. That is, they express ultimate reality, which is another term for religion.

Myths are expressed in symbols and images which reach us less at the cognitive level than at the level of dreams and fantasy. Stanely Kubrick, creator of such memorable films as *Dr. Strangelove, 2001, A Space Odyssey* and *Barry Lyndon,* understands what is happening: "I think an audience watching a film or a play is in a state very similar to dreaming, and that the dramatic experience becomes a kind of controlled dream. . . . But the important point here is that the film communicates on a subconscious level, and the audience responds to the basic shape of the story on a subconscious level, as it responds to a dream."[6]

The image-symbol-fantasy level of communication is more powerful than the cognitive level because we find it more difficult to bring it up to a level of consciousness where we can analyze and talk about it in a verbal, linear, and thus relatively non-threatening way.

Of course television and films are not the only media which provide us with mythic symbols. Consider the fashion doll, which is big business today, with retail sales in excess of $3 million a year. The difference between fashion dolls and baby dolls, is, of course, that Barbie and Dawn and Crissy are designed to conform much more to the requirements of *Playboy* than playpen.

Barbie doll, the most successful, started out in the 1950's as a "teen fashion model," but it was in fact a master teacher of cultural values. Ruth Handler, Barbie's creator, writes in *Growing Up with Barbie: A Guide for Mothers* (itself an interesting teacher's guide): "Barbie can be used as a communicator" to guide the little girl through "the world of manners and etiquette and grooming and hygiene—all somehow considered more important for girls." And she closes on this inspirational note: "The Barbie concept, in many ways, gives little girls a guide to better living." Or, as her husband, Elliott Handler, once blurted out in an unguarded moment: "You get hooked on one and then you have to buy the other. Buy the doll and then you buy the clothes."[7]

Images and myths engulf us from every direction—from Washington, from the churches, from the schools, and from mother—to name a few. But mass media advertising provides the overwhelming input. Leo Bogard in his book, "Strategy in Advertising," says: "Every day 4.2 billion advertising messages pour fourth from 1,754 daily newspapers, millions of others from 8,151 weeklies, and 1.4 billion more each day from 4,147 magazines and periodicals. There are 3,895 AM and 1,136 FM radio stations broadcasting an average of 730,000 commercials a day. And 770 television stations broadcast 100,000 commercials a day. Every day millions of people are confronted with 2,500,000 outdoor billboards, with 2,500,000 car cards and posters in buses, subways and commuter trains and with 51,300,000 direct mail pieces."[8]

In summary, society needs a commonality of assumptions about what *is* in order to remain a society. To do this the society constantly engages itself—the people—in a communication system of propaganda and censorship which strives to create and reinforce traditions, values and ends which are held in common by most of us. This communication system operates in all media, and at every level, at all times, but it is most effective in the mass media and at the level of symbols, images and fantasies which express commom myths about reality, identity, power, value and morality. Thus the culture, primarily through mass media, constantly communicates back to us, at the mythic level, the national religion of which we are all a part.

## The World of Mass Media

Now what are mass media telling us about who we are, what we can do and be, and what is of value? As we examine the media world, remember that we are looking for the symbolic meanings and the underlying myths which are far more important than any individual story line or message or content. We are looking for environment, functions, and context, and most important of all, for human relationships which define social roles and tell who has power, who is aggressor and who is victim.

For example consider who populates the television world today. For most Americans, this TV world becomes their world at least three hours a day, every day, throughout most of their lives. George Gerbner tells us that about half of all characters are married in TV-land, but among TV teachers, only 18% of the women and 20% of the men are married.[9] Furthermore, the women "find themselves, and a man" by leaving teaching. Failure in love and life is a requisite to teaching success. The problems of TV teachers are solved by their leaving the profession—not by towns raising taxes, building schools and giving higher salaries. TV *journalists,* on the other hand, are strong and honest. TV *scientists* are deceitful, cruel, dangerous; their research leads to murder in fully half the situations.

In the TV world two-thirds to three-fourths of important characters are male, American, middle class, unmarried, and in the prime of life. They are, incidentally, the people who really run the world.

Unlike real life, TV violence rarely occurs between people who know each other well; and most of it does not come about because of rage, hate, despair, panic, but from the businesslike pursuit of personal gain, power or duty. In fact, one-third of the violent people, according to Gerbner, could be considered "professionals" in the business of violence.

Marriage seems to shrink men and make them unfit for the unmarried, free-wheeling, powerful and violent parts. On the other hand, women appear to gain more power through marriage, and they lose some of their capacity for violence. Finally, dominant majority-type Americans are more than twice as likely as all "others" to commit lethal violence, and then live to reach a happy ending. In the symbolic shorthand of TV, the free and the strong kill in a good cause to begin with.

Thus there is an interesting tradeoff in the TV world. The price of being **good** (the teacher) is **impotence.** On the other hand, the price of having **power** (the scientist) is to be **evil,** unless you happen to be a powerful white American, in which case the end justifies the means and you are rewarded with the American image of happiness (in *To Catch a Thief* the thief always steals and always gets the girl—a different girl each episode).

But what do you do if you don't have any power? Let's take a medium which is not so familiar to most of us—the comic book, a powerful medium, principally among the semi-literate and disadvantaged youth who today have so little power that they face between 25% and 50% unemployment.

Frederick Leaman has conducted an informal study of the hidden message of comic books. He visited three drugstores in a large city and asked for their best-selling comics. From a group of 26 stories and 87 characters he constructed the comic-book world. It is a world of conflict and contest. It is predominantly young, white, and middle-majority. Of every ten characters, seven commit some crime. Killers represent 13% of the population.

But here is the underlying message: in more than half (54%) of the stories, the key to superstatus is the consumption of some *chemical substance* that can effect a drastic transformation. One out of every five characters uses drugs to seek super-power, super-intelligence, or eternal life. Furthermore, positive, active, violent characters use drugs the most. The heroes of the comic-book world comprise two-thirds of all drug takers. Only 17% of their antagonists—the villains—use drugs. But the role of the drug user is untainted by villainy. Heroes use drugs in good causes.[10]

Or consider the roles of blacks in Saturday morning television for children. According to Joyce Sprafkin, blacks occupy 40% of all human roles in record commercials, while in commercials for board games, less than 6% of the parts are assigned to blacks.[11] Black (and white) children are systematically being taught that blacks may be musical, but they don't engage in games that require thinking.

## Central Myths of Our Society

Of course we are dealing with a very complex society, and it would be impossible to detail all the images and symbols that go into creating its commonality. However, there are a few central myths and values from which most of the images and symbols spring.

1. According to sociologist Marie Augusta Neal, the major myth of our Western culture is the Social Darwinian theory initiated by Herbert Spencer that *the fittest survive.* This is the concept that between ethnic groups today there exist genetic differences large enough to justify programming for unequal natural capacities for responsible decision-making, specifically in the interest of the group one represents. Sister Marie points out that Social Darwinism dominates our policy-making regarding education, jobs, geographical residential allotments, provision for recreation, health services and the uses of human beings to carry on wars.

It is no accident that in Gerbner's TV violence profile,[12] lower class and non-white characters were especially victimization-prone, that they were more violent than their middle class counterparts, and that they paid a higher price for engaging in violence. Majority-type Americans were twice as likely as the minority types to commit lethal violence and then live and reach a happy ending. As Gerbner says, "In the symbolic shorthand of television drama, the free and the stronger kill in a good cause to begin with." Or, as our myth suggests, the fittest survive, and the fittest in our mass media are *not* lower-class, non-white Americans.

2. Another central myth is that *power, including decision-making, starts at the central core and moves out.* The political word is from Washington; the financial word is from New York. The very nature of the mass media convey the message that power moves from center out. While watching the television set one has the sense of being at the edge of a giant instantaneous network where a single person at the center pushes the right button and millions of us "out there" see what has been decided.

Of course, there are alternatives to the myth of power moving from the center to the edges. Our own Declaration of Independence proposed that government derives its power from the consent of the governed, in other words, that the flow of power should be from the periphery to the center. But the opposite model was much supportive of the needs of the industrial revolution and the rise of a major nation state, and today it is clearly essential to the maintenance of both centralized governmental bureaucracy and a capitalist economy.

In our society, people at the center make decisions about what the others need and what they get. Mass production means standardization so, for example, whether people want it or not, the items on the shelves of our supermarkets get more and more the same, while mass advertising convinces us that we are getting more and more diversity. The idea that people in the power center should plan for others extends from the corporate home offices, to the national church bureaucracies, to the social welfare agencies. The result is that corporate business wonders why they are so low in the credibility polls, church leaders wonder why they are losing their jobs and their budgets, and social workers wonder why the poor don't appreciate the plans that have been worked out for them.

And none of us can figure out what is wrong, because the myth of power from center-out seems the way it ought to be.

3. A third central myth is that *happiness consists of limitless material acquisition.* This has several corollaries. One is that *consumption is inherently good*—a concept driven home so often and so effectively by the advertising industry as to need no elaboration. Another is that *property, wealth and power are more important than people.* This is a more difficult concept to have to face. But we need only consider the vast following for Ronald Reagan's proposition that the Panama Canal is ours because we bought and paid for it, to see how far it has made its way into our consciousness. We did, after all, pay for the Canal Zone. The fact that today it results in depriving people of Panama of their human rights is regrettable, but a deal is a deal. Or recall the city riots in the late sixties. It was when looters started into the stores that the police started to kill. Both humans and property may be sacred, but in our mythology today property rights are just a little more sacred.

4. Then there is the central myth that *progress is an inherent good.* At one level this is symbolized by the word "new" and "improved" attached periodically to every old product. But the myth goes much deeper. In his reflections last year in *The New Yorker,* Lewis Mumford said that the "premise underlying this whole age, its capitalist as well as its socialist development, has been 'the doctrine of Progress.' "Progress,' " he said, "was a tractor that laid its own roadbed and left no permanent imprint of its own tracks, nor did it move toward an imaginable and humanly desirable destination. *'The going is the goal,'* " not because there was any inherent beauty or usefulness in going. Rather, to stop going, to stop wasting, to stop consuming more and more, quicker and quicker, to say at any given moment enough is enough, would spell immediate doom.[13]

5. Finally, we have the central myth that *there exists a free-flow of information.* Of course the whole import of this analysis is that instead of a genuine free-flow there is consistent, pervasive and effective propaganda and censorship. Such a view is resisted most of all by the men and women who dedicate their lives to reporting the news. But they are the very ones least able to judge the matter, for they were selected and trained by the system so they could be depended upon to cooperate within its assumptions and myths.

This is not to condemn the news men and women any more than

others of us who function uncritically within the system year in and year out. Nevertheless, when Walter Cronkite says, "And that's the way it is . . .", what he is summing up is mostly information our society needs and wants to hear that particular day.

Consider, for example, the recent flap when Roger Mudd, on the campaign trail with Ronald Reagan, filed a story on how the telenews for all three networks had covered Reagan that day. Reagan had said absolutely nothing new or newsworthy, and he had indeed talked before a total of only about 2,000 people at shopping centers. But that morning he appeared before the network cameras so each could have something to send back as the day's "news." Mudd's story about the manufacture of news was killed by Cronkite, because it reflected negatively on the profession. But when Cronkite's rejection itself began to be circulated around pressrooms of the nation, CBS decided to run the Mudd story on the morning news—so a small fraction of viewers saw it and CBS averted revelation of censorship which could have been even more harmful to its "free flow" image than the original story.

## Media and Values

And what are the values that the mass media communicate on behalf of our culture? They are also too numerous to analyze in detail, but we can mention the predominant ones. *Power* heads the list: *power over others, power over nature.* As Hannah Arendt points out, in today's media world it is not so much that power corrupts as that the aura of power, its glamorous trappings, attracts.[14] Close to power are the value of *wealth and property,* the idea that *everything can be purchased,* and that *consumption is an intrinsic good.* The values of *narcissism, immediate gratification* of wants, and *creature comforts,* follow close behind.

Thus the mass media tell us that *we are basically good,* that *happiness is the chief end of life,* and that *happiness consists in obtaining material goods.* The media transform the value of sexuality into *sex appeal;* the value of self-respect into *pride;* the value of will-to-live into *will-to-power.* They exacerbate acquisitiveness into *greed;* they deal with insecurity by generating more *anxiety.* They change the value of recreation into *competition* and the value of rest into *escape.* And perhaps worst of all, the media constrict our experience and substitute media-world for real-world so that we are becoming less and less able to make the fine value judgments that such a complex world requires.

In terms of the economic system, the media are a perfect handmaiden of capitalism. The high technology required for our current mass communication system, with its centralized control, its highly profitable experience, its capital-intensive nature, and its ability to reach every individual in the society, is perfectly suited for a massive production-consumption system that is equally centralized, profitable and capital-intensive. Our production-consumption system simply could not exist without a communication system that trains people to be knowledgeable, efficient and hard-working producers and equally knowledgeable, efficient and hard-working consumers. The fact that capitalism turns everything into a commodity is admirably suited to the propaganda system of the mass media, which also turns each member of the audience into a consumer.

In terms of the political system, the media, again reflecting the values of the society, give us politics by image. The whole media experience of Vietnam was guided by the necessity of a super-power to create for itself an image which would convince the world—and itself—that it was number one, the mightiest power on earth (our most important value). The experience of Watergate is also revealing. Several observers have pointed out that the public, its leaders and the media all were offended and shocked not by what the President and his men *did* but by the fact that they got caught—publicly, red-handed, in a way that simply could not be imaged away. After Watergate we see the immediate return to the old value system: those who were indicted and convicted have been overwhelmed with high offers from publishers, the press and television (and the campuses) to tell their stories. This simply drives home the point that our society demands "positive images," including even more lies and fabrications, in order to justify and mitigate the horror of the cover-up, to rehabilitate the criminals, and above all, to help restore through imagery the public's loss of confidence in the political systems.

## Christian Values

What is the Christian response to this value system? The answer is obvious and undeniable. Regardless of your theology, whether you are conservative, liberal or middle-of-the-road, and regardless of whether you believe the Bible word for word or demythologize it piece by piece—the whole weight of Christian history, thought and teaching stands diametrically opposed to the media world and its values I have just been describing.

Instead of power over individuals, the Bible calls for justice and righteousness (Amos 5:23-24); kindness and humility (Micah 6:6) and the correction of oppression (Is. 1:17). Instead of power over nature in order to consume and waste, the Genesis story affirms the value of man's guidance and transformation of nature, in harmony with the whole creation (Genesis). Instead of the value of wealth, Jesus tells the rich young ruler to sell all that he has, and he says that the value of wealth in terms of the Kingdom of God is about the same value of a rope in threading a needle (Matt. 19:17-22; Mk. 10:17-21; Lu. 18-18-23).

The value of property is especially interesting. Sister Marie Augusta Neal compares the classic definition of justice rooted in entitlement, that is, the protection of property already possessed, with Leviticus 25:1-29, where the property must always be returned every fifty years to the people who sit on the land. And

Jesus simply tells us to give our coat to the person who needs it (Matt. 5:40). As for the idea that money can purchase anything, there is the story of the wealthy man who built bigger barns, and Jesus asking, "What does a man gain by winning the whole world at the cost of his true self?" (Mk. 8:36). The values of narcissism, of immediate gratification and creature comforts are placed against Jesus' affirmation that if anyone wants to be a Christian that person must leave self behind, must take up his or her cross and follow Jesus' way (Matt. 16:24).

Against the myths that we are basically good, that happiness is the chief end of life and consists in obtaining material goods, there are arrayed the affirmations that man is susceptible to the sin of pride and will-to-power, that the chief end of life is to glorify God, and that happiness consists in creating the Kingdom of God within one's self and among his neighbors—that is, among all inhabitants of the earth.

It is easy to see how completely at variance these values are with those clearly underlying the whole process, context and environment created by the mass media today. In short, we find ourselves living in a society which is completely at odds with our professed religion.

What can we do about this situation that will make any difference?

## Media Education

Perhaps the first and the most difficult thing we have to do is spend a good deal more thought in understanding what the media are really saying to us and to help others understand it also. Media education involves much more than reading film reviews and the media sections of *Time*, the *New Yorker* and *Saturday Review*. It means quite literally having to be in this world but not of it.

Let me illustrate what this requires. In *The Broken Covenant*,[15] Robert Bellah describes the growing dominance in America from the middle of the 18th century of what William Blake called "single vision." Single vision is the scientific-technocratic view of the world that everything is amenable to logic and reason and that there is no need for the imaginative vision and perspective supplied by religion. But Blake called for "twofold vision," which adds to practical rationalism the awareness that there is always more than what appears, and that behind every literal fact there is a depth of meaning and implication. To Blake the cutting off of this depth of meaning was a kind of sleep or death.

I suggest that our society is today cultivating single vision, and that the desensitization and the dehumanization that we feel all around us is a kind of sleep or death of awareness and conscience, and that we must revive in people a habit of double vision that can identify myths and values underlying society and can evaluate them from a perspective that transcends the limitations of that society.

And I suggest that the best place to do this is in the myths and symbols found in the mass media. Here we see all the appeal of a practical, rational, well-organized society, and here we can see what it does to people. We can see how it drives the rich and powerful onward by preaching the rewards of success; how it motivates and channels the energies of the working millions by encouraging them to be good, to follow the rules, to do what is right, and to produce in order to consume; and how it teaches the poor and powerless who they are and that they had better stay that way or get hurt.

This media education, this twofold vision, is no good without a reference point that transcends the culture. For Christians this is almost too simple. The Bible makes it clear that God is on the side of the poor and powerless, and all that this implies. And if this is not sufficient, the lives of the faithful right up to Martin Luther King illustrate that this is where God wants his people to be.

But where will this analysis take place? We have already shown that media propaganda is most effective when it reaches people as individuals in the mass—such as watching TV. The place least hospitable to such propaganda is where people regularly meet face-to-face in small groups. And this is precisely where the church has its strength. For all its failings, the church remains one of the few places in society where people regularly come together on a face-to-face basis. Here is where media education can and must take place.

Such analysis is not easy. It is complex and it is threatening. It is far easier for church people to gather together to condemn a TV episode containing more sex or violence than usual—and thus miss the whole thrust of the analysis.

Media education would have a different focus with different groups. Among the poor it would aim at helping them define what the media says about them, and then to define their real problems and their real role in society—which could very well lead to action to get out of that role. Among the vast middle class workers and consumers it would help them understand the ways in which they are being manipulated to ends not their own, and to evaluate the satisfactions held out to them by the media in terms of their reality, and then to establish values independent of those of the media, and develop lifestyles which can achieve *their* goals instead of media goals. As this is happening, new myths, new symbols and images would develop which would move into competition with the old and which could help transform the society into one better suited to meet human needs.

## Social and Political Action

The third and final thing we can do is to engage in direct social and political action to change the structures of the media so that they will be more open and responsive to points of view which differ from the cultural norm. I am thinking of such action as testimony before the FCC, initiating and building political support for bills in Congress, and instigating court suits and developing stockholder action with broadcasters, advertising agencies and sponsors.

Again, if we choose only to develop programs where the primary criteria are those of the media industry, and if we cosy up to the industry in order to get whatever scraps of goodwill and time and space they are willing to offer, then we simply have gained access to the media at the loss of our own soul. The mass media surely constitute one of the most powerful of all institutions in society and if we believe that God is on the side of the poor and the powerless and of justice and love, then we have to be ready to challenge the pretensions of that power and to do battle with it. At the same time, we must do battle in love, not forgetting that many in positions of power accept their role as uncritically and even unknowingly as those who are powerless and accept their situation uncritically.

I therefore think it is wrong to attack the media as if they were being manipulated and mishandled by greedy people at the top. In reality, the media reflect our *own* greed and weaknesses far more than we care to admit or to analyze. This means that we can't solve the problems of TV by grouping spots or reducing the number of ads. Although this might help take out some of the irritation, it would do nothing about the fundamental media problem. The solution is much more radical: a change in the beliefs and assumptions and economic base of the entire society.

Our social and political problem is thus to change enough individuals to bring about a change in the social structures which will make it possible for even more individuals to change. Social action and personal persuasion are reciprocal, and we cannot afford to neglect either one.

In concluding, I do not want to leave the impression that we are simply doomed to be shaped wholly by inexorable social forces and that the situation is hopeless. Let me quote Reinhold Niebuhr, who observed: "Nothing that is worth doing can be achieved in our lifetime; therefore we must be saved by hope. Nothing which is true or beautiful or good makes complete sense in any immediate context of history; therefore we must be saved by faith. Nothing we do, however virtuous, can be accomplished alone; therefore we are saved by love."[17]

I suggest that our Christian theology is fundamentally at odds with the theology of our society, and the mass media happens to be the arena where the matter is going to be resolved. It is a unique opportunity. It will take clear thinking, hard work and a good deal of faith, hope and love. But it is our society and our lives at stake, and I can think of no more exciting challenge.

***REFERENCES***


1. Ellul, Jacques, *Propaganda,* Alfred A. Knopf, New York, 1965, p. 121.
2. Ibid., p. 9
3. Ibid., p. 17
4. Petersen, James, "Eyes Have They, But They See Not," a conversation with Rudolf Arnheim, *Psychology Today,* June 1972. p. 55
5. Moles, Abraham, "A Skylight Open to the Neighbourhood," in *Intermedia,* International Broadcast Institute, Feb. 1976. p. 6
6. Kubrick, Stanley, *Cultural Information Service,* Jan. '75, p. 12
7. Troy, Carol, "Little TV Doll, Who Made You?" *New York Magazine,* Dec. 1973, pp. 51-66
8. *Advertising Age,* November 21, 1973, p. 7
9. Gerbner, George. Address at International Communication Association, April 21, 1972.
10. Gerbner, George, "The Social Uses of Drug Abuse." Annenberg School of Communications, Philadelphia, Pennsylvania. Updated.
11. Sprafkin, Joyce, "Stereotypes on Television." Media Action Research Center, New York City.
12. Gerbner, George and Gross, Larry P., *Violence Profile No. 5,* Univ. Of Penna., Philadelphia 19104, June 1973.
13. Mumford, Leis, Quoted in Arendt, Hannah, "Home to Roost: A Bicentennial Address," *New York Review,* June 26, 1975, p. 3
14. Ibid., p. 4
15. Bellah, Robert, *The Broken Covenant,* Seabury Press, New York, 1975, p. 72 ff.
16. Fore, William F., "Communication: A Complex Task for the Church." *Christian Century,* July 9, 1975, pp. 653-654
17. Quoted by James Reston in "Carter, Politics and Religion," *International Herald Tribune,* May 3, 1976.

# WHAT IS MARC?

Media Action Research Center, Inc. is an independently incorporated, not-for-profit organization, its work made possible by grants and individual contributions. The center was established in 1974 with a grant from the United Methodist Church.

MARC'S purposes are to:

—Study the impact of TV on viewers through scientific research.

—Make available in understandable form information on what is known about TV's influence.

—Help viewers develop strategies for more intentional, selective, questioning approaches to viewing through workshops and other events.

—Help bring about positive changes in the television system.

MARC has conducted or participated in scientific studies on sexual behavior on TV, stereotyping of women and minorities, effects of pro-social viewing on children, a profile of Saturday morning children's programs, and content messages in Saturday morning commercials. MARC is a partner in the creation and effectiveness-testing of 30-second public service TV spots designed to teach children positive ways of dealing with conflict.

The widely used Television Awareness Training program was developed by MARC in cooperation with the Church of the Brethren, American Lutheran Church, and the United Methodist Church. T-A-T is a curriculum which helps persons become more aware of the messages and influence of the television experience, more creative in the use of TV, and work for a television system that better serves the needs of the public. T-A-T resource materials, in addition to this book, include nine films, a leader's training manual, and a design for study in settings ranging from one hour to major workshops of eight two-hour sessions.

T-A-T National Trainers conduct weekend workshops throughout the Unted States. Persons participating in the intensive training events have an opportunity to become accredited T-A-T Leaders, equipped with training resources for leading workshops and making presentations in their own areas.

Major T-A-T training events have also been held in Canada and Brazil. T-A-T Canada was developed by Canadian communication leaders as an independent organization that works in close cooperation with MARC. Leaders in Brazil have also developed their own T-A-T program.

Through an arrangement with the World Association of Christian Communicators (WACC) T-A-T is now available throughout the world, with groups and institutions in each country adapting the television curriculum to their own situation and needs.

MARC's latest project is the creation of a new church school curriculum, *Growing With Television: A Study of Biblical Values and the Television Experience.*

The curriculum focuses on television as our most common experience, something much more than an advertising or entertainment medium, a major transmitter of culture, a shaper of values, setter of lifestyles, a conditioner of the way we think, believe, and feel. Since so much of what television portrays of our culture is antithetical to human values, the course of study stresses the need of making viewing an intentional and evaluative experience. For church educators, television can provide yet another opportunity for self-development through examining and defining cultural values, personal values, and Biblical values.

The study includes five course levels of 12 sessions each—younger elementary, older elementary, junior high, senior high, and adult. The program is flexible and suitable for many different settings: church school, youth or adult group meetings, mid-week or vacation church schools, intergenerational events. The teachers' books are designed and written for use by relatively untrained leaders.

The core of the curriculum is television. Homework assignments are made each week to help create a home learning environment for each member of the household. It is here where most learning and discovery will occur. The group session is the beginning point, and the debriefing point, of the home viewing experience. This study can involve all members of the household and congregation in a growth, self-discovery, group learning process.

The new curriculum is being published (September 1980) by six different publishers under the coordination of the 10-denomination Cooperative Publishers Association.

For information about MARC and its activities write to Media Action Research Center, Inc., Suite 1370, 475 Riverside Drive, N.Y., N. Y. 10115, or call 212, 865-6690.

MARC is incorporated in the State of New York under Section 216 of the Education Law and Section 404 of the Not-For-Profit Corporation Law. It has been granted Sections 501 (c) (3) and 509 (a) (2) status by the United States Internal Revenue Service.

A contribution to MARC will help insure that television is better understood, that viewers are empowered to use it more creatively, and that the rights and needs of viewers are better represented when decisions are made.

## THE CREATORS OF T-A-T

NELSON PRICE is director of the Division of Public Media of United Methodist Communications, a national communications agency for The United Methodist Church. He directs the denomination's activities in news and public media, including press relations, radio and television programming, media advocacy activities, and film and cable television production and relationships. He is president of the Media Action Research Center, Inc. and serves on the boards of directors of the Religious Public Relations Council, Inc., the North American Broadcast Section of the World Association for Christian Communication, and the Communications Commission of the National Council of Churches. He is an advanced member of the International Transactional Analysis Association. Mr. Price has served as producer or executive producer for films, radio and television programs such as *A Fuzzy Tale,* and *Begin With Goodbye.* He was consultant to Westinghouse Broadcasting Company on a joint television series entitled *Six American Families* on PBS. He is a national trainer for T-A-T.

BEN LOGAN is a broadcast producer with United Methodist Communications, a member of the board of directors of Media Action Research Center, and a free-lance writer. He has written drama for television, written and produced TV and film documentaries, public service spots, and was producer of *Night Call,* the highly acclaimed nation-wide radio call-in program.

Mr. Logan is author of *The Land Remembers—the Story of a Farm and Its People,* Viking Press, 1975 (Avon paperback, 1976). He is the editor of a curriculum, *Growing With Television* (a study of Biblical values and the television experience), has been a magazine editor and a teacher of creative writing, and is active as a T-A-T trainer.

CAROLYN KAY LINDEKUGEL has been director of media education for the American Lutheran Church (A.L.C.). She is a graduate of Denison University, has an M.A. from the University of Texas and has done graduate work at Miami of Ohio and Immaculate Heart College.

Ms. Lindekugel is a member of American Women in Radio and Television (AWRT), and past chairperson of the North American Broadcast Section of the World Association for Christian Communication. As a staff member of the A.L.C. she has served on the Task Force of Men and Women in Church and Society, written church school curriculum and created multimedia productions. She is a national trainer for T-A-T.

STEWART M. HOOVER is a graduate student at the Annenberg School of Communication at the University of Pennsylvania, while he continues to be Consultant for Media Education and Advocacy for the Church of the Brethren. He is chairperson for the steering committee of the North American broadcast section of World Association for Christian Communication. He is executive producer of a 26-part radio series titled, *Think About It.* Mr. Hoover has written and lectured on various aspects of mass communication and society. He is a national trainer for T-A-T.

## COOPERATING ORGANIZATIONS

### AMERICAN LUTHERAN CHURCH

Harry Souders
1568 Eustis Street
St. Paul MN 53108
(612) 645-9173

### CHRISTIAN CHURCH (DISCIPLES OF CHRIST)

Rev. Fred Erickson
P. O. Box 1986
Indianapolis, IN 46206
(317) 353-1491

### REFORMED CHURCH IN AMERICA

Rev. Peter Paulsen
107 Spanish Village Center
Dallas, TX 75248
(214) 386-0395

### UNITED CHURCH OF CHRIST

Ralph Jennings
289 Park Avenue, South
105 Madison Avenue
New York, N. Y. 10016

*(Continued on next page)*

COOPERATING ORGANIZATIONS (Continued)

**CHURCH OF THE BRETHREN**
Stewart Hoover
1450 Dundee Avenue
Elgin, IL 60120
(312) 742-6100

**PRESBYTERIAN CHURCH IN THE US**
Belle Miller McMaster
341 Ponce de Leon Avenue, NE
Atlanta, GA 30308
(404) 873-1531

**PROTESTANT EPISCOPAL CHURCH**
Sonia Francis
815 Second Avenue
New York, N. Y.10017
(212) 867-8400

**UNITED METHODIST CHURCH**
Nelson Price
475 Riverside Drive, Suite 1370
New York, N. Y. 10115
(212) 663-8900

**WORLD ASSOCIATION FOR CHRISTIAN COMMUNICATION (WACC)**
Thelma Awori
122 King's Road
London SW3 4TR
England

**T-A-T CANADA**
Rev. Keith Woollard
85 St. Clair East
Toronto, M4T 1M8
Canada

## NATIONAL TRAINERS

HELEN HALLUM BURNS

GEORGE C. CONKLIN, JR.

STEWART M. HOOVER

RICHARD P. JAMESON

PATRICIA KOWALSKI, OSM

BEN T. LOGAN

CAROLYN LINDEKUGEL MANLOVE

JOHN D. MULLER

NELSON PRICE

DIANE ZIMMERMAN UMBLE

SHIRLEY KORITNIK, SCL

# BIBLIOGRAPHY

## Books


1. Barnouw, Erik. *The Tube of Plenty.* New York: Oxford University Press, 1975.
   "The best single-volume history of radio and TV in this country," said John Leonard.
2. Barnouw, Erik, *The Sponsor: Notes on a Modern Potentate.* New York: Oxford University Press, 1978.
   An expert's description of the role of the sponsor in governing program choices and public craving for products.
3. Barnouw, Erik. *The Image Empire:* A History of Broadcasting in the United States from 1953, Vol. III. New York: Oxford University Press, 1970.
   Part three of the classic work. Thorough, readable, and academic accounting of the growth of the television system from 1953-1970.
4. Borgman, Paul. *TV Friend or Foe? A Parents' Handbook.* Elgin, IL: David C. Cook Publishing Co., 1979.
   A Christian's analysis of TV, and positive roles parents can have in their children's viewing experiences.
5. Brown, Les. *Keeping Your Eye on Television.* New York: Pilgrim Press, 1979.
   The author asks, Is the public interested in the public interest? He discusses the public's rights and responsibilities in broadcasting. Introduction is by Dr. Everett Parker.
6. Brown, Les. *The New York Times Encyclopedia of Television.* New York: Times Books, 1977.
   A single standard reference book for television that encyclopedically covers the entire field, past and present. Upwards of 10,000 entries about all areas including programs, technology, public policy, legal cases, and more.
7. Brown, Les. *Television: The Business Behind the Box.* New York: Harcourt, Brace and Jovanovich, 1971.
   A basic accounting of how economic priorities govern the construction of TV fare. Numerous vignettes of stars' and executives' behaviors.
8. Carnegie Commission on the Future of Public Broadcasting. *A Public Trust.* New York: Bantam Books, 1979.
   This report of the Carnegie Commission includes chapters on the rise of public broadcasting, programs and services, funding, public radio, the new telecommunications environment and more.
9. Cline, Victor B., editor. *Where Do You Draw the Line?* Provo, Utah: Brigham Young University Press, 1974.
   A book which raises questions about explicit images in the media. How far should freedom go? Where is the line of acceptability?
10. Cole, Barry, and Oettinger, Mal. *Reluctant Regulators—The FCC and the Broadcast Audience.* Reading, MA: Addison-Wesley Publishing Co., 1978.
    A comprehensive dissection of the workings and non-workings of FCC. Both a discouraging and enlightening review marred only by lack of notes and index.
11. Comstock, George, et. al. *Television and Human Behavior.* New York: Columbia University Press, 1978.
    A compendium of research findings including viewing patterns, programming trends, effects of violence, news and much more.
12. Fore, William F. *Image and Impact.* New York: Friendship Press, 1970
    The influence of mass media, especially TV, on American values and perceptions of people.
13. Kaye, Evelyn. *The ACT Guide to Children's Television.* Boston: Beacon Press, 1979.
    A parents' guide to understanding children and TV.
14. Lesser, Gerald, *Children and Television—Lessons from Sesame Street.* New York: Random House, 1974.
    A report on how *Sesame Street* started and what researchers learned about children's responses as the program was being developed. Lesser headed the advisory team.
15. Liebert, Robert M., Neale, John M., Davidson, Emily S. *The Early Window—Effects of Television on Children and Youth.* New York: Pergamon Press, Inc., 1973.
    A succinct explanation of how children's behaviors are affected via observational learning.
16. Mander, Jerry. *Four Arguments for the Elimination of Television.* New York: William Morrow, 1978.
    A thorough discussion of what Mander deems the "unreformable aspects" of television and their effects on human beings.
17. Mankiewicz, Frank and Swerdlow, Joel. *Remote Control: Television and the Manipulation of American Life.* New York: Quadrangle/The New York Times Book Co., 1978.

Lively, entertaining, anecdotal—this is a highly readable commentary on contemporary TV. Good bibliography but limited use of citations will frustrate the scholar.

18. Melody, William. *Children's Television: The Economics of Exploitation.* New Haven and London: Yale University Press, 1973.
    The author lays out the money-facts concerning the construction of children's TV and reasons for programming decisions.
19. Muggeridge, Malcolm. *Christ and the Media.* Grand Rapids, Mich.: Wm. B. Eerdmans Publishing C., 1977.
    Author discusses his conviction that the media have had an extremely negative effect on civilization.
20. Phelan, John M. *Media World: Programming the Public.* New York: Seabury Press, 1977
    Presents a strong case that the communications industry is dehumanizing persons while we allow this to happen. Thoughtful essays.
21. Powers, Ron. *The Newscasters: The News Business as Show Business.* New York: St. Martin's Press, 1977.
    Vibrant critique covering the spectrum of quality and qualities in TV news with lots of interviews of celebrities and journalists. Helpful for understanding news business, its limitations, excesses and strengths.
22. Schiller, Herbert I. *Communication and Cultural Domination.* New York: International Arts and Sciences Press, Inc., 1976.
    A look at how communications is used as a source of power within and between nations, the relation of that to cultural domination and some observations on processes for resisting such domination.
23. Schiller, Herbert I. *The Mind Managers.* Boston: Beacon Press, 1974.
    The story of manipulation by media—some of the effects and why they happened.
24. Schrank, Jeffrey, *TV Action Book.* Evanston: McDougal, Littell & Co., 1974.
    A highly readable text/workbook explaining how the TV system works and how to become an agent of change.
25. Schwartz, Tony. *The Responsive Chord.* New York: Anchor Press/Doubleday, 1973.
    Philosophy and instructions on how to elicit the responses of eye and ear via electronic media.
26. Shanks, Bob. *The Cool Fire—How to Make It In Television.* New York: Vintage/Random House, 1977.
27. United States Commission on Civil Rights. *Window Dressing on the Set: an Update.* Washington: U. S. Government Printing Office, 1979.
    A report of stereotypic portrayals of women and minorities on TV, with recommendations for change.
28. Ward, Scott, Wackman, Daniel B., and Wartella, Ellen. *How Children Learn to Buy.* Beverly Hills, CA.:
    The authors describe how children become consumers in the tutelage of television.
29. Winn, Marie. *The Plug-In Drug—Television and the Family.* New York: Viking Press, 1977.
    Strong assertion that the effects of TV are interfering with learning abilities of children and destroying family relationships.

## Newsletters

—*access* Magazine, $24 per year for 26 issues, from National Citizens' Committee for Broadcasting, 1530 "P" Street, N. W. Washington, D. C. 20005

—*Arts in Context,* including "The Televisionary." 423 West 43rd St., New York, N. Y. 10036. Twelve issues a year, $15 ($10 renewal). Provides advanced listings and reviews of outstanding TV programs.

—*Cultural Information Service,* P. O. Box 92, New York, N. Y. 10016. A bi-weekly review of the arts/media designed for leaders in education and religion, as well as occasional special discussion guides of current television and film resources.

—*Mass Media Newsletter,* 2116 N. Charles Street, Baltimore, MD 21218. Twenty-two issues a year, $10. Provides advanced listings and reviews of outstanding TV programs.

—*Media and Values,* published four times a year by the National Sisters Communications Service, 1962 South Shenandoah, Los Angeles, CA 90034. Reviews and articles on the media from a value perspective.

—*Re:act,* published four times a year in magazine format by ACT, 46 Austin Street, Newtonville, MA 02160. Articles on child and TV and updates on proceedings at FTC and FCC regarding various broadcast issues.

## Directories of Television Advertisers

—*The Target List,* NCG, P. O. Box 1039, Palo Alto, CA 94302. $2.00. Directory lists major television advertisers, their products and addresses.

—*Television Sponsors Cross-Reference Directory,* Everglades Publishing Co., P. O. Drawer Q, Everglades, FL 33929. $4.95 single issue, $13.80 a year for four issues. Lists 4,000 name-brand consumer products with addresses of parent and subsidiary companies. The most complete list and service of its kind available.

## Teacher's Guides

*Teacher's Guides to Television,* 699 Madison Avenue, New York, N. Y. 10021. $5.00 for two semesters, Fall and Spring. Lists TV programs of special educational value and provides teacher's guides for selected programs.

—*Prime-Time School Television.* 120 South LaSalle Street, Chicago, IL 60603. Written with the teacher in mind, PTST's guides deal with a variety of social studies issues that can be addressed using commercial television programs for discussions in the classroom.

## Audiovisual Resources

—*It's As Easy As Selling Candy To A Baby.* This film shows the effects of TV on the health and buying habits of young children, specifically data on effects of sugar consumption. This 15 minute color film comes from Action for Children's Television, 46 Austin Street, Newtonville, MA 02160.

—*Kids For Sale.* This newest release from Action for Children's Television looks at the discrepancies between what children's television is and what it could be, and explores the pervasive effects of television commercials, stereotyping, and violence on young viewers. This 20 minute film is suitable for parent, professional, and PTA audiences. Rental, $30; sale, $285. From Action for Children's Television, 46 Austin Street, Newtonville, MA 02160.

—*Seconds To Play.* This behind-the-scenes look at the televising of a football game points up the manipulation which makes it more than just a sporting event. In 16 mm. color, this 28 minute film is available from Films Incorporated, 1144 Wilmette Ave., Wilmette, IL 60091.

—*Six Billion $$$ Sell: A Child's Guide To TV Commercials.* A 15 minute film which uses clips from TV commercials, animation, and an original pop theme song to teach children about the techniques used by advertisers. Appropriate for grades three through eight. Rental, $25; sale, $220. From Consumer Reports Films, Box XA-35, 256 Washington Street, Mount Vernon, NY 10550.

—*The Television Newsman.* This 28 minute, color film shows a day in the life of an "eye-witness" news reporter. Some insights, but we do not see hassles on the screen. By Pyramid Films, Box 1048, Santa Monica, CA 90406. Rental $35.

—*The 30-Second Dream.* This 15 minute film, a fascinating montage of some of the best TV commercials around, explores the fantasy world of television advertising, where intimacy, vitality and success are "only a purchase away." Suitable for junior high school through adult audiences. Rental, $25; sale, $250. From Mass Media Ministries, 2116 N. Charles St., Baltimore, MD 21218.

—*TV: The Anonymous Teacher.* Designed for parents and teachers, this 15 minute film provides a rare opportunity to watch children react as they watch TV. Includes interviews with psychologists and child specialists on the influence of television. Rental, $20; sale, $225. From Mass Media Ministries, 2116 N. Charles St., Baltimore, MD 21218.

# INDEX